Advances in Astrobiology and Biogeophysics

For further volumes:
http://www.springer.com/series/5118

Advances in Astrobiology and Biogeophysics

springer.com

This series aims to report new developments in research and teaching in the inter-disciplinary fields of astrobiology and biogeophysics. This encompasses all aspects of research into the origins of life – from the creation of matter to the emergence of complex life forms – and the study of both structure and evolution of planetary eco-systems under a given set of astro-and geophysical parameters. The methods considered can be of theoretical, computational, experimental and observational nature. Preference will be given to proposals where the manuscript puts particular emphasis on the overall readability in view of the broad spectrum of scientific backgrounds involved in astrobiology and biogeophysics.

The type of material considered for publication includes:

- Topical monographs

- Lectures on a new field, or presenting a new angle on a classical field

- Suitably edited research reports

- Compilations of selected papers from meetings that are devoted to specific topics

The timeliness of a manuscript is more important than its form which may be unfinished or tentative. Publication in this new series is thus intended as a service to the international scientific community in that the publisher, Springer-Verlag, offers global promotion and distribution of documents which otherwise have a restricted readership. Once published and copyrighted, they can be documented in the scientific literature.

Series Editors

Dr. André Brack
Centre de Biophysique Moléculaire
CNRS, Rue Charles Sadron
45071 Orléans, Cedex 2, France
Brack@cnrs-orleans.fr

Dr. Gerda Horneck
DLR, FF-ME
Radiation Biology
Linder Höhe
51147 Köln, Germany
Gerda.Horneck@dlr.de

Dr. Lisa Kaltenegger
Max-Planck-Institut für Astronomie
Koenigstuhl 17
69117 Heidelberg
Germany

Dr. Christopher P. McKay
NASA Ames Research Center
Moffet Field
CA 94035, USA

Prof. Dr. H. Stan-Lotter
Institut für Genetik und Allgemeine
Biologie
Universität Salzburg
Hellbrunnerstr. 34
5020 Salzburg, Austria

Douglas A. Vakoch
Editor

Astrobiology, History, and Society

Life Beyond Earth and the Impact of Discovery

 Springer

Editor
Douglas A. Vakoch
SETI Institute
Mountain View
CA, USA

and

California Institute of Integral Studies
San Francisco
CA, USA

ISSN 1610-8957   ISSN 1613-1851   (electronic)
ISBN 978-3-642-35982-8   ISBN 978-3-642-35983-5   (eBook)
DOI 10.1007/978-3-642-35983-5
Springer Heidelberg New York Dordrecht London

Library of Congress Control Number: 2013936532

*To Tom Pierson,
for his steadfast leadership of research
and education
at the forefront of astrobiology*

Foreword

The burgeoning field of astrobiology poses three basic questions: What is the origin and distribution of life in the universe? Are we alone? What is life's future on Earth and beyond? Astrobiology is remarkable not only for its scientific significance but also because of its relevance to longstanding philosophical and existential issues. Who has *not* at one time or another asked how did we get here, are we alone, and what will become of us? Astrobiology's human side, long evident in philosophy, history, and theology, is making inroads in the social sciences. Today it is safe to say it has two agendas: a science agenda and a societal agenda. The broad umbrella of astrobiology provides a common frame of reference and promotes dialogue among people whose backgrounds and interests give them different perspectives. Astrobiology stresses interdisciplinary and multidisciplinary collaboration at a time when research is becoming increasingly specialized, and, as the anthropologist Ben Finney pointed out long ago, may represent a step towards the unification of science.

Researchers pursuing the science agenda explore microcosm and macrocosm: their interests range from molecules to galaxies. In addition to peering through microscopes and telescopes they analyze samples of air, earth, and water; hunt for life that thrives under extreme conditions; track environmental changes including global warming; crack open meteorites to hunt for extraterrestrial fossils or biological specimens; search for signs of extraterrestrial intelligence; and identify and track asteroids that could collide with Earth. Researchers following the societal agenda also explore microcosm and macrocosm, ranging from individual cognition to the qualities of entire societies and cultures. They seek ways to minimize the roles of anthropocentrism and ethnocentrism in our thinking about extraterrestrial life, turn to history for analogies to possible astrobiological discoveries, look to cross-cultural data from societies ancient and modern to gain insights into human diversity and the outcomes of cross-cultural encounters, develop scenarios, sample public opinion, and conduct experiments.

The history sections of this book advance both the scientific and societal agendas, while the remaining chapters focus on the societal agenda. This book's focus on the detection of extraterrestrial life and its consequences has a direct bearing on astrobiology's first two great questions (What is the origin and distribution of life in the universe? Are we alone?) and a clear relationship to the third (What is the future of life on Earth and beyond)?

What a difference a few decades can make in a millennia-old quest to understand our place in the universe! Apart from some interest in flying saucers shared, no doubt, with many other adolescent American males in the 1950s, I recall no special interest in the search for life beyond Earth. More or less, I fell into astrobiology by accident. In the late 1970s, I began collaborating with Mary M. Connors of NASA Ames Research Center on human requirements for post-Apollo space missions. In the 1980s we were still hard at it, and by then, under the tutelage of John Billingham, Mary had become involved in the cultural aspects of the Search for Extraterrestrial Intelligence (SETI). An unsung SETI pioneer, Mary shared two of her white papers on the topic (duly referenced in my publications) and, as luck would have it, on one of our meeting days she introduced me to the psychologist Donald Norman whose report to Congress included a few paragraphs on SETI. My new interest was confirmed at a meeting of the Society for Applied Anthropology, where I became entranced with the ideas of Ben Finney.

When I began writing about SETI in the late 1980s there was very little for a psychologist to go on: Mary's papers, some history and philosophy books, two anthropology books, one psychology book, and musings on the part of bright people who were not trained in social science methodology. One figure that attests to progress in astrobiology is that whereas in 1990 not one exoplanet had been discovered, on 16 July 2012 the Extrasolar Planets Encyclopaedia proclaimed a veritable jackpot of confirmed planets: 777. Progress on the societal agenda is more difficult to prove, but suffice it to say that my original half shelf of books and single crate of reprints has now overflowed my home library and is taking up about half the floor in a garage-sized rented storage unit.

Astrobiology, History, and Society is a fresh and important addition to the literature. Within these pages we find breathtaking overviews and fine-grained analysis of specific episodes and events. We find chapters that shed new light on the earliest phases of the extraterrestrial life debate and bring the debate up to today. Writers with decades of experience are joined by newcomers who apply new talent and introduce new views. We find popular and contrarian ideas, consensus and controversy, old saws, and Facebook. Of particular importance we find substantial new material on the discovery of non-intelligent extraterrestrial life. For too long interest in extraterrestrial intelligence has obscured the many profound and practical implications of finding non-intelligent life within our solar system. Another powerful prevailing theme is the role and limitations of analogies.

Astrobiology, History, and Society deserves to make a mark on its debut and then earn a place in the reader's permanent library. Let us hope that the future will see the publication of many additional worthwhile books. Such is by no means assured. We live in an age of rapidly advancing science and technology. We also live in an age marked by the widespread denial of scientific findings (such as evolution and global warming), disdain for education (manifested in increasing class sizes and hopeless debt for college students), and political coalitions that seek to cut funding for science and suppress supposedly "liberal" scientific ideas.

The battle against enlightenment cannot be won in the arena of science, so it is fought in the courts, in election campaigns, on talk radio, and in blogs. The danger is real, and the outcome uncertain.

Albert A. Harrison
Professor Emeritus
Department of Psychology
University of California
Davis

Preface

Astrobiologists must continually struggle with the "N = 1 Problem." Thus far, we know only one example of a planet that bears life: Earth. Consequently, we search for life beyond Earth by drawing analogies to "life as we know it." Similarly, we can contemplate possible habitats for extraterrestrial life by examining the varied habitats of life on Earth, where "extremophiles" thrive in environments ranging from deep-sea hydrothermal vents to high altitude mountaintops, from frozen Antarctic polar deserts to the cores of nuclear reactors. This volume grapples with the challenges of discovering life beyond Earth on the basis of life *on* Earth, placing the contemporary search for extraterrestrial life in larger historical, cultural, and scientific contexts. *Astrobiology, History, and Society* examines the history of the idea of extraterrestrial life through a combination of broad overviews and in-depth case studies, while also exploring the varied societal dimensions of discovering life beyond Earth.

There has been a notable lack of consensus about the defining characteristics of good analogies for making inferences about extraterrestrial life. As we see repeatedly in this volume, even the most direct observations must be interpreted—sometimes leading different investigators to radically divergent conclusions. Consider, for example, late eighteenth and early nineteenth-century theories of the habitability of the Sun and the Moon, as seen through the work of William Herschel and Georg Wilhelm Friedrich Hegel. Both Herschel and Hegel allowed for other planets of our solar system to be inhabited due to their shared characteristics with the Earth, but they came to opposite conclusions about the habitability of the Sun and the Moon. Even when they agreed on important criteria for comparisons, they largely relied on different evidence and gave differing weight to disanalogies that contradicted their conclusions.

When Herschel looked at the Sun, he saw both clouds and landmasses. Seeing sunspots as patches of the Sun's solid surface visible through its otherwise bright layer of clouds, Herschel argued the Sun has both solidity and atmosphere, providing a favorable environment for life. So too, Herschel reasoned, is the Moon much like the Earth. Both are massive bodies with mountains and valleys, whose inhabitants would experience environmental changes from season to season, as well as from day to night. The laws of gravity apply equally on these two bodies, Herschel noted, and the Moon has its own satellite in the sky above: the Earth.

The disanalogies Herschel saw between the Moon and Earth—the former lacking oceans, atmosphere, and rain-bearing cloud cover—did not dissuade him from imagining the Moon to be inhabited; the differences would merely yield selenites who are markedly different from humans.

In Hegel's cosmology, the Sun is made of light, while the Moon is made of fire. Only planets have solid cores, he reasoned. Moreover, neither the Sun nor the Moon has clouds, argued Hegel, indicating they contain no air or water. This lack of meteorological processes for Hegel indicated the absence of a dynamic system capable of supporting life on either the Sun or Moon. In contrast, the Earth and other planets are solid bodies with atmospheres, which are also alike in their motions, rotating around their own axes while also revolving around the Sun. This reflects a complex combination of motions fitting for an inhabited world, in Hegel's view. The Sun merely rotates about its own axis, but revolves around nothing else—making it unlike the Earth, and thus uninhabited.

While modern astrobiologists do not count on finding life on either the Sun or the Moon, the case for other planetary bodies is less clear-cut. Consider, for example, the history of efforts to determine whether Mars is a life-bearing planet, as seen through the intricate interplay of advancing scientific understanding and increasing technological capabilities. Viewed as a "wandering star" by the ancients who observed the night sky with the naked eye, over the millennia Mars was reconceived as one of several planets circling the Sun, along with Earth. This heliocentric model of our solar system became widely accepted in the seventeenth century, supported by increasingly accurate measurements of the motions of heavenly bodies as well as concepts of physics that challenged the distinct division between celestial and terrestrial realms, as illustrations later in this book remind us (Figs. 1.2 and 2.1). At the turn of the twentieth century astronomers sketched quite divergent maps of the surface of the red planet, giving rise to a debate about whether the apparent lines were natural features, canals manifesting the work of Martian engineers, or merely optical illusions (Fig. 1.5). As the twentieth century progressed, larger telescopes brought the surface of Mars closer to our comprehension, while the same instruments allowed spectroscopic studies of the Martian atmosphere, leading to a better understanding of its chemical composition (Fig. 7.2).

Viewed as a twin planet to Earth, in recent centuries Mars has often been seen as the other planet in our solar system most likely to be habitable. Contemporary astrobiologists draw analogies between present-day Earth and the red planet in an earlier era, when water flowed freely, as these scientists note parallels between the riverbeds cut into the Martian surface and similar features in satellite photos of Earth's surface. NASA has "followed the water" in its exploration of Mars, hoping that the link between water and life we see on Earth today can provide a guide for finding extraterrestrial life. If some day scientists find conclusive evidence of life beyond Earth—even in the form of long extinct microbes on our neighboring planet—this discovery would profoundly affect our understanding of our place in the universe.

Prior to the advent of the Space Age, humankind made its most direct contact with Mars only by serendipitous discoveries of meteorites, ejected from their home world by impacts so powerful they could break free of Martian gravity and make their way to Earth across the distances of interplanetary space. In recent years, the most notable example has been meteorite ALH 84001, which was formed on Mars about four billion years ago, when liquid water may have flowed on the planet, and eventually arrived on Earth. An examination of its interior with electron microscopy showed structures that were reported in 1996 as potentially being fossilized microbes—a view now rejected by a majority of scientists (Fig. 7.4).

As humankind has begun exploring other planets via spacecraft, remote sensing of planetary features has been complemented in some missions by onboard laboratories. After NASA's two Viking landers reached the surface of Mars in 1976 and began sending data back to Earth (Fig. 7.3), results that initially seemed to indicate the presence of Martian biochemistry were later largely seen as being caused by inorganic chemical processes. In 2008, when the Phoenix lander discovered perchlorates on the Martian surface, once again a minority of scientists argued this discovery provided additional support that the Viking experiments had detected life; the Viking experiments that showed no evidence of organics when soil samples were heated might be due to their reactions with perchlorates, it was argued, not because the organics were absent. The spectroscopic analysis of Mars, once conducted solely from Earth-based observatories, entered a new phase in August 2012, when Curiosity, the Mars Science Laboratory Rover, began examining the composition of rocks by studying the spectra they emit upon being bombarded by a series of laser pulses (Fig. 1).

History provides important lessons about the need for astrobiologists to be open to reinterpreting data as more is learned. To return to the mid-twentieth-century spectroscopic analysis of Mars mentioned above, the data initially interpreted as evidence for Martian vegetation was explained a half dozen years later by the same experimenter as being due to a form of water (deuterated hydrogren) in the Earth's atmosphere. In other cases, beliefs once held are not so easily let go of—at least by the scientists initially making the claims. In another example from roughly the same time period, early efforts to use the astrometric method of planet detection led to claims that Barnard's star was accompanied by a planet 1.6 times the size of Jupiter (Fig. 7.5). It took decades before this claim was demonstrated conclusively to be spurious.

An examination of such episodes can provide insights into the nature of scientific discovery, and this book provides a series of in-depth case studies of interest to astrobiologists and historians of science alike. The episodes detailed in this volume also allow the broader public to appreciate better the incremental nature of scientific progress, where interpretations of data can reverse as new observations and insights become available. The more clearly scientists and other scholars can articulate the ambiguities and uncertainties involved in the normal course of doing science, the better we can help prepare people from all walks of life as they follow the latest reports about efforts to find life beyond Earth.

Throughout the history of the search for extraterrestrial life, we have seen tantalizing suggestions of the impact that its discovery might have. For several days in 1835, people around the world thought intelligence had been directly observed on the Moon's surface, as the New York *Sun* recounted the latest discoveries of

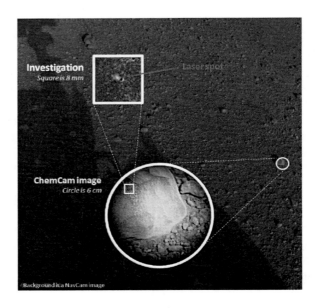

Fig. 1 Curiosity, NASA's Mars Science Laboratory, conducted the first spectroscopic laser test of a rock on an extraterrestrial planet on August 19, 2012, less than two weeks after this rover landed on the surface of Mars. The *widest* context view in this composite image was taken by Curiosity's Navigation Camera (NavCam). The *circular* inset, taken by the Chemistry and Camera instrument (ChemCam), shows a portion of the fist-sized rock named Coronation, which was hit by 30 laser pulses, each lasting five one-billionths of a second and delivering more than a million watts of power. By using ChemCam's three spectrometers to examine the glowing plasma from the laser-excited atoms in the rock, scientists were able to study Coronation's chemical composition. The *square* insert at the *top* of the image shows a higher magnification of the area repeatedly targeted by the laser. Image credit: NASA/JPL-Caltech/LANL/CNES/IRAP

renowned astronomer John Herschel using a 24-foot diameter telescope at the Cape of Good Hope in South Africa. Among the most astounding discoveries reported was the existence of lunar "bat people," or *Vespertilio-homo* (Fig. 1.4). In truth, this series in a daily newspaper was intended by its author as a satire about overly enthusiastic advocates of the plurality of worlds. As with Orson Welles' radio broadcast of *War of the Worlds* just over a century later, an unsuspecting public took these media accounts as true.

Might such historical incidents provide illuminating analogies for an actual detection of extraterrestrial life? As the chapters in this volume show, the analogies we might first think of are not necessarily the most useful. When historian Bruce Mazlish sought an appropriate analogy for understanding the development of the space program, he bypassed one obvious alternative—the Age of Exploration—and opted instead to draw an analogy to the development of the railroad (Yerxa and Mazlish 2009, 61). Similarly, scientists involved in the Search for Extraterrestrial Intelligence (SETI) have looked beyond voyages of exploration for analogies of contact with other civilizations, arguing that scenarios for contact at interstellar distances via radio transmissions are quite different from the direct physical contact involved in European expeditions to the New World. A more apt analogy, they argue, acknowledges the possibility of unidirectional transmission of knowledge across generations, akin to the transmission of ancient Greek writings through the ages, safeguarded and translated by medieval Islamic scholars, and eventually contributing knowledge critical for Europe's Renaissance (Finney and Bentley 1998).

Could such analogies help us anticipate the future? An early proponent of using encounters between *Homo sapiens* and Neanderthals as an analogue for human—extraterrestrial contact also suggested that the best analogies are discovered by starting with multiple analogies, then systematically eliminating the less favorable ones (Ascher 1961, 322–323; Ascher and Ascher 1963, 308). Indeed, forecasters have shown that experts who make judgments based on multiple analogies—identifying the single best analogy from the possibilities—can markedly improve their predictions (Green and Armstrong 2007). Some social scientists value prediction, not necessarily as a means to say with certainty what will happen in the future, but to identify some possibilities that are more likely to occur than others. For example, sociologists and psychologists have identified widespread patterns in people's attitudes about life beyond Earth, such as the tendency for more religious people to be more skeptical about the existence of extraterrestrial life (Bainbridge 2011; Pettinico 2011).

In contrast, historians have been loathe to attempt to predict the future through analogies with the past. Familiar with the vicissitudes of historical events and mindful of the multiple societal forces involved, historians warn against efforts to anticipate the future. As the chapters in the final section of this volume highlight, historians and anthropologists tend to be cautious about predicting the precise reactions of humankind to the possible discovery that we are not alone. The contingencies of historical circumstance and cultural context, as well as the difficulty of anticipating the precise nature of contact, make such anticipations inherently imprecise, they argue.

By definition, analogies make use of inexact comparisons, which necessarily limit our certitude about the conclusions we draw. Nevertheless, the use of analogies can be illuminating—often in unexpected ways. To return to Mazlish's *The Railroad and the Space Program*, while that project was limited in its ability to provide specific, concrete predictions about the impact of space exploration, the comparison proved especially fruitful in providing new perspectives on the development of railroads (Fischer 1970, 257). Similarly, in this volume we see ways that analogies involving extraterrestrial life can help us better understand life on Earth—even if we are left uncertain about whether life exists *beyond* Earth. For example, an examination of the analogies that have been drawn between terrestrial evolution and possible mechanisms of biological change on other worlds has highlighted key elements of evolutionary theory. By studying the history of the interrelationship of evolutionary theory and estimates of the likelihood of life beyond Earth, we can better chart the varied receptions of evolutionary theory within diverse scientific specialties in the twentieth century.

Other chapters in this volume argue that we may make significant advances by seeking relevant heuristics, rather than constructing grand theories of contact. Though we may not be able to predict the specific responses that people will have upon discovering life beyond Earth, analogies can provide us with insights into how we understand life and intelligence that is radically "other." Our views of "other life" are influenced not only by science, but also by science fiction and popular culture. We may not be able to anticipate the precise reactions of Earth's diverse religious traditions to discovering life beyond Earth, but even here, historical examinations of theological reflections on the plurality of worlds may help understand the range of possible responses within traditions (Vakoch 2000). The complete reactions of the populace may be impossible to anticipate with accuracy, but astrobiologists who understand how the news will be disseminated in a world of Twitter and Facebook will be better prepared to communicate clearly and effectively about scientific results that may not be as conclusive as the public would wish. And as we are reminded repeatedly in this book, we must be wary of anticipating the global response to the discovery of extraterrestrial life by relying solely on studies of Europeans and North Americans. Even the means we use to conduct a valid study may vary from nation to nation, with structured questionnaires asking about the participants' beliefs coming across as difficult to understand and complete in some cultures. Alternative approaches, such as ethnographic studies and in-depth interviews, provide critical opportunities for researchers to build trust, while yielding rich understandings of cultural context.

Astrobiology, *History*, *and Society* is divided into three parts, with each part beginning with an overview essay that includes a preview of the other chapters in that part. The first part covers the plurality of worlds debate from antiquity through the nineteenth century, while part II covers the extraterrestrial life debate from the twentieth century to the present. The final part examines the societal impact of discovering life beyond Earth. Throughout the book, authors draw links between their own chapters and those of other contributors, emphasizing the interconnections between the various strands of the history and societal impact of the search for extraterrestrial life.

Contemporary astrobiologists seek answers to questions previously asked, albeit in different form, by theologians and natural philosophers. As estimates of the prevalence of extraterrestrial life have swung like a pendulum through the ages, thoughtful individuals have contemplated the significance of competing alternatives: that life exists only on Earth, or that it is also found elsewhere in the universe. Astrobiologists search with increasingly sophisticated instruments, using the methods of modern science in place of pure reason or revelation, seeking evidence of both habitable environments and organisms that have evolved independently from life as we know it. By understanding the historical and social context of the modern search for life beyond Earth, we can better appreciate the cultural, ideological, and scientific factors that make it plausible at one time to imagine the universe brimming with life, and mere decades before or after, to see the cosmos as devoid of life, except for that found on Earth. By gaining a greater perspective on the nature and significance of this search as a whole, we are better prepared to sort through the many challenging and sometimes ambiguous questions facing today's astrobiologists.

Douglas A. Vakoch

References

Ascher, Robert. 1961. "Analogy in Archaeological Interpretation." *Southwestern Journal of Anthropology* 17 (4): 317–325.

Ascher, Robert, and Marcia Ascher. 1963. "Interstellar Communication and Human Evolution." *In Interstellar Communication: A Collection of Reprints and Original Contributions*, ed. A. G. W. Cameron, 306–308. New York: W. A. Benjamin, Inc.

Bainbridge, William Sims. 2011. "Cultural Beliefs about Extraterrestrials: A Questionnaire Study." In *Civilizations Beyond Earth: Extraterrestrial Life and Society*, ed. Douglas A. Vakoch and Albert A. Harrison, 118–140. New York: Berghahn Books.

Finney, Ben, and Jerry Bentley. 1998. "A Tale of Two Analogues: Learning at a Distance from the Ancient Greeks and Maya and the Problem of Deciphering Extraterrestrial Radio Transmissions." *Acta Astronautica* 42 (10-12): 691–696.

Fischer, David Hackett. 1970. *Historians' Fallacies: Toward a Logic of Historical Thought*. New York: Harper & Row.

Green, Kesten C., and J. Scott Armstrong. 2007. "Structured Analogies for Forecasting." *International Journal of Forecasting* 23: 365–376.

Pettinico, George. 2011. "American Attitudes about Life beyond Earth: Beliefs, Concerns, and the Role of Education and Religion in Shaping Public Perceptions." In *Civilizations Beyond Earth: Extraterrestrial Life and Society*, ed. Douglas A. Vakoch and Albert A. Harrison, 102-116. New York: Berghahn Books.

Vakoch, Douglas A. 2000. "Roman Catholic Views of Extraterrestrial Intelligence: Anticipating the Future by Examining the Past." In *If SETI Succeeds: The Impact of High Information Contact*, ed. Allen Tough, 165-174. Bellevue, WA: Foundation For the Future.

Yerxa, Donald A., and Bruce Mazlish. 2009. "From Psychohistory to New Global History: An Interview with Bruce Mazlish." In *Recent Themes in World History and the History of the West: Historians in Conversation*, ed. Donald A. Yerxa, 60-66. Columbia, SC: University of South Carolina Press.

Acknowledgments

To the contributors of chapters that appear in this volume, I especially appreciate the depth and innovation of the research they share here. These authors deserve special thanks for engaging one another in dialogue during the process of writing their works, as reflected in the numerous cross-references throughout the volume. I am grateful to Karen Anderson for so capably recording our thoughts during an informal meeting where many of us gathered to identify common themes across our chapters. I am pleased that many longtime colleagues involved in astrobiology worldwide were able to participate in this project, along with other researchers relatively new to the field.

Over the past 15 years, many colleagues from the SETI Institute have shared with me their insights into astrobiology, as well as the ways we can best communicate this work to the broader public. I especially thank Molly Bentley, Anu Bhagat, John Billingham, Edna DeVore, Frank Drake, Sophie Essen, Andrew Fraknoi, John Gertz, Gerry Harp, Jane Jordan, Ly Ly, Chris Munson, Chris Neller, Tom Pierson, Karen Randall, Jon Richards, Pierre Schwob, Seth Shostak, and Jill Tarter.

More recently, I warmly acknowledge the administration, faculty, staff, and students of the California Institute of Integral Studies (CIIS), especially for support from Joseph Subbiondo, Judie Wexler, and Tanya Wilkinson. The work of editing this volume was made possible through a generous sabbatical leave from my other academic responsibilities at CIIS in the spring of 2012. In addition, I thank Harry and Joyce Letaw, as well as Jamie Baswell, for their intellectual and financial contributions to promoting space exploration.

I thank the NASA Astrobiology Science Conference 2012 co-chairs Loren Williams and Eric Gaucher, as well as local organizing committee members Sue Winters and Elizabeth Wagganer, for their support of conference sessions in which several of the chapters in this volume first appeared in nascent form.

For encouraging me to develop this project and then shepherding the book through the editorial process at Springer, I am indebted to Ramon Khanna. Also at Springer, Tamara Schineller and Charlotte Fladt have my gratitude for helping to move the book swiftly and efficiently into production. My appreciation for faithfully overseeing all aspects of copyediting, layout, and indexing goes to S.A. Shine David of Scientific Publishing Services.

Finally and most importantly, to my wife Julie Bayless, I am grateful in more ways that I can or will share here. Thank you, forever.

Douglas A. Vakoch

Contents

8   The Creator of Astrobotany, Gavriil Adrianovich Tikhov 175

Danielle Briot

9   Life Beyond Earth and the Evolutionary Synthesis 187

Douglas A. Vakoch

**13 Cultural Resources and Cognitive Frames: Keys to an
Anthropological Approach to Prediction**   259

Ian Lowrie

14 The Detection of Extraterrestrial Life: Are We Ready?   271

Klara Anna Capova

**15 Impact of Extraterrestrial Life Discovery for Third World Societies:
Anthropological and Public Health Considerations**   283

M. Margaret Weigel and Kathryn Coe

18 Christianity's Response to the Discovery of Extraterrestrial Intelligent Life: Insights from Science and Religion and the Sociology of Religion................................... 329

Constance M. Bertka

19 Would the Discovery of ETI Provoke a Religious Crisis? 341

Ted Peters

Part I
The Early Extraterrestrial Life Debate

Chapter 1
The Extraterrestrial Life Debate from Antiquity to 1900

Michael J. Crowe and Matthew F. Dowd

Abstract This chapter provides an overview of the Western historical debate regarding extraterrestrial life from antiquity to the beginning of the twentieth century. Though schools of thought in antiquity differed on whether extraterrestrial life existed, by the Middle Ages, the Aristotelian worldview of a unified, finite cosmos without extraterrestrials was most influential, though there were such dissenters as Nicholas of Cusa. That would change as the Copernican revolution progressed. Scholars such as Bruno, Kepler, Galileo, and Descartes would argue for a Copernican system of a moving Earth. Cartesian and Newtonian physics would eventually lead to a view of the universe in which the Earth was one of many planets in one of many solar systems extended in space. As this cosmological model was developing, so too were notions of extraterrestrial life. Popular and scientific writings, such as those by Fontenelle and Huygens, led to a reversal of fortunes for extraterrestrials, who by the end of the century were gaining recognition. From 1700 to 1800, many leading thinkers discussed extraterrestrial intelligent beings. In doing so, they relied heavily on arguments from analogy and such broad principles and ideas as the Copernican Principle, the Principle of Plenitude, and the Great Chain of Being. Physical evidence for the existence of extraterrestrials was minimal, and was always indirect, such as the sighting of polar caps on Mars, suggesting similarities between Earth and other places in the universe. Nonetheless, the eighteenth century saw writers from a wide variety of genres—science, philosophy, theology, literature—speculate widely on extraterrestrials. In the latter half of the century, increasing research in stellar astronomy would be carried out, heavily overlapping with an interest in extraterrestrial life. By the end of the eighteenth century, belief in intelligent beings on solar system planets was nearly universal and certainly more common than it would be by 1900, or even today. Moreover, natural theology led to most religious thinkers being comfortable with extraterrestrials, at least until 1793

M. J. Crowe (✉) · M. F. Dowd
University of Notre Dame, South Bend, IN, USA
e-mail: crowe.1@nd.edu

D. A. Vakoch (ed.), *Astrobiology, History, and Society*, Advances in Astrobiology
and Biogeophysics, DOI: 10.1007/978-3-642-35983-5_1,
© Springer-Verlag Berlin Heidelberg 2013

when Thomas Paine vigorously argued that although belief in extraterrestrial intelligence was compatible with belief in God, it was irreconcilable with belief in God becoming incarnate and redeeming Earth's sinful inhabitants. In fact, some scientific analyses, such as Newton's determination of the comparative masses and densities of planets, as well as the application of the emerging recognition of the inverse square law for light and heat radiation, might well have led scientists to question whether all planets are fully habitable. Criticism would become more prevalent throughout the nineteenth century, and especially after 1860, following such events as the "Moon Hoax" and Whewell's critique of belief in extraterrestrials. Skepticism about reliance on arguments from analogy and on such broad metaphysical principles as the Principle of Plenitude also led scientists to be cautious about claims for higher forms of life elsewhere in the universe. At the start of the twentieth century, the controversy over the canals of Mars further dampened enthusiasm for extraterrestrials. By 1915 astronomers had largely rejected belief in higher forms of life anywhere in our solar system and were skeptical about the island universe theory.

1.1 Introduction

The door that inadvertently let in the aliens was opened in 1543. It was in that year that Copernicus published his *De revolutionibus*, the treatise in which he posited a universe centered upon the Sun, about which the Earth and the other planets revolved. This counterintuitive notion of a moving Earth was initially restricted to a finite universe, though one large enough to account for the lack of observed parallax of the stars. Eventually, however, the bounds of the universe would be pushed away from the Sun, our solar system would be relegated to an unremarkable corner of this universe, and all the stars of the sky would become suns in their own right.

In this chapter, we will provide an overview of the history of the extraterrestrial life debate in the Western intellectual context prior to 1900. It is based in large part on two long and fully referenced books by Michael Crowe published in 1986 and in 2008[1]and on a course the two of us have co-taught over the last three years titled "The Extraterrestrial Life Debate: A Historical Perspective." More complete discussion of the themes and authors in this essay can be found in those books.

1.2 Before the Eighteenth Century

Prior to Copernicus, a worldview heavily informed by the work of Aristotle (384–322 B.C.) held sway in Europe. In that worldview, the Earth occupied the middle of the universe. The terrestrial region alone saw life, death, birth, decay, coming

---

[1] Providing full references for all the quotations by and information about the over one hundred authors discussed in the essay would be cumbersome. This has led to our decision to reference most of the quotations and some of the information to these two widely accessible books.

to be, and passing away. Above the sphere of the Moon, which surrounded the terrestrial region, no change occurred except for locomotion—movement in space. In that region, all things were eternal and unchanging, moving forever in cycles that remained inviolate. There was no place except on Earth where life of a terrestrial nature could be found. There was, literally, no place for aliens. By the end of the seventeenth century, however, extraterrestrials were understood to populate an extended universe, a universe teeming with life in much the same way as the Earth was filled with a wide variety of living creatures.

1.2.1 The Ancient and Medieval Periods

It was not always the case that aliens had no place in the universe prior to Copernicus's revolution. Two schools of thought in the ancient world came to opposite conclusions regarding life outside the confines of the Earth. The Atomists, typified by Leucippus (fl. 480 B.C.) and Democritus (d. 361 B.C.), as well as the later Epicurus (d. ca. 270 B.C.) and Lucretius (99–55 B.C.), believed in a cosmos that was infinitely large and had an infinite amount of matter that interacted in random patterns. Periodically, chance collisions could cause a world to form, leading to life, such as the world in which we live. Throughout this infinite space, similar chance events would inevitably lead to other worlds. Thus there was "life out there," but not life with which we would be likely to interact. The "world" we see, that is, the Earth below and the heavens above, did not inevitably hold life, but similar systems surely exist out in the cosmos.

The Aristotelian worldview, on the other hand, understood the physical cosmos to be fully constituted by the objects visible to us. This finite, bounded system, described above, encompassed all that is. No other world could exist, and thus no other life could exist beyond that which we find around us. This latter view became the basis for the European view of the physical cosmos in the Middle Ages. Combined with the mathematical astronomy of Ptolemy, the learned view of the physical world was of a finite cosmos, a unified system in which other inhabitable worlds simply did not fit.

Of course, scholars of the Middle Ages were creative thinkers, and so they did speculate about how our cosmos might in fact be different from this view. Thomas Aquinas (1224–1274), for example, considered in his *Summa Theologica* (part 1, question 47, article 3) the question of whether there could be more than one Aristotelian-type cosmos; he concluded in the negative (Crowe 2008, 18–20). Nicole Oresme (1325–1382) too considered this question, in his *Le livre de ciel et du monde*, speculating whether worlds might exist sequentially in time, or whether multiple worlds might be arrayed throughout space or one within another (Crowe 2008, 22–26). Although Oresme ultimately did not suggest the abandonment of an Aristotelian worldview, he did argue that

> God can and could in His omnipotence make another world besides this one or several like or unlike it. Nor will Aristotle or anyone else be able to prove completely the contrary. But, of course, there has never been nor will there be more than one corporeal world (Crowe 2008, 26).

John Buridan (ca. 1295–1358) had come to a similar conclusion—namely, that God could have made other worlds, even if logically we ought not expect them to exist—though via different arguments (Dick 1982, 29–30).

Such writings demonstrate that medieval authors were analyzing and probing the claims of Aristotle. Even though for the most part the Aristotelian worldview was retained, various aspects of it were questioned. The question of the eternity of the world, for example, was a source of wide-ranging arguments (Dales 1990). Even the motion of the Earth was discussed prior to Copernicus's revolutionary book: Nicholas of Cusa (1401–1461), for example, stated in his *Of Learned Ignorance* that the Earth moves. Moreover, he claimed that the Earth was not so distinct a region as Aristotle claimed, and that in fact the rest of the universe shared attributes of the earthly realm, including the presence of living beings (Crowe 2008, 27–34).

1.2.2 The Copernican Cosmological Shift

Despite the medieval questioning of various aspects of Aristotelian cosmology, however, the overwhelming majority of scholars would have shared the understanding that the universe was basically as Aristotle described it. Thus, in the middle of the sixteenth century, Copernicus found himself with a cosmos that was bounded with a spherical heavens with the Earth in the middle. But the system was not elegant, and longstanding problems with the calendar had yet to be solved. In an effort to address these issues, Copernicus proposed a physical system that placed the Sun in the middle of the cosmos, put the Earth revolving about it, as well as rotating about its own axis, and retained the bounded sphere of the stars.

This system thus made the Earth similar to five other planets: the Earth revolved around the Sun, as did Mercury, Venus, Mars, Jupiter, and Saturn. Setting aside the apparent problems of a terrestrial physics on a swiftly moving Earth for later scholars, Copernicus's cosmology brought about massive shifts in the understanding of the physical nature of the universe. Moving the Earth away from the center of the universe was not by itself of the greatest significance for the theory of extraterrestrial life. It was instead the breaking of the celestial sphere.[2]

As mentioned above, there was no place for aliens in Aristotle's universe. Placing the terrestrial region, the sphere of change, into motion might have opened up other places for aliens, namely, turning other planets into additional regions of change. Indeed, some early heliocentrists considered such schemes, as discussed below. But Copernicus's finite universe, with the Earth occupying a homocentric sphere somewhere between the Sun and the celestial sphere, did not persist for an extended period of time as a physical theory. Instead, the universe would burst open into a multitude of suns and accompanying planets, creating abodes for extraterrestrials throughout the now much larger universe.

---

[2] See Chap. 2 of this volume, "Early Modern ET, Reflexive Telescopics, and Their Relevance Today," by Dennis Danielson for complementary analysis of the changes brought about by the Copernican revolution.

1.2.3 Two Underlying Principles

With the walls of the celestial sphere sundered, and the basic problem that scholars didn't have the ability to directly access through observation—beyond the developing telescope technology—the vast majority of the universe, certain assumptions were made about the nature of the universe. Though not contemporarily labeled as such, we can see two important principles that operated implicitly in many scientists in the following centuries, even up to our current situation: the Copernican Principle and the Principle of Plenitude.

The Copernican Principle, also called the Principle of Mediocrity, was not one to which Copernicus himself ascribed. This principle holds that everywhere else in the universe is basically similar to Earth; alternatively, one can say that the Earth is in no way particularly special. In a weak sense, the scientific process assumes something similar: that what we observe here on Earth, and the laws of nature that we determine from them, will operate similarly elsewhere. But in the extraterrestrial life debate, this principle leads to the assumption that, because life is present here, it will exist in other places. How similar those places must be to Earth will be different for different authors and in different eras. That the other place is a planet circling a star will be enough for some. For others, as we shall see, various considerations such as the average temperature of the planet, the presence of water and an atmosphere, and a rocky surface, for example, will become more important, refining the Copernican Principle to a more stringent restriction on the expectation of life.

The Principle of Plenitude is related to, indeed, overlaps, the Copernican Principle. Frequently leading from theological assumptions, the Principle of Plenitude suggests that the universe ought to be as rich as possible. Because living things best demonstrate the richness that nature can produce, the principle implies that life will be plentiful throughout the universe. This was often based upon the claim that God values life and thus would widely populate the universe, not restrict life to the relatively miniscule environment of the Earth. The theological basis for this principle is explicit in many authors, but is not necessary for the principle, as it is also operative at times in authors who reject a divine principle underlying the physical universe; an appeal to "Nature," for example, can often substitute for a divine force responsible for life.

1.2.4 The Seventeenth Century

The Copernican system was not embraced swiftly, at least not as a physical system.[3] But a few important writers and scientists of the early seventeenth century did adopt some form of the system, particularly a moving Earth orbiting the Sun. By the end of the century, especially after the work of Descartes and Newton gave

---

[3] Westman (1980) identifies only ten Copernicans of the sixteenth century, at 136n6. But see Westman (1975) for ways in which the Copernican system was used as a calculational device without adherence to the physical system.

scientific validity to a heliocentric solar system and an extended universe, the cosmological change initiated by Copernicus had overwhelmingly changed our view of the universe.

An early adopter of heliocentrism was Giordano Bruno (1548–1600). Bruno was a controversial figure in his own day, and remains so today.[4] He lived a somewhat nomadic lifestyle, in part because of his propensity to anger colleagues and local officials, making a prudent withdrawal advisable. His interests were wide ranging: theology, philosophy, cosmology, magic, and mnemonics. He became a Dominican in 1566, but left the order in 1576 under a cloud of suspicion. His life ended after a trial in which he was found guilty of heresy and was burned at the stake. Controversy over his life and the reasons for his condemnation and execution remains, largely because many of the trial documents are lost.[5]

His role in the extraterrestrial life debate centers around his claims that the universe is infinite in extent and filled with life. He defends such a picture of the universe based on its being a truer reflection of the nature and grandeur of God than the bounded Aristotelian universe. A number of his writings touch on the subject. Especially significant are his *La cena de la ceneri* (*The Ash Wednesday Supper*; 1584), *De l' infinito universo e mondi* (*On the Infinite Universe and Worlds*; 1584), *De la causa, principio et uno* (*On Cause, Origin, and Unity*; 1584), and *De immenso et innumerabilibus* (*On the Immense and Innumerable*; 1591). Convinced of the Copernican claim that the Earth orbits the Sun, and decidedly anti-Aristotelian, Bruno posited that the universe is vastly extended in space and that it is filled with systems of stars and planets. In fact, he is the first to claim that the stars are objects like our Sun, and that they are all orbited by planets, just as in Earth's system. Moreover, consistently with the Principle of Plentitude, he argued for an animistic universe, in which life not only populates the stars and planets but in fact characterizes the celestial objects themselves (Crowe 1986, 10).

Two revolutionary astronomers of the seventeenth century, still heralded today as among the most important astronomers in history, were also Copernicans: Galileo Galilei (1564–1642) and Johannes Kepler (1571–1630). Galileo did not write a great deal about extraterrestrials, but he did have a bit to say. In his *Letter on Sunspots* of 1613, Galileo rejected as "false and damnable the view of those who would put inhabitants on Jupiter, Venus, Saturn, and the moon" (Crowe 2008, 52). He reiterates his rejection of extraterrestrials in a letter of 1616 to Giacomo Muti, in which he argues that life cannot exist on the Moon for not only does the Moon lack certain vital elements, namely, earth and water, each part of it also undergoes a cycle of fifteen days of unrelenting sunshine followed by fifteen days of darkness, which would make it impossible for life to subsist there (Crowe 2008, 52–53).

---

[4] For more on Bruno's life and thought, see Yates (1964) and Singer (1950). See also Rowland (2008), but note the critical review of Gregory (2012).

[5] For an account in English of the trial, including what materials we do and do not possess, see Finocchiaro (2002).

Kepler, on the other hand, reached the opposite conclusion. Curiously, he did not argue for an extended universe of the sort Bruno envisioned, and in fact Kepler had strong reasons for maintaining a bounded universe, as such a universe, in his view, better reflected a universe created by a Trinitarian God (Kozhamthadam 1994, 16–18, 29–34) and admitted of a clever geometric proof of heliocentrism via the Platonic solids (see Crowe 2008, 53–55). Nonetheless, Kepler was willing to populate other regions of the cosmos with life. In a 1610 response to Galileo's *Starry Messenger*, Kepler argued that the telescopic discovery of the moons of Jupiter, invisible to the unaided terrestrial eye, demonstrate that God provided those moons for the benefit of the inhabitants of Jupiter. He went on to argue that the Earth has the best place in the cosmos for observing because of its central position among the bodies orbiting the Sun. So whereas Kepler saw humanity in the most privileged position in the finite cosmos, we are not the sole place in which life resides (Crowe 2008, 59–64).

Bruno's idea of an unbounded universe remained suspect due to its association with a person tried and sentenced to death for heresy. In the hands of a leading philosopher of the seventeenth century, however, the idea became more palatable. René Descartes (1596–1650) proposed in his *Principles of Philosophy* (1644) a system in which matter is extended throughout space. Running up against one another were vortices, whirlpools of ethereal matter, with stars in their center and planets being carried along in the eddies of the moving plenum (see Fig. 1.1). Just as our planet had become one planet circling the Sun, so did our solar system become one of many such systems (Dick 1982, 106–112). Descartes himself did not, however, boldly assert whether extraterrestrials exist, stating in a letter of 1647, "although I do not at all infer … that there would be intelligent creatures in the stars or elsewhere, I also do not see that there would be any reason by which to prove that there were not" (Crowe 1986, 16). Still hedging, he made a slightly more positive statement in a letter to Frans Burman where he wrote, "An infinite number of other creatures far superior to us may exist elsewhere" (Hennessey 1999, 37).

Another cosmological system appeared near the end of the century that would eventually displace the Cartesian universe: the Newtonian system. Isaac Newton, in his *Philosophiae naturalis principia mathematica* of 1687, also known by the abbreviated title of the *Principia*, proposed a system in which "forces," rather than a plenum, were responsible for planetary systems and the movement of their constituent objects. Moreover, Netwon's system united terrestrial and celestial physics into a unified whole. The impact of the Newtonian understanding of nature can hardly be overstated, and it would become the dominant view of the physical cosmos. It too posited an extended universe with numerous star-planet systems. Though Newton did not make strong positive statements about the existence of extraterrestrials, the system could be sympathetic to them, as we shall see below.

Numerous authors of the seventeenth century worried at the problem of extraterrestrials. Tackling the issue from the Cartesian point of view were authors such as Henry Regius (1598–1679), Jacques Rohault (1620–1675), Henry More (1614–1687), and Pierre Borel (ca. 1620–1671), among others, some arguing for extraterrestrials, some against (Crowe 1986, 16–18; Dick 1982, 112–120). Pierre

Fig. 1.1 A diagram of the
Cartesian vortices from
Descartes' *Le Monde* (1664)

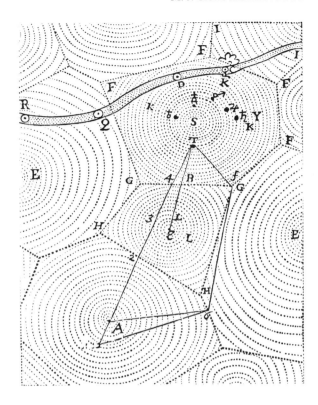

Gassendi (1592–1655) and Otto von Guericke (1602–1686) applied the Principle
of Plentitude to the question of extraterrestrials existing on other planets in our
solar system as well around other stars (Crowe 1986, 17–18; Hennessey 1999,
38–39). Throughout the latter half of the century, then, extraterrestrials were gaining intellectual traction.

1.2.5 Reaching a Wider Audience

Undoubtedly one of the most engaging and influential authors to bring the question of extraterrestrial life into the limelight was Bernard le Bovier de Fontenelle
(1657–1757). Fontenelle wrote deliberately for a wide, popular audience—including women—when he composed his *Entretiens sur la pluralité des mondes*
(*Conversations on the Plurality of Worlds*), published in 1686. The book was quickly
translated from its original French into a number of languages, including Danish,
Dutch, English, German, Greek, Italian, Polish, Russian, Spanish, and Swedish. The
book presents a dialogue between a philosopher and a curious marquise, in which the
philosopher presents Copernican and Cartesian ideas to this intellectually able woman.

While doing so, Fontenelle also incorporates a great deal of discussion about extraterrestrials. Not only does he assert their existence, he also discusses the characteristics they must have, which differ based on their environment. The inhabitants of Mercury, for example, "are so full of Fire, that they are absolutely mad." Moreover, they are used to a much greater intensity of light and heat than any inhabitant of Earth. But, the philosopher assures the marquise, they would be perfectly at home in their own environment, even one that we would find unbearable, because "Nature never gives life to any Creature, but where that Creature may live; then thro' Custom, and ignorance of a better Life, those people may live happily" (Crowe 2008, 79–80).

Fontenelle was cognizant of the appeal of the work and its musing on extraterrestrials. In an addition made to the book in 1687, he added the following summary of important arguments of the work:

> [1] the similarities of the planets to the earth which is inhabited; [2] the impossibility of imagining any other use for which they were made; [3] the fecundity and magnificence of nature; [4] the consideration she seems to show for the needs of their inhabitants as having given moons to planets distant from the sun, and more moons to those more remote; and [5] that which is very important—all that which can be said on one side and nothing in the other (Crowe 2008, 73).

Note that we can see both the Copernican Principle and the Principle of Plentitude at work here. The Earth and our solar system are taken as typical representatives of what the rest of the universe must be like, and, due to the fecundity of nature, we can expect other places to be filled with life with characteristics appropriate to the circumstances in which they find themselves.

Another author of the late seventeenth century also had a great influence on the extraterrestrial life debate, this time a highly esteemed scientist: Christiaan Huygens (1629–1695). Published posthumously in 1698, his *Cosmotheoros*, or, as the English version has it, *The Celestial Worlds Discover'd: or, Conjectures Concerning the Inhabitants, Plants and Productions of the Worlds in the Planets*, was an extended treatise sprinkled with pieces of scientific information, which provide a springboard for conjectures about the inhabitants of other planets. He too makes use of the Copernican Principle. In some cases, he seems to be defending the principle. For example, in discussing whether Jupiter has water, he asserts that such a claim "is made not improbable by the late Observations" that the darker areas of the planet change, implying the presence of clouds on the planet. He thus has tried to demonstrate that the assumption of water's presence is reasonable. But for the most part, Huygens simply assumes the principle is true, making a number of assertions that seem unwarranted. In the case of water on Jupiter, for example, he discusses how, due to the planet's distance from the Sun, the water on Jupiter must in some way be different from that on Earth such that it can maintain its liquid form in colder temperatures (Crowe 2008, 91–92). He makes similar arguments about animals, in this case arguing that just as on Earth we find different animals in different regions of the planet, so too will we find different animals on different planets. Eventually, Huygens comes to the conclusion that the planets house rational creatures in some degree similar to us, both in the senses by which they learn about the world and the sciences they must have developed.

1.2.6 Extraterrestrials and Religion at the End of the Seventeenth Century

Throughout the medieval and early modern periods, natural philosophy, what we most closely identify with science, was linked with considerations of God's relationship to the physical world. Up to this point, however, we have not concentrated our discussion on the religious implications and arguments surrounding the idea of extraterrestrial life. But it had played a role in the debate, as the brief mentions above regarding Bruno and Kepler, for example, make plain. Also significant is that, again using those two authors as examples, religious arguments could be used for different purposes. Bruno posited that God's nature suggests an extensive universe filled with life, whereas Kepler envisioned a harmonious, bounded universe reflective of God's order. As extraterrestrials became a more common assertion, the religious implications of their existence also had to be considered.

An early example of the way extraterrestrials could be used in the religious context is revealed in the writings of Richard Bentley (1662–1742). Bentley was commissioned in 1692 to deliver a series of sermons under funding from Robert Boyle given for the purpose of "proving the Christian religion" (Crowe 2008, 115). A devotee of Isaac Newton, Bentley incorporated the Newtonian system into these sermons, which meant he had to address the ways in which this system, far different from the traditional Aristotelian universe, was compatible with and indeed supportive of the Christian religion.

In the eighth sermon of the series, Bentley discussed the issue of extraterrestrial life. Repeating a claim similar to Kepler's, Bentley argues that recent telescopic discoveries of celestial objects invisible to the unaided human eye demonstrate that portions of the universe are not made for the benefit of humans. Because God does not act in vain, those objects must have been made for the purpose of other beings.

> It remains therefore, that all Bodies were formed for the sake of Intelligent Minds: and as the Earth was principally designed for the Being and Service and Contemplation of Men; why may not all other Planets be created for the like Uses, each for their own Inhabitants which have Life and Understanding? (Crowe 2008, 117–118).

Moreover, recognizing that different planets might have very different environments, Bentley asserts that God would create those extraterrestrials so that they were of a nature proper to their own situation, not unlike the claim Fontenelle had made.

By the end of the seventeenth century, then, the size and nature of the universe had changed drastically from what it had been: from a bounded, geocentric universe in which life was restricted to the terrestrial region to a vastly extended universe of multiple solar systems that, despite a lack of observational evidence, were assumed to teem with planets and life, similar to ours in some ways but perhaps significantly different in the details. Extraterrestrials populated the new universe, and intellectuals of the following centuries would expend a great deal of effort to say more about them.

1.3 The Extraterrestrial Life Debate in the Eighteenth Century

Let us begin with a broad and important claim: Nearly half the leading intellectuals of the eighteenth century (sometimes labeled the Enlightenment) discussed extraterrestrial life issues in their writings. Our basis for this claim is an examination done about a decade ago of eight anthologies of Enlightenment thought, taking the list of authors included in each as an indication of the editor's judgment as to who were the leading Enlightenment thinkers. This study (Crowe 2008, xvii–xviii) found that at least 41% of the authors included by each of the eight editors met this criterion; moreover, in five of the eight cases, the percentage exceeds 53%. This test seems adequate to justify the claim that not only many, but, in fact, close to a majority of Enlightenment intellectuals actively engaged in a debate that many persons currently assume first arose in their own lifetimes. It is noteworthy that the percentages found in this study are all lower bounds, which can only increase as future research finds extraterrestrials in the pages of other authors in these anthologies. Moreover, it is significant that relatively few scientists and no astronomers appear in these anthologies; had at least the latter been included, the percentages would have risen significantly.

The materials that follow not only demonstrate how successfully extraterrestrials have propagated, but also their extraordinary adaptability. Hundreds of authors have found ways to adapt them to their thought and writings. So pliable is the doctrine of a plurality of worlds, or pluralism, that almost any author can create extraterrestrials suited to his or her system. Fiction writers beset by such limitations as characters with a single head or civilizations with only two sexes, or frustrated by the small scale of merely terrestrial catastrophes, have long since learned that extraterrestrials can rescue their writings from such restrictions. Writers from almost every era, and nearly all the authors discussed in this survey, have succeeded in creating extraterrestrials suited to their needs.

1.3.1 The Period from 1700 to 1750

During the first half of the eighteenth century, numerous authors discussed the question of a plurality of worlds, most embracing the positive side. Space limitations prohibit mentioning all, so we will focus on illustrative cases. Among these, one who especially effectively illustrates the tendencies of the time is Rev. William Derham (1657–1735), who was a key figure in linking pluralism with Newtonian science and with natural theology, as well as dissociating it from the astronomy of the previous century. While serving in 1714 as chaplain to the future King George II, Derham published his *Astro-Theology, or A Demonstration of the Being and Attributes of God from a Survey of the Heavens*, a volume that by 1777 had attained fourteen English and six German editions. As the title suggests, its goal was to illustrate the power and goodness of God through astronomy, which fits very well with pluralism. Simultaneously, it championed the Newtonian system. Pluralism permeates the book, adding to its attractiveness and religious appeal. Derham's indebtedness to Huygens's *Cosmotheoros* did not preclude his arguing against the Dutchman's waterless and lifeless Moon. A particularly important feature of the book is that it delineates three world systems, these being the "Ptolemaick," which he rejects, the "Copernican," which he accepts, but only as a precursor to his "New Systeme." This third system includes the Copernican, but goes beyond it by the supposition that "there are many other Systemes of *Suns* and *Planets*, besides that in which we have our residence; namely, that every Fixt Star is a sun and encompassed with a Systeme of Planets, both Primary and Secondary as well as ours" (Crowe 2008, 121) (see Fig. 1.2). One of his arguments for this system is that it "is far the most magnificent of any; and worthy of an infinite CREATOR" (Crowe 2008, 123). That Derham should be seen more as symbol than as the source of the increasing acceptance of pluralism in early eighteenth century is shown by mention of Thomas Burnet (1635–1715), John Ray (1628–1705), and Nehemiah Grew (1641–1712), all of whom in the three decades before Derham's *Astro-Theology* had endorsed pluralism in physicotheological treatises.

Various universities taught the doctrine of a plurality of worlds, for example, Oxford. At that esteemed establishment, students learned of pluralism from David Gregory (1661–1708), the Salvilian professor of astronomy, as well as from his eventual successor in that chair, John Keill (1671–1721). Whereas Gregory was

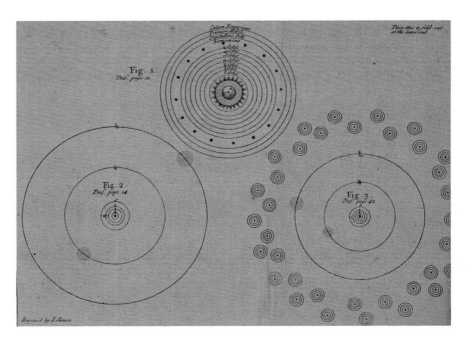

Fig. 1.2 Derham's diagram of his third system. Reproduced from the original held by the Department of Special Collections of the University Libraries of Notre Dame

noncommittal in his *Astronomiae physicae et geometricae elementa* (1702), Keill was outspoken, claiming in his *Introductio ad veram astronomiam* (1718) that it is "no ways probable" that God would create stars without placing planets around them "to be nourished, animated, and refreshed with the Heat and Light of these Suns" (Crowe 1986, 30–31). Oxford's Savilian professor of geometry at this time was Edmond Halley (1656–1742), who in 1720 became Astronomer Royal. Not only did Halley endorse pluralism, he argued from his statement that all planets "are with Reason suppos'd habitable" to the possibility that habitable globes exist beneath the Earth's surface. He had proposed subterranean globes to explain apparent shifts in the Earth's magnetic poles, but was delighted to add concerning their habitation: "Thus I have shew'd a possibility of a much more ample Creation, than has hitherto been imagin'd" (Crowe 1986, 31).

At Cambridge University William Whiston (1667–1752), who in 1702 succeeded Newton as Lucasian professor of mathematics, repeatedly advocated pluralism. As early as his *New Theory of the Earth* (1696), Whiston urged that other planets and planetary systems have inhabitants subject to moral trials. Two decades later in his *Astronomical Principles of Religion* Whiston extended his pluralism by proposing denizens dwelling in the interiors of the sun, planets, and comets; moreover, he posited "Not wholly Incorporeal, but Invisible Beings" living in planetary atmospheres (Crowe 1986, 31). In the intervening years, Whiston advanced the Newtonian system by publishing his Cambridge astronomical lectures wherein he identified stars as suns, but in 1710 he lost the Lucasian chair because of charges of religious heterodoxy, in particular, of Arianism. Nonetheless, he continued to lecture on astronomy at various locations and to spread the pluralist doctrine.

One such location was Button's coffeehouse in London, where Whiston's lectures in 1715 introduced Alexander Pope (1688–1744) to the new pluralist universe (Crowe 1986, 31). Two decades were to pass before Pope presented pluralism in his *Essay on Man*, but from the time of its publication, wherever or in whatever language that poem was read, Pope's message that the "proper study of mankind" must include consideration of extraterrestrials was brought before the public. Pope's poem sets out this point in the following lines, which had such appeal to Thomas Wright and Immanuel Kant that they quoted them when in the 1750s they published their theories of the universe:

> He, who thro' vast immensity can pierce,
> See worlds on worlds compose one universe,
> Observe how system into system runs,
> What other planets circle other suns,
> What vary'd Being peoples ev'ry star,
> May tell why Heav'n has made us as we are
>   (Crowe 1986, 31).

Of all eighteenth-century English poems using scientific materials, only one rivaled and possibly surpassed Pope's *Essay on Man* in popularity. This was *Night Thoughts*, published between 1742 and 1745 by the already sixtyish rector of Welwyn, Edward Young (1683–1765). Translated into French, German, Italian, Magyar, Portuguese, Spanish, and Swedish, it was, according to one of Young's biographers, "for more than a hundred years … more frequently reprinted than

probably any other book of the eighteenth century" (Crowe 1986, 84). Young divided
Night Thoughts into nine "Nights," in the last of which he makes a final attempt
to reform the libertine Lorenzo by means of "A Moral Survey of the Nocturnal
Heavens." Whereas other physicotheological poets were finding in flora and fauna
the finest field for proving the Deity, Young above all seeks Him in the celestial:

> Devotion! Daughter of Astronomy!
> An undevout astronomer is mad.
> True; all things speak a God; but in the small
> Men trace out Him; in great, He seizes man…
>   (Crowe 2008, 200).

Young's universe was thoroughly pluralistic, being constructed, if not from
Derham, at least on Derham's "New Systeme":

> One sun by day, by night ten thousand shine,
> And light us deep into the Deity…
>   (Crowe 2008, 200).

This pluralistic universe raises a host of questions, scientific and religious, concerning the extraterrestrials encountered late in "Night Ninth":

> What 'er your nature, this is past dispute,
> Far other life you live, far other tongue
> You talk, far other thought, perhaps, you think,
> Than man. How various are the works of God?
> But say, what thought? Is Reason here Inthroned,
> And absolute? Or Sense in arms against her?
> Have you two lights? Or need you no reveal'd?….
> And had your Eden an abstemious Eve?….
> Or if your mother fell, are you redeemed?….
> Is this your final residence? If not,
> Change you your scene, translated? or by death?
> And if by death; what death? Know you disease?
>   (Crowe 2008, 200–201).

A final excellent illustration of the debate about extraterrestrials in the first half
of the eighteenth century involves Russia. Intent to drag mother Russia into the
modern world, Peter the Great (1672–1725) encouraged translation of various
European books into Russian. The need for this is evident in the fact that before
1717 no published exposition in Russian of the Copernican system existed. This
situation was remedied not by one of the classic volumes of Copernicus, Galileo,
Kepler, or Newton, but rather by Peter having Jacob Bruce prepare a translation of
Huygens's *Cosmotheoros*. After Bruce completed his translation in the mid-1710s,
Peter the Great ordered Mikhail Avramov to publish it. Avramov upon reading the
treatise found it a work of "Satanic perfidy" (Crowe 1986, 158). Decades later
he described his dilemma in a letter to the more traditional Empress Elizabeth:
"concealing his godless, frenzied, and atheistic heart, Bruce praised the book by
the delirious author—Kristofer Huiens, … pretending that it was very clever and
wholesome for the educating of all the people, … and with such habitual and
godless flattery deceived the sovereign." Avramov was not however duped; as he

explained: "I examined this book which was contrary to God in all ways, and with my heart quaking and my soul overawed, I fell before the mother of God with the sobbing of bitter tears, frightened to publish and frightened not to publish" (Crowe 1986, 158). The daring decision he made was to subvert the Tsar's order by printing only thirty copies of the book and making efforts to conceal these. Such was the entry of pluralism into Russia, a country that for some decades was the most active in radio searches for extraterrestrials.

By 1750, pluralism had been championed by an array of authors, including some of the most prominent figures of the age. Presented with exceptional appeal by Fontenelle, given legitimacy in scientific circles by Huygens and Newton, reconciled to religion by Bentley and Derham, set to poetry by Pope and Young, integrated into philosophical systems by Berkeley and Leibniz, taught in textbooks by Wolff and in taverns by Whiston, the idea of a plurality of worlds was winning international acceptance. This transformation was revolutionary. Nevertheless, from the perspective of the present its foundation was frail, resting as it did on analogical arguments of dubious force, on metaphysical principles such as the Principle of Plenitude and Copernican Principle, and on a scattering of astronomical observations. However that may be, the era of the extraterrestrial had begun and would continue even unto the present.

1.3.2 The Sidereal Revolution of the 1750–1800 Period

The single most important eighteenth-century development in astronomy took place in the 1750–1800 period and centered on four authors. This revolutionary development was nothing less than the founding of stellar astronomy. Moreover, it will be suggested that these four authors—Thomas Wright, Immanuel Kant, Johann Lambert, and William Herschel—were not only heavily committed to extraterrestrial life ideas, but also that in a number of ways these ideas were significantly involved in their contributions to stellar astronomy.

The sidereal revolution, second in significance for astronomy only to the Copernican Revolution, opened astronomy to the regions beyond our solar system and culminated in the 1920s. During its founding period it encompassed three conclusions: (1) that the Milky Way is an optical effect arising from the scattering of millions of stars over a roughly planar area, (2) that these stars form a giant disk-shaped structure many light years in diameter, and (3) that many of the thousands of nebulous patches seen in the heavens are galaxies comparable in magnitude to our own Milky Way galaxy.

The chief publication of Thomas Wright (1711–1786), an Englishman with no university training, is his *An Original Theory or New Hypothesis of the Universe* (1750), a volume rich in both illustrations from Wright's hand and speculations from his active imagination. Both reflect his desire not only to work out the physical structure of the universe but also to integrate it with the spiritual. In the course of his book, Wright attains the first of the three fundamental notions of sidereal astronomy, but—despite decades during which he was credited with the second—he failed to attain the disk structure, although he does propose three wrong models, the last of which he formulated late in his life. This consists in the idea that the Milky Way's appearance is due to the overflow of a ring of volcanoes inside the starry vault. Regarding the idea that the nebulae are island universes, some attribute it to him, whereas others dispute this (Crowe 1986, 45–47). A careful reading of Wright's volume, which is filled with mention of extraterrestrials, suggests this his enthusiasm for them played a major role in this romp through the sidereal realm. Among dozens of quotations he takes from such authors as Bruno, Fontenelle, Newton, Huygens, Derham, and Pope supporting the plurality of worlds, Wright describes this idea as having "ever been the concurrent Notion of the Learned of all Nations" (Crowe 1986, 43). His pious purposes appear when his calculation that possibly 170,000,000 inhabited globes exist within "our finite view" spurs him (along with the Copernican Principle) to reflect: "In this great Celestial creation, the Catastrophy of a World, such as ours, or even the total Dissolution of a System of Worlds, may possibly be no more to the great Author of Nature, than the most common Accident in Life with us" (Crowe 2008, 135). This deistic remark in a book containing no mention of Christ stimulates Wright to comment: "This Idea has something so cheerful [!] in it, that I own I can never look upon the Stars without wondering why the whole World does not become Astronomers" (Crowe 2008, 136).

Immanuel Kant (1724–1824), although standing in maturity barely five feet two and never weighing more than a hundred pounds, is now recognized as a giant of thought who boldly explored the depths of the human mind and the heights of the heavens. Trained at Königsberg in Leibnizian and Wolffian thought, exposed simultaneously to Newtonianism, so devoted to Lucretius's *De rerum natura* that he memorized long passages, the young Kant was well prepared to publish in 1755 his *Allgemeine Naturgeschichte und Theorie des Himmels*, the first book to set out all three claims constitutive of sidereal astronomy. Nevertheless, the situation is more complex in that Kant had read a review in a German journal of Wright's 1750 volume and was led by the review to credit Wright with the disk theory of the Milky Way, which as mentioned above Wright had never attained. The review did not, however, mention Wright's speculation that nebulae may be other universes, a conceptual leap Kant himself attained. And not least, Kant lays out in his volume an early version of the nebular hypothesis, which was part of his concern for working out cosmic evolution.

Kant's book is in three parts, the above developments being set out in the first two along with some ideas of extraterrestrial life. For example, in part two, he turns to the Principle of Plenitude:

> Now, it would be senseless to set Godhead in motion with an infinitely small part of his creative ability and to imagine his infinite force, the wealth of a true inexhaustibility of nature and worlds to be inactive and locked up in an eternal absence of exercise. Is it not much more proper, or to say better, is it not necessary to represent the very essence of creation as it ought to be, to be a witness of that power which can be measured by no yardstick? For this reason the field of the manifestation of divine attributes is just as infinite as these themselves are (Crowe 2008, 139).

In the third part, centered on the solar system, Kant makes clear the extent of his commitment to extraterrestrials. He effectively sets out its fundamental purpose and methodology by describing part three as "an essay on a comparison, based on the analogies of nature, between the inhabitants of the various planets" (Crowe 2008, 140). Relying on various assumptions, including the Great Chain of Being, and the idea that one can infer the nature of a planet's inhabitants from the type of matter dominant on the planet, Kant asserts that the planetarians *become more excellent and perfect in proportion to the distance of their habitats from the sun*" (Crowe 2008, 145). Kant's Mercurians and Venusians are consequently dullards, whereas his earthlings occupy "exactly the middle rung … on the ladder of beings" and his Jovians and Saturnians are greatly superior beings. As he states: "From one side we saw thinking creatures among whom a man from Greenland or a Hottentot would be a Newton, and on the other side some others who would admire him as [if he were] an ape" (Crowe 2008, 148).

As a treatise on extraterrestrials, Kant's book is rather remarkable: almost never before and rarely since did an author advocate such widespread life in both the solar and sidereal systems. The book reveals that the philosopher, later so famous for his critiques of speculative systems, wrote not without having experienced their siren call. That Kant later realized that he had allowed his imagination to roam too far in the 1750s is indicated by the fact that when in 1791 J. F. Gensichen prepared an edition of Kant's book to accompany the German translation of three astronomical papers by William Herschel, he noted that Kant had forbidden him to include any materials beyond his fifth chapter because of their excessively hypothetical character.

The third pioneer of sidereal astronomy was Johann Heinrich Lambert (1728–1777), whom Kant in 1765 described as "the greatest genius in Germany" (Crowe 1986, 56). He was a brilliant mathematician, physicist, and philosopher, who was the first person to discern the structure of the Milky Way. Although Lambert did not publish the disk theory until 1761 when he brought out his *Cosmologische Briefe uber die Einrichtung des Weltbaues*, he had attained that idea in 1749. As Lambert recounted to Kant in 1765, his *Cosmological Letters* had its origin one evening in 1749 when "from my window I looked at the starry sky, especially the Milky Way. I wrote down on a quarto sheet the idea that occurred to me then, that the Milky Way could be viewed as an ecliptic of the fixed stars, and it was that note I had before me when I wrote the *Letters* in 1760" (Crowe 1986, 56). It is a complex question whether Lambert attained the notion of nebulae as other universes, but it is indisputable that this book was as richly stocked with extraterrestrials as those of Wright and Kant. Lambert in 1765 stated to Frederick the Great that the public considers his *Letters* as a second volume "to the one on plurality of worlds by Fontenelle" and an Italian physics professor, Rev. Giuseppe Toaldo, praised it as the most beautiful of pluralist books (Crowe 1986, 56).

If one compares the books of Kant and Lambert, one finds that only the former was concerned with cosmogony, whereas Lambert unlike Kant was above all interested in comets and their inhabitants. A knowledgeable Newtonian, Lambert was aware that comets move either in elliptical orbits or in nonrecurring parabolic or hyperbolic paths. Lambert incorporates the latter paths into his system by proposing that such comets pass from one system to another and are populated by astronomically interested extraterrestrials blessed with an atmosphere that by expanding at certain times preserves livable conditions on these comets. Lambert's passion for populating the millions of comets in his system as well as his overall pluralism derived from the teleological approach everywhere evident in his volume. He also relies heavily on the Principle of Plenitude and a version of the Copernican Principle. Nonetheless, unlike Kant, Lambert favored a finite universe.

William Herschel at one point described a version of Lambert's book as "full of the most fantastic imaginations" (Crowe 1986, 57). Nonetheless, far more frequently than Kant, Lambert labels particular claims as speculative and attempts to specify the degree of credibility each deserved. His book (in some cases in a condensed form) had by 1801 been translated into French, Russian, and English. One suspects that the interest in Lambert's book may have derived more from its ideas about extraterrestrials than from its pioneering views in stellar astronomy. Ironically these three pioneers of stellar astronomy may have had no more concern about stars than their contemporaries; it was rather inhabited planetary systems that interested them and that they sought to arrange into systems. The Milky Way for them was not primarily a giant array of glowing globes, but rather a visible collection of sources of heat and light serving the myriads of beings living on the admittedly unobservable planets of the stellar systems. Kant's dull Mercurians and super Saturnians and Lambert's cometary astronomers moving from one solar system to another are now seen as bizarre companions to the more durable doctrines developed in their books. Nevertheless, eighteenth-century readers may have had the opposite view.

Among the four pioneers of sidereal astronomy, Sir William Herschel (1738–1822) ranks first in importance; indeed, he is arguably the most influential astronomer of modern times. When his career is looked at broadly enough to include his role in the extraterrestrial life debate, he emerges not primarily as a tireless telescopic technician and model empiricist seeking to learn what his extraordinary telescopes would teach him. Rather it appears (1) that Herschel was at various times a speculatively inclined celestial naturalist, quixotically caught up in a quest for evidence of extraterrestrials; (2) that many of his efforts make most sense when seen as attempts to transform pluralism from being a delight of poets, a doctrine of metaphysicians, and a dogma of physicotheologians into a demonstration of astronomers; and (3) that pluralism was a core component in Herschel's research program and as such influenced a number of his astronomical endeavors, especially but not exclusively in his formative years.

Born in Hanover, Herschel lacked a university education but, after immigrating to England in the 1750s, established himself as a musician, composer, and music teacher, working in various locations, especially in Bath, where he settled with his sister Caroline, who also contributed to astronomy. Until well into his forties, Herschel worked as a professional musician. One of the key factors that led Herschel to turn from music to astronomy occurred in 1773, when he purchased and began reading *Astronomy Explained upon Sir Isaac Newton's Principles* by James Ferguson (1710–1776), a popular itinerant science writer, whose book scholars believe significantly influenced Herschel. Ferguson's book teaches some elementary astronomy and shows much enthusiasm for a plurality of worlds.

Ferguson's universe includes "Thousands of thousands of Suns ... attended by ten thousand times ten thousand Worlds ... peopled with myriads of intelligent beings, formed for endless progression in perfection and felicity" (Crowe 1986, 60). Ferguson focuses chiefly on the solar system, concerning which he notes that the outer planets are provided with extra moons, and in the case of Saturn with a ring, to illuminate the nights of their inhabitants. He stresses that "by the assistance of telescopes we observe the Moon to be full of high mountains, large valleys, and deep cavities. These similarities leave us no room to doubt but that all the Planets and Moons in the System are designed as commodious habitations for creatures endowed with capacities of knowing and adoring their beneficent Creator" (Crowe 2008, 173). Influenced by such readings, Herschel began to construct telescopes and to observe the heavens. The abilities he showed in this undertaking were recognized in 1781 when he became world famous for discovering the planet Uranus.

Striking evidence in Herschel's unpublished manuscripts shows that Herschel may well have thought at that time that he had made an even more important discovery five years earlier. Among the earliest of Herschel's lunar observations is that dated 28 May 1776, when with a newly acquired telescope, he reports that while observing the Moon,

> I was struck with the appearance of something I had never observed before, which I ascribed to the power and distinctness of my Instrument, but which perhaps may be an optical fallacy—I believed to perceive something which I immediately took to be *growing substances*. I will not call them Trees as from their size they can hardly come under that denomination, or if I do, it must be understood in that extended signification so as to take in any size how great soever.... My attention was chiefly directed to Mare humorum, and this I now believe to be a forest, this word being also taken in its proper extended signification as consisting of such large *growing substances*. (Crowe 2008, 177–178).

Herschel's ambivalence about these observations led him in late 1778 to compose a new analysis, which shows that by then Herschel believed he had evidence not only of forests but also of lunar towns. Suspecting that lunar craters may be the towns of the lunarians, he remarks:

> Now if we could discover any new erection it is evident an exact list of those Towns that are already built will be necessary. But this is no easy undertaking to make out, and will require the observation of many a careful Astronomer and the most capital Instruments that can be had. However this is what I will begin (Crowe 1986, 65).

Herschel continued his search for evidence of lunar life at least until 1783, finding "roads," a "city" and much else, but never secured observations that he deemed conclusive enough for publication. It is clear from his writings that he believed all the planets are inhabited; for example, he refers periodically to the inhabitants of such planets as Mars and Uranus, and eventually published two papers (treated later) advocating life on the Sun.

With this as background, we can turn to Herschel's very important contributions to sidereal astronomy. One of Herschel's most important achievements was his observation of nebular patches, then called nebulae. When Herschel began observing them regularly around 1782, just over a hundred were known, these having been listed by Charles Messier, who was one of the few astronomers of that period interested in them. In fact, Messier's chief interest was comets. What led to Messier's interest in nebulae was that they resemble comets in appearance, which led Messier to make a catalogue of nebulae so that astronomers would not mistake them for comets. Herschel, typically using giant reflecting telescopes that he himself had constructed, managed over the course of his career to discover nearly 2,500 more nebulae. What seems to have led him to this effort was that by 1784 he had formulated the disk theory of the Milky Way and proceeded from this to claim that the nebulae are other universes comparable to the Milky Way. It is within this context that one should view the exclamation of the poetess Fanny Burney who in 1786 visited Herschel: "he has discovered fifteen hundred universes! How many more he may find who can conjecture?" (Crowe 1986, 67). It is true that by 1791 Herschel had backed away somewhat from the island universe theory because he had concluded that some nebulae consist of a "shining fluid" rather than being a cluster of a vast number of stars. Such studies by Herschel served to launch stellar astronomy, which many decades later came to be the most important area of astronomy. As in the case of other founders of sidereal astronomy, it appears that the claim that inhabited planets orbit stars spurred interest in the stellar regions. In Herschel's case, it seems very possible that in building some of his telescopes, he was initially motivated more by hope of detecting evidence of extraterrestrials than by an interest in observing nebulae, in which objects his contemporaries took little interest.

In the eighteenth century, publications in stellar astronomy represented perhaps five percent of the astronomical literature. It should not be assumed that other leading astronomers were not continuing to advance astronomy in other areas. Nor should it be assumed that they were not concerned about extraterrestrials. Johann Elert Bode (1747–1826), director of the Berlin Observatory and editor of one of the leading astronomical journals of the period published so widely on extraterrestrials that his position has been labeled "un panpopulationnisme cosmique" (Crowe 1986, 73). Active especially on the observational level, for example, searching for lunarians was Johannes Schröter (discussed later). One indication of his passion for sighting terrestrial features elsewhere in the solar system is his report of sighting mountains on the rings of Saturn (Crowe 1986, 72). The leading mathematical astronomer of the period, Pierre Simon Laplace (1747–1827), also came out forcefully for pluralism as did his French contemporary Jérôme Lalande (1732–1827).

1.3.3 At the End of the Eighteenth Century

During the period from 1600 to 1800, pluralism, viewed in relation to the scientific community, passed through a remarkable transformation. Many seventeenth-century scientists—Galileo, Descartes, Newton (in his published writings)—had viewed it as a speculation at the limits of legitimate science. During the eighteenth century, however, an international array of astronomers, including Wright, Lambert, Herschel, Bode, Schröter, Laplace, and Lalande, embraced extraterrestrials; in fact, for some pluralism formed an integral part of their research program.

Moreover, while scientific journals published pluralist papers, astronomical texts and university courses regularly treated this topic. In short, pluralism made extraordinary progress with practitioners of the most ancient science.

If, however, we examine eighteenth-century pluralism from the perspective of the present, it is clear that its conjectural component was large. Broad analogy more than detailed astronomy, physicotheology more than physics, and teleology more than telescopes had been used to erect a vast edifice on what nineteenth-century scientists gradually found to be a frail foundation. Although by the eighteenth century, the medieval cosmos with its crystalline spheres, angelic planetary movers, and associated metaphysical and mythical elements had been discredited, it should not be forgotten that the cosmos championed by many Enlightenment figures was not free of comparable associations. Carl Becker did not discuss pluralism when in his *Heavenly City of the Eighteenth-Century Philosophers* he maintained that the Enlightenment "*philosophes* demolished the Heavenly City of St. Augustine only to rebuild it with more up-to-date materials" (Crowe 1986, 161), but he might have found in such authors as Wright, Kant, Lambert, and Herschel evidence to support a parallel claim for the heavens of Enlightenment astronomy. Crystalline spheres had disappeared, but Halley's inhabited subterranean spheres and Herschel's cool solar nucleus became available. Disputes about hierarchies of angels may have been abandoned, but a number of authors debated whether the superbeings of the solar system live on the outer planets, inner planets, or on the Sun itself. Moreover, although some authors analyzed the question of other worlds in largely scientific terms, the majority, explicitly or implicitly, invoked religious or metaphysical considerations.

In the case of science, so also in regard to religion: serious difficulties lay just beneath the surface. The very success of the natural theological enterprise, whether practiced by Christians or deists, tended to emphasize "Nature's God" while downplaying the idea of an incarnated redeemer. Structures of insects or solar systems may evidence God's existence, but they are mute as to a messiah. Furthermore, pluralist physicotheology set off in even starker relief the radical nature of the Christians' claim. Why would the God of all worlds select an insignificant planet for his most remarkable actions? In short, whereas by the 1790s pluralism had reached a rapprochement with theism, tensions with Christianity had not as yet been fully faced. This is at least the conclusion suggested by the sensation that resulted when in the 1790s Thomas Paine launched a vigorous attack on Christianity on a pluralist basis.

In 1793, the rapprochement worked out between extraterrestrials and many religious writers began to shatter as thousands of people read a book written by Thomas Paine (1737–1809). Entitling his book *The Age of Reason*, Paine argued that astronomical science had made it impossible for any thinking person to accept the central Christian notions of a divine incarnation and redeemer. In his book, Paine recounts that James Ferguson, a popular and pious lecturer on astronomy, had convinced him that a good and generous God must have populated the Moon and planets. When Paine confronted Christianity with this astronomical claim, he became a deist, that is, a person accepting a remote, impersonal God, but denying such central Christian doctrines as Christ's incarnation and redemption. In his book Paine argues that although the existence of intelligent life only on the Earth is not a specific Christian doctrine, it is nonetheless "so worked up therewith from ... the story of Eve and the apple, and the counterpart of that story—the death of the Son of God, that to believe otherwise ... renders the Christian system of faith at once little and ridiculous" (Crowe 2008, 224). Paine presses the same point in even stronger language by asking:

> From whence ... could arise the ... strange conceit that the Almighty ... should ... come to die in our world because, they say, one man and one woman had eaten an apple! And, on the other hand, are we to suppose that every world in the boundless creation had an Eve, an apple, a serpent, and a redeemer? In this case, the person who is irreverently called the Son of God, and sometimes God himself, would have nothing else to do than to travel from world to world, in an endless succession of death, with scarcely a momentary interval of life (Crowe 2008, 229).

Paine's conclusion was stark: either reject belief in extraterrestrial life—a doctrine that he claimed had been established by astronomy—or reject Christianity.

Paine's *Age of Reason* attracted an immense readership both in Britain, where 60,000 copies of it were printed, and in America, where a single Philadelphia bookshop sold over 15,000 copies. It also generated more than fifty published responses, some explicitly opposing Paine's extraterrestrial life attack on Christianity.

1.4 The Extraterrestrial Life Debate in the First Half of the Nineteenth Century

1.4.1 Some Considerations from Science

As of 1800, the level of belief in extraterrestrial intelligences was higher than it had ever been before or than it would ever be again. Educated Europeans and Americans believed that wherever celestial objects might be, there also must be extraterrestrial subjects. It is true that not a shred of satisfactory direct empirical evidence confirmed this conviction, but powerful principles—chiefly the Copernican Principle and Principle of Plenitude—supported this belief. So also, it was thought, did such observations as lunar mountains and whitish Martian polar caps. Not only itinerant lectures but also prestigious university professors and eloquent preachers shared this exciting news with their audiences. It is true that Paine's *Age of Reason* (1793) caused some ripples, but chiefly it set Christian religious writers to searching for ways to reconcile this exciting message with traditional Christian belief in an incarnated redeemer.

On one level at least, the widespread confidence shown by early nineteenth-century intellectuals in the existence of extraterrestrials is surprising. In particular, by the early 1800s various scientific results were available, some of which scientists had attained more than a century earlier, that presented, one would think, serious difficulties for those wishing to populate the planets. For example, already in the late seventeen century, Newton in his *Principia* had used his law of universal gravitation to determine the relative masses and densities of the Sun, Earth, Jupiter, and Saturn and also the relative force of gravitation on the surface of each of these objects. (His calculation depended on the fact that each of these bodies has a body in orbit around it.) His results revealed immense diversity among these objects. For example, he found that it would take 169,282 Earth masses to match the Sun's mass. Similarly, it would take 3,021 Saturns or 1,067 Jupiters to match the Sun's mass. This suggested, one would think, the problematic character of claims that these giant bodies are analogous to the Earth and hence probably also inhabited. Even more striking was Newton's calculation that if one assigns a density of 100 to the Sun, the density of Jupiter is 94.5, Saturn 67, and Earth a whopping 400. Even more problematic is that if an earthling weighs 435 units, a Saturnian would weigh 529, a Jupiterian 943, and a solarian 10,000 units. This indicated that were a human transported to the Sun, the human's weight would increase by a factor of 235.[6]

If one turns from gravitational considerations to optical and thermal ones, the situation becomes more complex historically, but no less striking. Consider first of all light, especially the fundamental photometric result that the amount of light radiating from a point source decreases according to the inverse square law. Thus if a piece of paper is placed one unit from a point source of light radiating in all directions, then moving the paper twice as far away leads to its receiving only one fourth as much light per unit area at that distance. At ten units distance, the amount of light drops to 1/100 of the original intensity. The inverse square law for light propagation was well known at least from 1720 when Pierre Bouguer published experiments demonstrating its correctness with Johann Lambert in 1760 adding further evidence (Mach n.d., 13–17). This makes it easily possible to determine that Saturn, being about 9.5 times farther from the Sun than the Earth, must receive about 90 times less light per unit surface area than our Earth. This would seem to present a problem for proponents of Jupiterians; instead Christian Wolff writing in 1735 saw it as an opportunity. Setting the distance of Jupiter as 26/5 times farther from the Sun than the Earth and using the inverse square law and calculating what pupil diameter would be necessary in the eye of a Jupiterian to see as much as we do and assuming that pupil diameter would enable him to calculate the height of a typical Jupiterian, Wolff calculated that Jupiterians must be extremely tall (Crowe 1986, 30). Some were impressed by this calculation, but Voltaire viewed it as so absurd that he satirized it in his famous *Micromégas* (1752) (Crowe 1986, 120–121).

Regarding the amount of heat each planet receives from the Sun, the situation was still more complicated, partly because the nature of heat was still under discussion in 1800, around which time William Herschel discovered infrared rays. It does seem correct to say that by the middle of the nineteenth century scientists were fully aware that the amount of heat from our Sun reaching any planet would, like the amount of light, drop off at a rate governed by the inverse square law.

Severe as these problems may seem to us, proponents of inhabited planets rarely took them seriously, or at most (say in the case of light) viewed them as an explanation of why God had provided Saturn with a ring.[7] One example of an

---

[6] For Newton's text and a full explanation, see Crowe (2006, 194–99).

[7] James Ferguson and Emanuel Swedenborg and a number of scientists made this claim. See Crowe (2008, 172, 218).

author who used these considerations against extraterrestrials was Thomas Young, who, commenting in 1807 on William Herschel's arguments for life on the Sun, suggested that the Sun's great mass would make human-sized solarians weigh over two tons (Crowe 1986, 168).

In the 1830s, arguably the most prominent astronomer on our planet was John Herschel (1792–1871). In 1833, he published his *Treatise on Astronomy*, which was the most respected presentation of astronomy available in English. Regarding the heat/light problem discussed above, he states: "The intensity of solar radiation is nearly seven times greater on Mercury than on the earth, and on Uranus 330 times less; the proportion between these two extremes being that of upwards of 2000 to one" (J. Herschel 1833, 277–288). Regarding the gravity issue, he declared that "the intensity of gravity, or its efficacy in ... repressing animal activity on Jupiter is nearly three times that on Earth, on Mars not more than one third, and on the four smaller planets probably not more than one twentieth; giving a scale of which the extremes are in the proportion of sixty to one" (Herschel 1833, 273). Regarding the density issue, he states that the density of Saturn is about one eighth of the Earth's, "so that it must consist of materials not much heavier than cork." All this does not lead him to back away from extraterrestrials but to remark on "what immense diversity must we not admit in the conditions of that great problem, the maintenance of animal and intellectual existence and happiness, which seems ... to form an unceasing and worthy object of the exercise of the Benevolence and Wisdom which presides over all" (Herschel 1833, 273). As we shall see, twenty years later, one of Herschel's contemporaries analyzed these factors in quite a different way.

1.4.2 Some Prominent Philosophers

A telling illustration of the overconfidence many had in the existence of extraterrestrials comes from consideration of two prominent philosophers known for their empiricist orientation, who nevertheless wrote about extraterrestrials in a very speculative manner—without apparently realizing it. The philosophers were Sir John Herschel and Auguste Comte.

In 1833, Herschel, as noted previously, published his *Treatise on Astronomy*. At that time, British intellectuals held Herschel in high esteem not only as a scientist but also as an expert on scientific method, the subject of his *Preliminary Discourse on the Study of Natural Philosophy* (1830). In its opening paragraph, he urges the prospective scientist to "strengthen himself ... for the unprejudiced admission of any conclusion which shall appear to be supported by careful observation and logical argument, even should it prove of a nature adverse to notions he may have previously formed for himself ... without examination, on the credit of others. Such an effort is, in fact, a commencement of that intellectual discipline which forms one of the most important ends of all science" (Crowe 2008, 238). In his second paragraph, he stresses that this restraint is especially necessary in astronomy, because "Almost all its conclusions stand in open and striking contradiction with those of superficial and vulgar observation, and with what appears to every one, until he has understood and weighed the proofs to the contrary, the most positive evidence of his senses" (Crowe 2008, 239). Shortly thereafter, he offers the following example: "The planets, which appear only as stars somewhat brighter than the rest, are to (the astronomer) spacious, elaborate, and habitable worlds; several of them vastly greater and far more curiously furnished than the earth he inhabits ... and the stars themselves properly so called ... are to him suns of various and transcendent glory—effulgent centres of life and light to myriads of unseen worlds" (Crowe 2008, 239). The actual situation was very different: Herschel's claims regarding planets being "spacious, elaborate, and habitable worlds" and stars being "effulgent centres of life and light to myriads of unseen worlds" was hardly empirical information revealed by telescopes nor were these beliefs "adverse to (popular) notions." In fact, such were among the popular notions of the day.

The case of Auguste Comte (1798–1857), famous as the founder of positivism, is comparable. In the second volume of his *Cours de philosophie positive* (1830–1842), Comte sets out a systematic analysis on empiricist grounds of the methodology of astronomy, the only science that Comte believed had as yet transcended the theological and metaphysical stages of development (Comte 1968, 2, 1). Despite its advanced state, Comte prescribes further purification for astronomy. For example, he warns astronomers against excessive speculation concerning the heavenly bodies: "we will never by any means be able to study their chemical composition or their mineralogical structure" (Crowe 2008, 312). The imprudence of this pronouncement, coming but a few decades before the development of spectroscopy, has often been noted. So has Comte's lack of insight in rejecting sidereal astronomy. What has not been noted is that Comte's strictures are not extended to the most highly speculative area of astronomy, the question of extraterrestrials. Yet one finds in the same discussion Comte's assertion that "If, what is highly probable, the planets provided with an atmosphere, as Mercury, Venus, Jupiter, etc. are in fact inhabited, we can regard those inhabitants as being in some fashion our fellow citizens, since from what is a sort of common fatherland, there ought of necessity result a certain community of thought and even of interests, whereas the inhabitants of other solar systems will be complete strangers" (Crowe 2008, 316).

Comte carries his critique of astronomy even farther in his *Systeme de politique positive*, the first volume of which appeared in 1851, five years after the discovery of the planet Neptune. In that volume Comte criticizes astronomers for their interest in remote regions of the solar system, lamenting the "mad infatuation ... which some years ago possessed, not only the public, but even the whole group of Western astronomers, on the subject of the alleged discovery (of Neptune), which, even if real, would not be of interest to anyone except the inhabitants of Uranus" (Crowe 2008, 317). One cannot but be struck by the inconsistency of Comte in banishing the stars and Neptune from astronomy, while embracing, without supporting arguments, the existence of Uranians. The explanation of the latter anomaly may lie in his use of pluralism in criticizing theology. Comte excoriates in this context those who see astronomy as allied with religion, "as if the famous verse 'The Heavens declare the glory of God' had preserved its meaning. It is, however, certain that all true science is in radical and necessary opposition to all theology." Moreover, he adds that for those familiar with the true philosophy of astronomy, "the heavens declare no other glory than that of Hipparchus, of Kepler, of Newton, and of all those who have cooperated in the establishment of laws" (Crowe 2008, 317). In particular, what has shown the unacceptability of theology is the realization that the Earth, rather than being the center of the universe, is only a secondary body circling the Sun "of which the inhabitants have entirely as much reason to claim a monopoly of the solar system, which is itself almost imperceptible in the universe" (Crowe 2008, 318). If it is correct that Comte's enthusiasm for extraterrestrials arose from a belief that their existence invalidated any theology in which humanity has a primacy, then it appears legitimate to suggest that Comte's own astronomy may not have transcended even the anti-theological stage.

Repeatedly in the history of the extraterrestrial life debate one encounters arguments based on analogies. It is interesting that various philosophers addressed this issue, sometimes directly in regard to extraterrestrial life issues. The first of these was Étienne de Condillac in his *La logique* (1780), who distinguished between different sorts of analogy, for example, those of resemblance and those where cause and effect are involved (Crowe 1986, 137). John Stuart Mill (1807–1873) in his famous *System of Logic* (1843) provided a much fuller discussion, centered on the question of life on the Moon and planets. In the course of his extended discussion, two of the most important points Mill makes are these: (1) Regarding the planets, he remarks "when we consider how immeasurably multitudinous are those of their properties which we are entirely ignorant of, compared with the few which we know, we can attach but trifling weight to any considerations of resemblances in which known elements bear so inconsiderable a proportion to the unknown" (Crowe 1986, 231), and (2) regarding analogies in general, he cautions that their chief value lies "in suggesting experiments or observations that may lead to positive conclusions" (Crowe 1986, 231). In more modern parlance, Mill's suggestion is that analogies are primarily involved in the logic of discovery rather than the logic of verification. A very perceptive remark came in 1885 from Charles Sanders Peirce (1839–1914), who remarked "There is no greater nor more frequent mistake in practical logic than to suppose that things that resemble one another strongly in some respects are any more likely for that to be alike in others," adding that "any two things resemble one another just as strongly as any two others, if recondite resemblances be admitted" (Crowe 1986, 552).

1.4.3 Religious Issues

During the first half of the nineteenth century, religious issues arose repeatedly in the extraterrestrial life debate. It should not, however, be assumed that the nature of the debate took the form of religious writers resisting belief in extraterrestrials.

The situation was far more complicated. By 1800, numerous authors had concluded that belief in God was not only reconcilable with the existence of extraterrestrials, but that belief in an omnipotent and omnibeneficent God implies life elsewhere. The Principle of Plenitude had not diminished in popularity or power. Authors of books on natural theology regularly featured extraterrestrials. On the other hand, in the period after Paine's *Age of Reason* (1793), tensions were felt in regard to Christianity with its notions of a divine incarnation and redemption. To put it differently, tensions arose in some cases between natural theology and revealed theology.

1.4.3.1   Three Authors in the Paine Tradition

Early in his education, the famous romantic poet Percy Bysshe Shelley (1792–1822) became convinced of the existence of extraterrestrials; in fact, one Shelley scholar has argued that it was astronomy that led Shelley to forsake Christianity (Evans 1954, 69). This fits with one of Shelley's early poems, *Queen Mab* (privately printed 1813), where he employed extraterrestrials against central Christian claims. At one point in his poem, he refers to "Innumerable systems," adding that the plurality of worlds makes it

> impossible to believe that the Spirit that pervades this infinite machine begat a son upon the body of a Jewish woman; or is angered at the consequences of that necessity, which is a synonym of itself. All this miserable tale of the Devil, and Eve, and an Intercessor, with the childish mummeries of the God of the Jews, is irreconcilable with the knowledge of the stars. The works of His fingers have born witness against Him (Crowe 2008, 234).

Shelley greatly expanded this argument in a satirical essay, "On the Devil, and Devils," which he wrote later in his short life but which was sufficiently shocking that it was withheld from publication for many years (Crowe 2008, 235–239).

Examination of the early diaries of John Adams (1735–1826), second president of the United States, indicate that already by 1756 he had forsaken traditional Christianity. As he explained in his diary:

> Astronomers tell us … that not only all the Planets and Satellites in our Solar System, but all the unnumbered Worlds that revolve round the fixt Starrs are inhabited…. If this is the Case all Mankind are no more in comparison of the whole rational Creation of God, than a point to the Orbit of Saturn. Perhaps all these different Ranks of Rational Beings have in a greater or less Degree, committed moral Wickedness. If so, I ask a Calvinist [Christian], whether he will subscribe to this Alternitive [*sic*], "either God almighty must assume the respective shapes of all these different Species, and suffer the Penalties of their Crimes, in their Stead, or else all these Being[s] must be consigned to everlasting Perdition?" (Crowe 2008, 207–208).

A letter Adams wrote on 22 January 1825 suggests that Adams retained this position throughout his life. Writing to his friend and successor as U.S. president, Thomas Jefferson, who was then seeking to hire faculty for the University of Virginia, Adams warned Jefferson against hiring European professors: "They all believe that great Principle which has produced this boundless universe, Newton's universe and Herschell's [*sic*] universe, came down to this little ball, to be spit upon by the Jews. And until this awful blasphemy is got rid of, there never will be any liberal science in the world" (Crowe 2008, 208).

The third religious figure in the Paine tradition is the famous American Transcendentalist Ralph Waldo Emerson (1803–1882). One of the best-known events in American literary and religious history occurred in September 1832, when Emerson resigned his pulpit because he could no longer reconcile his religious views with administering the Lord's Supper. A plausible case can be made that what triggered this event was linked to the idea of a plurality of worlds. On 27 May 1832 Emerson delivered a sermon titled "Astronomy," the theme of which appears in a rhetorical question in his diary for 23 May 1832, where he asked, given modern astronomy, "Who can be a Calvinist or who an Atheist[?]" (Crowe 2008, 318). In other words, astronomy in effect supports belief in God but refutes belief in the Incarnation and Redemption. To some extent, Emerson summarizes his position by announcing:

> And finally, what is the effect upon the doctrine of the New Testament which these contemplations produce? It is not contradiction but correction. It is not denial but purification. It proves the sublime doctrine of One God, whose offspring we all are and whose care we all are. On the other hand, it throws into the shade all temporary, all indifferent, all local provisions. Here is neither tithe nor priest nor Jerusalem nor Mount Gerizim. Here is no mystic sacrifice, no atoning blood (Crowe 2008, 322).

1.4.3.2 Some Authors in the Christian Tradition

Numerous Christian authors in the period after Paine responded to his polemics. Three of the most successful were Timothy Dwight, Thomas Chalmers, and Thomas Dick.

Timothy Dwight (1752–1817) served as president of Yale University from 1795 until his death. One of Dwight's goals as Yale president was to combat deism, to which end he prepared a series of 173 sermons, which he repeated every four years lest any undergraduate miss his message. In these sermons, Dwight not only urged students to good actions but also marshaled extraterrestrials on behalf of his evangelical urgings. For example, in his fifth sermon, Dwight states that God called into existence "the countless multitude of Worlds [which] he stored, and adorned, with a rich and unceasing variety of beauty and magnificence, and with the most suitable means of virtue and happiness" (Crowe 1986, 175). In his next sermon, Dwight calls Yale students to repentance by asking them: "How different will be the appearance, which pride, ambition, and avarice, sloth, lust, and intemperance, will wear in the sight of God, in the sight of the assembled universe" (Crowe 1986, 176). Lunarians were also employed. Dwight informed those hearing his seventh sermon that although many astronomical investigations had shown that the Moon lacks an atmosphere, nonetheless "it is most rationally concluded, that intelligent beings in great multitudes inhabit her lucid regions, being probably far better and happier than ourselves" (Crowe 1986, 176).

Late in the series of sermons Dwight turns to problems raised by Paine's polemics, resolving them by the suggestion that a rebellion from God occurred only among angels and among terrestrials. In particular, "The first [rebellion] was perpetrated by the highest [i.e., the angelic], the second by the lowest [i.e., human] order of intelligent creatures. These two are with high probability the only instances, in which the Ruler of all things was disobeyed by his rational subjects" (Crowe 1986, 177). Thus Dwight was combating Paine's objection by declaring that humanity is the only race in the universe that fell into sin and required redemption. This bold response to Paine made for powerful preaching; in fact, Dwight's sermons were so effective that in a number of years as many as a third of Yale's graduates entered the ministry.

Ideas of extraterrestrial life played an even larger role in the evangelical movement in Scotland, where Thomas Chalmers (1780–1847) was not only the leading evangelical but also the most prominent Scottish religious figure of his day. Chalmers's rise to fame began with a series of sermons he delivered in Glasgow in 1815. In these sermons, Chalmers mixed evangelical piety with extraterrestrial themes similar to those of Dwight, thereby delighting hundreds who waited hours to experience his eloquence. His sermons, when published as *Astronomical Discourses on the Christian Revelation*, went through dozens of editions in both Britain and America.

Even more energetic about employing extraterrestrials in the service of religion was another Scotsman, Thomas Dick (1774–1857). From his observatory near Dundee, Dick deluged America with books blending ideas of extraterrestrial life with various religious themes. He edified readers of his first book, *The Christian Philosopher* (1823), by stating that the wisdom of God is shown by our Sun being placed at just such a distance as best to benefit us. Dick hastens, however, to add that the Sun's position does not prevent other planets from being happily inhabited by beings appropriately formed for their varying distances from the Sun. We learn from this book that rational beings dwell not only on all the planets but also on the Moon and Sun. For example, Dick states that God placed within the immense body of the Sun "a number of worlds ... and peopled them with intelligent beings" (Crowe 1986, 196). Turning to the Moon, he predicts that "direct proofs" of the Moon's habitability will be forthcoming, supplementing this by appendices in which he discusses whether the observations of the German astronomers Schröter and Gruithuisen provide such proofs. Dick, moreover, boldly claims that the existence of extraterrestrial life "is more than once asserted in Scripture" (Crowe 1986, 197).

Dick presents similar ideas in his *Philosophy of Religion* (1826) and his *Philosophy of a Future State* (1828). In the former book, he asserts that "the grand principles of morality ... are not to be viewed as confined merely to the inhabitants of our globe, but extend to all intelligent beings ... throughout the vast universe [in which] *there is but one religion*" (Crowe 1986, 197). In the latter book, he calculates that 2,400,000,000 inhabited worlds exist in the visible creation. In his *Celestial Scenery* (1836), he provides a table of the population of each planet, including even the ring, and the edge of the ring, of Saturn! It has been reproduced here in Fig. 1.3.

SUMMARY VIEW OF THE SOLAR SYSTEM. 401

	Square Miles.	Population.	Solid Contents.
Mercury	32,000,000	8,960,000,000	17,157,324,800
Venus..............	191,134,944	53,500,000,000	248,475,427,200
Mars	55,417,824	15,500,000,000	38,792,000,000
Vesta	229,000	64,000,000	10,035,000
Juno	6,380,000	1,786,000,000	1,515,250,000
Ceres	8,285,580	2,319,962,400	2,242,630,320
Pallas.............	14,000,000	4,000,000,000	4,900,000,000
Jupiter	24,884,000,000	6,967,520,000,000	368,283,200,000,000
Saturn	19,600,000,000	5,488,000,000,000	261,326,800,000,000
Outer ring of Saturn.	9,058,803,600 ⎫		
Inner ring.........	19,791,561,636 ⎬ 8,141,963,826,080		1,442,518,261,800
Edges of the rings ..	228,077,000 ⎭		
Uranus	3,848,460,000	1,077,568,800,000	22,437,804,620,000
The Moon.........	15,000,000	4,200,000,000	5,455,000,000
Satellites of Jupiter .	95,000,000	26,673,000,000	45,693,970,126
Satellites of Saturn..	197,920,800	55,417,324,000	98,960,400,000
Satellites of Uranus .	169,646,400	47,500,992,000	84,823,200,000
Amount........	78,195,916,784	21,894,974,404,480	654,038,348,119,246

Fig. 1.3 Thomas Dick's population figures for the solar system from *Celestial Scenery* (1836)

1.4.3.3 Two New Religions Involving Extraterrestrials

Consideration of Dwight, Chalmers, and Dick suggests how deeply ideas of extra-terrestrial life had entered into religious thought in the nineteenth century. The same point can be made even more forcefully by noting two very prominent religious figures who not only founded major religious denominations but also provided these new religions with scriptures incorporating extraterrestrials. These persons were Ellen G. White (1827–1915), the prophetess of the Seventh-day Adventist Church, and Joseph Smith (1805–1844), founder of the Church of Jesus Christ of Latter-day Saints (or Mormon Church).

Ellen G. White during the 1840s became involved with the Millerite movement, which had predicted that Christ's second coming was imminent. By November 1846, she had begun to experience visions involving extraterrestrial beings. Regarding Saturn, she reported: "The inhabitants are a tall, majestic people.... Sin has never entered here" (Crowe 2008, 329). Further visions came in 1849, such visions convincing her associates that she possessed special gifts. By the early 1860s, she and her associates founded a new denomination, which they designated the Seventh-day Adventist Church. For it, White provided a theology incorporating extraterrestrials, including the doctrines that sin occurred only on earth and that correspondingly Christ came only to our planet. As she wrote in one of her books, *The Story of Patriarchs and Prophets*, "It was the marvel of all the universe that Christ should humble himself to save fallen man. That he who has passed from star to star, from world to world, superintending all ... [took] upon himself human nature, was a mystery which the sinless intelligences of other worlds desired to understand" (Crowe 2008, 331). This theology not only provided a way around Paine's dilemma, it also presented a remarkable cosmic conception that seems to have enhanced the attractiveness of this new religion. Ellen White's denomination has continued to grow and in fact to spread throughout the world, with current membership, according to the church's website, over seventeen million.

The second case of a new religion embracing extraterrestrials is no less remarkable. Joseph Smith provided his Latter-day Saints not only with the *Book of Mormon* but also with a number of other scriptures including *Doctrine and Covenants* and *The Pearl of Great Price*. In both these texts, Smith advocates the idea that the universe contains a vast number of inhabited worlds. In the later work, God is presented as revealing that "I can stretch forth mine hands and hold all the creations which I have made; and … among all the workmanship of mine hands there has not been so great wickedness as among thy brethren" (Crowe 2008, 326). Two other pluralist notions advocated by Smith are that some inhabited worlds have already passed away and that new inhabited worlds will arise. The great emphasis Mormons place on this doctrine may have played a significant role in the spread of the Mormon community, which, according to the church's website, now numbers over fourteen million members.

1.4.4 The Sun and Moon as Special Cases

In discussions of the habitability of celestial bodies, the Sun and Moon are not only special cases, but also provide illuminating instances of the intensity of interest in believing in extraterrestrials. It takes a robust commitment to extraterrestrials to conclude that they inhabit our Sun, which seems overly warm, or the Moon, which seems inhospitable for life because stars rapidly disappear behind the Moon indicating the absence of a lunar atmosphere.

1.4.4.1 The Sun

William Herschel certainly possessed a sufficiently robust commitment to extraterrestrials to populate our Sun—and thereby the millions of stars in the universe. An incident reported in the 1787 issue of the *Gentleman's Magazine* helps put Herschel's publications in this regard in perspective. A certain Dr. Elliot was brought to trial in London for having set fire to a lady's cloak by firing pistols near it. Insanity was the plea made for Elliot, in support of which a Dr. Simmons recounted examples of Elliot's bizarre behavior, especially his having submitted a paper to the Royal Society arguing for the Sun's habitability (Manning 1993).

This incident leads us to wonder what may have been the reaction among readers of the Royal Society's *Philosophical Transactions* when in 1795 and 1801 they encountered papers in which Herschel theorized that the Sun consists of a cool, solid, dark, spherical interior above which floats an opaque layer of clouds. In 1795,

Herschel suggested that separate rays carry heat and light and that heat rays generate a rise in temperature only when in contact with special material (W. Herschel 1912, 1:470–84). In 1801, Herschel expanded the theory by proposing two exterior layers, the upper consisting of the glowing matter, the lower being a reflecting shield keeping the inner surface cool. As Herschel states: "The sun … appears to be nothing else than a very eminent, large, and lucid planet, evidently the first, or in strictness of speaking, the only primary one of our system …. Its similarities to the other globes of the solar system … leads us to suppose that it is most probably … inhabited … by beings whose organs are adapted to the peculiar circumstances of that vast globe" (W. Herschel 1912, 2:479). Herschel argues for his solarians by suggesting that terrestrial life flourishes in a variety of situations and by arguing that terrestrials who deny life to the Sun have no more logic on their side than inhabitants of a planetary satellite who deny life to the primary around which they revolve. Such arguments seem to support E. S. Holden's statement that Herschel's views on solar and lunar life "rest more on a metaphysical than a scientific basis" (Crowe 1986, 67).

Holden's conclusion needs, however, to be qualified in one important way, which helps explain why the premier astronomer of that period favored such a strange theory. Although as early as 1780 Herschel had considered a form of this solar model (W. Herschel 1912, 1:xcvi), he had between then and 1795 accumulated astronomical evidences that, when viewed in terms of his strong belief in the plurality of worlds doctrine, substantially increased the attractiveness of this model. In particular, during this period Herschel's stellar researches had led him to observe what he describes in his 1795 solar paper as "very compressed clusters of stars." He goes on to argue that stars in such clusters will be too tightly packed to accommodate inhabited planets. This did not lead Herschel to abandon the region as a home for extraterrestrials; rather it led him to conclude that the stars themselves must be "very capital, lucid, primary planets" so structured as to allow habitation (W. Herschel 1912, 1:483). Thus Herschel had found a way to save these stars from being "mere useless brilliant points" (W. Herschel 1912, 1:484). That Herschel's solar theory was no passing fancy in his thought is shown by his having elaborated it further in his 1801 paper in which he refers to the Sun as "a most magnificent habitable globe" (W. Herschel 1912, 2:147) and by his 1814 description of stars as "so many opaque, habitable, planetary globes" (W. Herschel 1912, 2:529). However bizarre Herschel's solar theory may seem, there is good evidence that it persisted as the preferred theory of the Sun until the 1850s (Meadows 1970, 6; Crowe 2011, 172–173).

Among advocates of solarians in the first half of the nineteenth century, some knew almost nothing of astronomy and mathematics. Such a charge cannot, however, be brought against Carl Friedrich Gauss (1777–1855), recognized as possibly the most brilliant mathematician of all time. Moreover, Gauss by profession was professor of astronomy at the University of Göttingen and director of its observatory. We know Gauss's views regarding extraterrestrials partly from his writings, but also from other sources, for example, records kept by his Göttingen colleague Rudolf Wagner (1805–1864), of conversations with the great mathematician. Wagner's records show that Gauss had adopted the doctrine that after death our souls take on new material forms on other cosmic bodies, including even the Sun. That Gauss held such an extreme idea is also evidenced in the biography of Gauss written immediately after his death by Wolfgang Sartorius von Waltershausen. This intimate friend revealed that Gauss

> held order and conscious life on the Sun and planets to be very probable and occasionally called attention to the action of gravity on the surface of heavenly bodies as bearing preeminently on this question. Considering the universal nature of matter, there could exist on the sun with its 28-fold greater gravity only very tiny creatures ... whereas our bodies would be crushed (Crowe 1986, 208).

1.4.4.2 The Moon

The special significance of our Moon for the debate derives from its nearness. No celestial body provided more promising hunting grounds for extraterrestrials than the Moon, as numerous nineteenth-century astronomers realized. In 1802, Johannes Schröter (1745–1816) published the second and final volume of his *Selenotopographische Fragmente*, the most ambitious telescopic investigation of the Moon that had ever been undertaken. In the first volume, Schröter had stated that he was "*fully convinced that every celestial body may be so arranged physically by the Creator as to be filled with living creatures ... praising the power and goodness of God*" (Crowe 1986, 70–72). He backed up this claim for the Moon by numerous observations he made with telescopes that in size rivaled even those of William Herschel. More of the same followed in his second volume, fifteen sections of which he devoted to his observations relevant to selenites, as he called the lunar inhabitants (Crowe 1986, 70–73).

Although some of Schröter's contemporaries sought to cast doubts on his extravagant claims, others sought to surpass him. The most striking case is the German astronomer Franz von Paula Gruithuisen (1774–1852), who during his lifetime published 177 astronomical papers and edited three astronomically oriented journals. He would perhaps rank as his most important paper his 1824 "Discovery of Many Distinct Traces of Lunar Inhabitants, Especially of One of Their Colossal Buildings." On the one hand, this long paper, in which he reports observing cities, forts, a temple, and animal trails on the Moon may have helped his career, in that two years later he became the professor of astronomy at Munich University. In the longer run, however, this paper discredited him and made him an object of ridicule, Gauss, for example, complaining of Gruithuisen's "mad chatter" (Crowe 1986, 204).[8]

The British public learned about the remarkable results attained by Gruithuisen partly from an essay that Thomas Dick published on lunar observation and lunar life in which, although he expressed skepticism about Gruithuisen's 1824 claims and wrongly attributed them to another German astronomer (whose name he misspelled), nonetheless held out great hopes for the detection of evidence of lunar

---

[8] For further information on Gruithuisen and on the intense debate on lunar life, see Crowe (1986, 202–8), and Sheehan and Dobbins (2001, 49–118).

life, which he amplified in later publications. For example, Dick discusses observations made by William Herschel of bright points on the Moon. Dick comments:

> Certain luminous spots, which have been occasionally seen on the dark side of the moon, seem to demonstrate that fire exists in this planet. Dr. Herschel and several other astronomers suppose, that they are volcanoes in a state of eruption. It would be a more pleasing idea, and perhaps as nearly corresponding to fact, to suppose, that these phenomena are owing to some occasional splendid illuminations, produced by the lunar inhabitants, during their long nights (Crowe 2008, 262–263).

Although more could be said about ideas of and searches for life on the Moon during this period, let us conclude by recounting an extraordinary event that occurred in 1835, which illustrates the degree to which the public had come to believe in extraterrestrial life. In that year, Richard Locke (1800–1871), a writer with the New York *Sun* newspaper, created a sensation by publishing a series of articles reporting that intelligent beings had been telescopically detected on the Moon (for a depiction of those beings, see Fig. 1.4). The noteworthy feature of this event is that nearly everyone believed Locke's report, even though substantial evidence had already been gathered to show that the Moon lacks an atmosphere. Locke's articles won him a place in the history of journalism as the author of what is now called "The Great Moon Hoax." Our claim is that a careful analysis of this incident shows that this was *not* a hoax. Locke had a much more serious intent: he was writing satire, a satire that nearly all his contemporaries missed. The reason is that for decades they had believed excessive claims for extraterrestrials, including selenites and solarians, made by astronomers, philosophers, professors, religious

Fig. 1.4 An 1835 lithograph picturing the lunarians

writers, and the public press. When in 1852 William Griggs edited a new edition of Locke's Moon writings, he stressed Locke's satiric intent:

> we have the assurance of the author, in a letter published some years since, in the *New World*, that it was written expressly to satirize the unwarranted and extravagant anticipations upon this subject, that had been first excited by a prurient coterie of German astronomers, and then aggravated almost to the point of lunacy itself … by the religio-scientific rhapsodies of Dr. Dick (Crowe 2008, 294).

Griggs added in regard to Dick that

> it would be difficult to name a writer who, with sincere piety, much information … and the best intentions, has done greater injury, at once, to the cause of rational religion and inductive science, by the fanatical, fanciful, and illegitimate manner in which he has attempted to force each into the service of the other (Crowe 2008, 294).

Professor James Secord has located Locke's letter, which fully supports Griggs's claim (Crowe 2008, 294–296) We would, perhaps, dispute somewhat Griggs's claim about the difficulty of locating authors who rival Dick in the degree to which their enthusiasm overwhelmed their discretion. In fact, a few of the authors analyzed in this section may be proposed as competitors. We would also suggest—admittedly it would be difficult to prove—that Locke's 1835 satire had a sobering effect on the extraterrestrial life debate.[9]

1.5 The Extraterrestrial Life Debate in the Second Half of the Nineteenth Century

As of 1850, the belief that extraterrestrial intelligent beings populate the planets and moons not only of our solar system but of other systems as well, remained widespread, at least in Europe, the United States, and other areas where there was a high level of learning. Soon after 1850, scientists found themselves entering the debate equipped not only with telescopes of greatly increased size, but also with powerful spectroscopic methods, which methods were central in astronomy coming to include astrophysics and astrochemistry. Moreover, scientists in the quest for extraterrestrial life had already gained the fundamental insights of geology and were soon to have access to the theory of evolution by means of natural selection. The tensions that pluralism had presented for religion had by 1850 somewhat subsided; in fact, it was widely assumed that belief in extraterrestrial life was not only compatible but also supportive of religion. Bright as prospects might seem for major advances in establishing the legitimacy of belief in extraterrestrial intelligent life, we shall see that by about 1910, various authors had gradually succeeded in driving intelligent extraterrestrials certainly from the solar system and possibly even from the universe.

---

[9] For recent analyses of the "Moon Hoax" see Sheehan and Dobbins (2001, Chap. 7), and Goodman (2008).

A major campaign in the battle against the extraterrestrials began in 1853, led by a warrior who earlier had eloquently endorsed extraterrestrials (Crowe 2008, 300–307) and who in 1853 came forth, not under his own name, but cloaked in anonymity. His campaign, it was later learned, was launched not only from Cambridge University, not only from its most prominent college, Trinity College (which had been the college of John Wilkins, Isaac Newton, and Richard Bentley), but also from the heart of Trinity, the Master's Lodge, where dwelt Rev. William Whewell. As one might suspect, most terrestrial religious figures and prominent scientists rallied around the extraterrestrials in opposing this attack. They won this *battle*, but gradually lost the *war*.

By the late 1870s, extraterrestrials had been expelled from nearly every planet of the solar system except Earth and Mars. Beginning in 1877, a major counteroffensive began on behalf of the Martians, coming from Milan, Italy, and Flagstaff, Arizona. Even this effort on behalf of the extraterrestrials had by 1915 failed. And after the Martians had been driven from our system, the extraterrestrials of our solar system managed only a few small and fruitless forays. The remaining sections of this survey are on one level an account of this crusade against the extraterrestrials of our solar system.

1.5.1 William Whewell

In 1850, British scholars recognized Rev. William Whewell (1794–1866) as one of the wonders of his age. He had made significant contributions not only to a number of sciences, but also to the history and philosophy of science. He wrote on ethics, economics, architecture, tidal theory, and a diversity of other areas too numerous to mention. He was well known for a treatise he published in 1833 on science and religion titled *Astronomy and General Physics Considered with Reference to Natural Theology*, in which he befriended extraterrestrials. Yet in 1853, he attacked them in his *Of the Plurality of Worlds: An Essay*. Various authors have offered explanations of this reversal (including Crowe 1986, Chap. 6), but a discussion of these would take us too far afield. Probably the least popular but most prophetic claim in Whewell's book was that that all other planets of our solar system lack intelligent life. In fact, he was the first person to glimpse the solar system of the twenty-first century, a system lacking higher forms of life except on Earth. Taking up astronomical data long available but largely neglected, Whewell argued that solar radiation, given the inverse square law of solar light propagation, must make the inner planets excessively hot and the outer planets excessively cold (Crowe 2008, 345). Moreover, he took Newton's determination that the density of Jupiter is about four times lower than the density of our Earth, to indicate that Jupiter may well be a planet without a solid surface (Crowe 2008, 345–349). Similarly, he used observations of the Magellanic Clouds made by John Herschel to suggest that the nebular patches that some claimed were other universes comparable to our Milky Way could not be composed of vast numbers of stars, which created serious difficulties for the island universe theory (Crowe 2008, 344). The breadth of Whewell's analysis is suggested by the fact that he used the best available information on the age of the Earth to critique those who maintained that God's efforts would have been wasted had the vast universe not contained plentiful intelligent life. Against this, he maintained that the best scientific determinations of the age of the Earth were such as to reveal that throughout most of the vast age of our Earth, no intelligent beings were present. In other words, he argued that though the Earth may occupy an extraordinarily small area of space, we cannot infer from this that intelligent life must be widespread unless we are also willing to accept the idea that God's efforts were wasted because through vast periods of the past, the Earth lacked intelligent beings (Crowe 2008, 343–344).

Whewell's position in the debate has at times been dismissed because religious passages appear periodically in his writings on a plurality of worlds. Such dismissal in a number of ways seriously distorts his position. Because so many of his contemporaries had employed the Principle of (Divine) Plenitude in support of extraterrestrials and because pluralists writing in the tradition of natural theology had so frequently argued that if other celestial bodies were not inhabited, it would entail that God's efforts would have been wasted had the planets been bereft of life, in short, because so many authors had based their claims for extraterrestrials on religious ideas, Whewell could not but discuss theological issues. What is crucial to understand is that Whewell typically argued that theology and religion shed no light on issues regarding extraterrestrial life. In fact, it would be possible to argue that Whewell played a key role in *opposing* and *freeing* the extraterrestrial life debate from religiously based arguments, which were so prominent in the eighteenth and nineteenth century. Thus it would be nearer the truth to say that Whewell led the way in opposing the practice of basing of claims about extraterrestrials on religious ideas. Moreover, a careful reading of Whewell's writings will show that he was quite scrupulous about basing his anti-pluralist claims on scientific information, as illustrated in the prior paragraph. In fact, we have a direct statement from Whewell that shows that such was both his position and practice. In his *Dialogue on a Plurality of Worlds*, Whewell responded to various critics of his book. In replying to Critic Y, who according to Whewell had chastised him for building "the philosophy of your Essay on a religious basis [and taking] for granted the truths of Revealed Religion, and reason[ing] from them," Whewell stressed that "I do not reason in the way which you ascribe to me. I obtain my views of the physical universe from the acknowledged genuine sources: observation and calculation" (Whewell 2001, 54).

In the preface to his book, Whewell had suggested "It will be a curious, but not a very wonderful event, if it should now be deemed as blameable to doubt the existence of inhabitants of the Planets and Stars as, three centuries ago, it was held heretical to teach that doctrine" (Crowe 2008, 335). And this prediction was confirmed. Numerous authors who had come to believe in extraterrestrials on the basis of natural theology, especially the principle of God's plenitude, were deeply distressed, perhaps none more than the well-known Scottish scientist David Brewster (1781–1868), who responded with his *More Worlds than One: The Creed of the Philosopher and the Hope of the Christian* (1854), wherein he asserts that our Sun is "a domain so extensive, so blessed with eternal light, it is difficult to claim that it is not occupied by the highest order of intelligences" (Crowe 2008, 356). The debate, which in some ways set the stage of the Darwin debate later in the 1850s (Brooke 1977), ended up generating at least twenty other books and over fifty journal publications, some scientific, some religious, and most a blend of the two. Over two-thirds of these publications opposed Whewell (Crowe 1986, Chap. 7).

1.5.2 Richard Anthony Proctor and Camille Flammarion

The most interesting and influential response to Whewell's claims came (gradually) from Richard Anthony Proctor (1837–1888), a British astronomer and prolific expositor of that science. Proctor's first major success as an author came in 1870, by which time the development of spectroscopy was transforming astronomy into astrophysics and astrochemistry. In that year, Proctor published his *Other Worlds than Ours*, an immensely popular discussion of extraterrestrial life ideas, one theme of which was an analysis of the Whewell-Brewster debate. Although siding in many cases and ultimately with Brewster, Proctor jettisoned Jupiterians, precisely for the reasons that Whewell had indicated. Also, because William Huggins's spectroscopic work in the 1860s had shown that earlier observational claims that Orion consists of a vast number of stars, indicating that it may be an island universe, could not possibly be correct because Orion gives a bright line spectrum, Proctor showed hesitation at the island universe theory, an important component of the pluralist position. Whewell's analysis of Herschel's observations of the Magellanic Clouds also influenced him.

By 1875, Proctor had moved further in what he called a "Whewellite" direction. A key essay in this alteration was Proctor's "A New Theory of Life in Other Worlds" (1875). In this essay, he withdraws intelligent extraterrestrials not only from most planets of our solar system but also of other systems. Writing in this Darwinian period Proctor suggests that planets are evolving, that "Each planet, according to its dimensions, has a certain length of planetary life, the youth and age of which include the following eras:—a sunlike state; a state like that of Jupiter or Saturn, when much heat but little light is evolved; a condition like that of our earth; and lastly, the stage through which our moon is passing, which may be regarded as planetary decrepitude" (Crowe 2008, 402). Within this perspective, he admits that not only most planets, but also most solar systems are bereft of intelligent life. Then he adds:

> Have we then been led to the Whewellite theory that our earth is the sole abode of life? Far from it. For not only have we adopted a method of reasoning which teaches us to regard every planet in existence, every moon, every sun, every orb in fact in space, as having its period as the abode of life, but the very argument from probability which leads us to regard any given sun as not the centre of a scheme in which at this moment there is life, forces upon us the conclusion that among the millions on millions, nay, the millions of millions of suns which people space, millions have orbs circling round them which are at this present time the abode of living creatures (Crowe 2008, 403–404).

One wonders whether Whewell, dead nearly a decade by then, would have been pleased or rather would have commented: "Pluralism dies hard!"

Prolific and popular as Proctor was, he was outdone by his French counterpart, Camille Flammarion (1842–1925), who like Proctor established his reputation by publishing a very popular plurality of worlds volume (*La Pluralité des mondes habités*, 1862), which in later editions expanded to nearly ten times its original size. Rarely has the pluralist position had a more prolific and enthusiastic advocate, an advocacy that lasted even to the last year of Flammarion's long life. One of the most forceful features of the volume is Flammarion's stress on the prodigality of nature. He also wrote about ideas of extraterrestrials in such a way as to link them with transmigrational, agnostic, and existentialist ideas. His advocacy of the Martians at the turn of the century, carried out in two huge volumes, was legendary. Nevertheless, because he sided with Schiaparelli and Lowell rather than Maunder and Antoniadi, he ended up on the losing side.

1.5.3 The Sun and Moon

Solarians are gone. Their departure was far more than an event of local significance. When our Sun lost its inhabitants, so did every star in the universe. This raises a question: When and by whom were the solarians slaughtered? Their departure was not caused by a direct attack. It is true, as we have seen, that Newton raised serious problems for them as did Thomas Young in the first decade of the nineteenth century and also François Plisson in an 1847 volume skeptical of extraterrestrials (Crowe 1986, 168, 248–249). But what above all drove the solarians from the Sun was the progress of astrophysics. As more and more was learned about the Sun and stars, for example, their mechanisms and temperatures, it became not just difficult but nearly impossible to assume solarians. The modifier "nearly" has been added because in 1861, no less an authority than John Herschel publicly suggested that certain immense structures on the surface of the Sun that James Nasmyth reported observing might be "*organisms* of some peculiar and amazing kind" (Crowe 1986, 221). Although in the early decades of the nineteenth century, many saw solar life as plausible, by the last decades of the nineteenth century, such seemed impossible. For example, when in 1870 Richard Proctor published his *Other Worlds Than Ours*, he labeled life on the Sun "too *bizarre* [for] consideration" (Proctor 1870, 20). Some castles crumble as a result of rapid and direct attack; others fall vacant and over decades become uninhabitable. The latter fate befell the Sun and stars.

Selenites proved to be of hardier stock. In fact, they were sirens who drew various astronomers to them. It is true that from the 1860s on, the spectroscope made it far more difficult to argue for a lunar atmosphere. But it is also true that some selenographers, as those who map our Moon are called, had turned against the selenites. Writing in 1876, the lunar specialist Edmund Neison lamented the low level of interest in selenography, attributing it to the high quality lunar mapping done in the 1830s by Wilhelm Beer and J. H. Mäedler, suggesting that this was because Beer and Mäedler and their work had "finally solved … the great questions" in regard to the Moon; in particular, they had "demonstrated that the moon was to all intents an airless, waterless, lifeless, unchangeable desert" (Crowe 1986, 387). Selenography, however, was given new life after 1866 when Julius Schmidt (1825–1874), a respected Greek observer, reported that the lunar crater Linné had disappeared! According to Neison, this was seen as so exciting that "Nearly every astronomer was led to study the moon, and for months all the principal telescopes of Europe were turned upon our satellite. Moreover, many of our present amateurs were then for the first time led to purchase telescopes, and take up the study of astronomy" (Crowe 1986, 387). Of course what caused such excitement was the prospect of finding evidence of selenites. Reports of the disappearance of or changes in other lunar features followed, for example, in the Alpetagius, Plato, Messier, and Hyginus N regions. It has now been concluded the Schmidt's Linné observations were almost certainly spurious, but the important point is that, as Neison noted, many were drawn to astronomical observation on this basis. Moreover, selenites had a significant role in the erection of the Lick Observatory in 1888 with the largest refracting telescope that had ever been erected. Such at least was reported by the director of the observatory, Edward S. Holden, who revealed what led Lick, an uneducated millionaire, to donate $700,000 for the observatory. Lick originally planned to build, according to Holden, "a marble pyramid larger than CHEOPS on the shores of San Francisco Bay," but was dissuaded from this by the fear that it would be destroyed in a bombardment. Lick then hit upon the idea of an observatory; as Holden recorded, "The instruments were to be so large that new and striking discoveries were to follow inevitably, and, if possible, living beings on the surface of the moon were to be described, as a beginning" (Crowe 1986, 389–390).

Persons hoping for selenites found encouragement not only from observers but also from mathematical astronomers of distinguished reputation, such as Peter Andreas Hansen, who in 1856 published a paper arguing that a previously unaccounted for motion of the Moon might be explained were the Moon egg-shaped, the narrow end pointing to the Earth. In particular, he estimated that the Moon's center of figure is located about thirty-three miles nearer us than its center of mass. Hansen added that because of this, "one can no longer conclude that the [remote] hemisphere may not be endowed with an atmosphere, and that it has no vegetation and living beings" (Crowe 1986, 390). Various pluralists, including John Herschel, immediately jumped on this possibility to save the selenites. Their efforts ceased after 1868, when Simon Newcomb found a better way of explaining those motions of the Moon that had led to Hansen's calculation (Crowe 1986, 391).

1.5.4 Messages and Meteorites

In the latter half of the nineteenth century, there was widespread awareness of two ways in which the confirmation of life elsewhere could come suddenly: the reception of a message from an extraterrestrial source or the finding on Earth of some object from space that contains life from another planet. This section will survey these possibilities.

In 1866, Victor Meunier (1814–1903) suggested that lunarians may be trying to communicate with us. Encouraged by this, another Frenchman, Charles

Cros (1846–1888), in 1869 wrote an essay suggesting the use of parabolic mirrors to signal Mars or Venus, which essay was published in 1871 in *Cosmos*, a journal edited by Meunier. Cros also proposed that flashes reported by such astronomers as Messier and Schröter on Venus and Mars might be signals. In 1891, Flammarion announced that a woman had offered the Académie des sciences a prize of 100,000 francs to be awarded to the first person who communicates with the Martians. Flammarion encouraged the competition by assuring the public that Mars was the ideal object for the competition because "its intelligent races … are far superior to us" (Crowe 1986, 395). The offering of this prize sparked much controversy, which extended to England, where in 1892 Francis Galton proposed a method employing mirrors. Some labeled such discussions silliness, but others took them seriously, noting that flashes had at times been seen on the surface of Mars. In 1896, Galton published a now famous paper proposing a method of communicating with the Martians through the use of dots and dashes. In 1892, J. Norman Lockyer, prominent astrophysicist and founder of a leading scientific journal, *Nature*, actively joined the discussion. Later in 1896, Konstantin Tsiolkovskii (1857–1935), eventually famous as a rocket pioneer in Russia, followed up on Galton's proposals. As the century ended, A. Mercier and Flammarion kept these issues alive among the French.

In the early years of the nineteenth century, scientists attained an understanding that gave them hope that evidence of extraterrestrial life might literally fall from the skies at any time. The recognition that they had attained was that meteorites, rather than being ejecta from terrestrial volcanoes, actually came from outer space. This led them to examine meteorites in detail, especially determining whether they contain evidence of extraterrestrial life, possibly in fossilized form.

In the 1830s, interest in this approach grew especially intense, partly because on 12 November 1833, the most brilliant meteor shower in recorded history drew widespread interest. In 1834, J. J. Berzelius (1799–1848), the leading analytical chemist of the period, performed a chemical analysis of a carbonaceous chondrite that had fallen in the Alais region of France, concluding that he could not tell whether it contained carboniferous materials of extraterrestrial origin (Crowe 1986, 401–402).

By 1871 the consensus among scientists was that some meteorites contain organic materials, but that there was no direct, clear-cut evidence that any meteorites contain remains of an extraterrestrial life forms. In that year's British Association for the Advancement of Science meeting, William Thomson (1824–1907), later Lord Kelvin, brought his presidential address to a sensational conclusion by simultaneously discussing spontaneous generation, evolutionary theory, meteorites, and the plurality of worlds. Raising the question of the origin of terrestrial life, Thomson states that both "philosophical uniformitarianism" and Pasteur's experiments rule out spontaneous generation (Crowe 1986, 402). Given this, Thomson asks: "How, then, did life originate on earth?" The answer offered is that

> because we all confidently believe that there are at present, and have been from time immemorial, many worlds of life besides our own, we must regard it as probable in the highest degree that there are countless seed-bearing meteoric stones moving about through space…. The hypothesis that life originated on this earth through moss-grown fragments from the ruins of another world may seem wild and visionary; all I maintain is that it is not unscientific (Crowe 1986, 403).

Thomson was not the first to have proposed a meteoric origin of terrestrial life, but he was the most prestigious. The reaction to Thomson's theory was mixed, even after 1875 when the premier German physicist of the period, Hermann von Helmholtz (1821–1894), revealed that he had arrived at the same hypothesis slightly before Thomson and had presented it in a public lecture in spring 1871 at Cologne and Heidelberg. Contained in the lecture is his statement:

> Who can say whether the comets and meteors … may not scatter germs of life wherever a new world has reached the stage in which it is a suitable dwelling place for organic beings? We might, perhaps, consider such life to be allied to ours, at least in germ, however different the form it might assume in adapting itself to its new dwelling place (Crowe 1986, 405).

Despite this support, many scientists were hesitant to accept this theory, possibly because though rich in explanatory power, it was poor in falsifiability.

1.5.5 The Rise and Fall of the Island Universe Theory

At the beginning of the nineteenth century, the theory that the thousands of nebular patches observed by William Herschel are in fact universes comparable to our own Milky Way system had a measure of support, which was important for advocates of extraterrestrials because it was assumed that these universes would, like ours, be well stocked with life. Crucial to this claim was the question whether with improved telescopes the nebular patches could be resolved into individual stars, which would support the claim that they are distant universes rather than nearby patches of glowing gas. Many were resolved into individual stars, but giant Orion resisted such resolution until the mid-1840s when Lord Rosse in Ireland and William Cranch Bond at Harvard reported their successful resolution.

The theory, however, was beset by the difficulty that whereas the nebular patches, if other milky ways, should be seen as randomly distributed over the heavens, they tended to cluster toward the poles of our galaxy. Another problem had been pointed out by Whewell, who noted that in the Magellanic Clouds nebulae were seen as in many areas comparable in size to nearby stars, indicating that they could not be vast, very remote milky ways. A further difficulty arose when in the 1860s William Huggins found that the spectroscope indicated that the Orion nebula produced a bright-line spectrum, indicative of glowing gas, whereas if it were an island universe, it should generate a dark line spectrum indicative of stars, in other words, that the resolutions reported by Bond and Rosse were spurious. Richard Proctor also pressed the distribution problem, which led to the situation that by 1900 the island universe theory had been discredited. Agnes Clerke accurately expressed the consensus of astronomers when in the 1880s she wrote:

> There is no maintaining nebulae to be simply remote worlds of stars in the face of an agglomeration like the Nebecula Major [the Large Magellanic Cloud], containing in its (certainly capacious) bosom *both* stars and nebulae. Add the evidence of the spectroscope to the effect that a large proportion of these perplexing objects are gaseous, with the facts of their distribution telling of an intimate relation between the mode of their scattering and the lie of the Milky Way, and it becomes impossible to resist the conclusion that both nebular and stellar systems are parts of a single scheme (Clerke 1887, 456–457).

The island universe theory would be successfully resurrected decades later, but until then the positive side in the extraterrestrial life debate was deprived of one of its chief supports.

1.5.6 *Extraterrestrials and Evolutionary Theory*

Ideas of extraterrestrial life were associated in various ways with the growing significance in the last half of the nineteenth century of evolutionary ideas. Of course, evolutionary theory was linked even earlier (think of Kant and William Herschel) with astronomy and cosmogony. As John Brooke has stressed, William Whewell's *Essay* of 1853 raised many of the same issues as were debated after the publication of Darwin's *Origin of Species* in 1858 (Brooke 1977). Darwin himself never directly entered the extraterrestrial life debate, but it is possible that the idea of a plurality of worlds may have helped him to his theory. In particular, in the 1844 essay he drafted (but did not then publish) to formulate and explain his theory of natural selection, he commented "It is derogatory that the Creator of countless Universes should have made by individual acts of His will the myriads of creeping parasites and worms, which since the earliest dawn of life have swarmed over the land" (Crowe 2008, 372). It is also clear that the naturalistic approach associated with evolutionary theory influenced debates about extraterrestrials, in both a pro-pluralist and at other times in an anti-pluralist direction. It is especially striking that the co-discoverer of the theory of evolution by natural selection, Alfred Russel Wallace (1823–1913), emerged near the end of the nineteenth century as not only a critic of pluralism but also to publish in 1904 an appendix to his *Man's Place in the Universe* (1903), an appendix that lays out for the first time the type of argument based on evolutionary theory against extraterrestrials (Crowe 2008, 427–437) that has been taken up by many evolutionary theorists in the last thirty years, for example, by Ernst Mayr, Simon Conway Morris, William Burger, and Jared Diamond.

1.5.7 *The Controversy over the Canals of Mars*

As of 1877, Mars, which seen from Earth is so small that a teacup a half mile distant will cover it, was in the dramatic position of appearing to be the last, best hope for life elsewhere in our planetary system. In the period from 1877 to about 1910, developments occurred that led to dozens of books, hundreds of telescopes, thousands of articles, and millions of people being focused on whether intelligent beings, possibly desperately struggling to survive, conceivably seeking to signal us, roam its surface. Various earlier astronomers had attempted to draw a map of Mars, including its polar caps, and had determined its period of rotation. Then in 1877, Giovanni Schiaparelli (1835–1910) at Brera Observatory in Milan reports sighting numerous "canali" on Mars. Other astronomers at first had difficulty confirming his observations, but gradually over the next decades many succeeded in sighting them. During the next Mars opposition in 1879, Schiaparelli reported sighting a doubling of some canals, or what he called a gemination. By 1890, a significant portion of the astronomical community accepted the canal observations, but with little agreement as to their cause, one of the theories of the canals being that Martians, because they were very short on water, constructed them as irrigation devices.

In 1893, Percival Lowell (1855–1916), a wealthy Harvard-educated businessman and orientalist, returned from Japan to enter the canal controversy. He founded, funded, directed, and served as chief astronomer and publicist for the Lowell Observatory, which he erected in Flagstaff, Arizona, and equipped with a 24-inch refractor. Lowell announced his intention to launch an "investigation into the conditions of life in other worlds" and prophesized that "there is strong reason to believe that we are on the eve of pretty definite discovery in the matter" (Crowe 1986, 508). For a chronology of this and other major developments in the canal controversy, see Table 1.1.

In 1894, W. W. Campbell at Lick Observatory attempted to reconfirm earlier spectroscopic reports by William Huggins and other astronomers of water vapor in Mars's atmosphere. Campbell "fails," but concludes that if the Martian atmosphere contains water vapor, its quantity must be below what could be detected by available instrumentation. In the same year, E. W. Maunder (1851–1928) of Greenwich Observatory published a paper questioning the canal observations. Having noted that adjacent sunspots at times appear as a continuous line, he suggests that the canals may be optical illusions arising from the tendency of the eye to integrate fine detail below the limits of vision. Maunder asserted regarding Mars: "We cannot assume that what we are able to discern is really the ultimate structure of the body we are examining" (Crowe 1986, 500).

Certainly by 1896, the canal controversy had spread through the educated world, with many advocates for and against the Martians. For example, in that year Campbell reviewed Lowell's *Mars* for the journal *Science*, accusing Lowell of taking "the popular side of the most popular scientific question afloat" (Crowe 2008, 484). He argued that Lowell has neither adequate observations nor mastery of the Mars literature. On the other hand, Agnes Clerke in the *Edinburgh Review* described the canals "as among the least questionable, though perhaps the very strangest of planetary phenomena" (Crowe 1986, 513). In the same year, a Greek astronomer living in Paris, Eugène Antoniadi (1870–1944), became director of the British Astronomical Society's Mars section and for the next two decades edits its reports. Although at first an advocate of the canals, by the late 1890s he becomes skeptical, moved especially by the arguments made by E. W. Maunder. One indication of the intensity of interest in Mars at the end of the nineteenth century is the publication in 1899 of H. G. Wells's *War of the Worlds*, one of the most popular works of science fiction of all time.

In the first decade of the twentieth century, various developments occured that culminate in the astronomical community concluding that the canal claims are untrustworthy, even though there was agreement that under certain conditions some observers could see thin lines on Mars. Among these developments were tests in which subjects in classrooms were instructed to draw what they saw in a test diagram, which showed spots but no thin lines. The result was that the test

Table 1.1 Highlights of observational evidence for Martian canals

Observer	Date	Observatory	Instrument	Method	Result/remarks
Schiaparelli	1877	Brera (Milan)	8 in.	Visual	First observation of system of *canali*
Green	1877	Madeira	13-in. reflector	Visual	Shaded area boundaries; no canals
Hall/Harkness	1877	U.S. Naval	26 in.	Visual	Moons of Mars found but no canals
Schiaparelli	1879/1880	Brera (Italy)	8 in.	Visual	First report of double canals
Maunder	1882	Greenwich	28 in.	Visual	Some canals
Perrotin/ Thollon	1886 1888	Nice (France)	15 in. 30 in.	Visual	Many canals; first confirmation of double canals
Holden/ Keeler et al.	1888	Lick	36 in.	Visual	Some canals but no doubles
Pickering/ Douglass	1892	Harvard Arequippa (Peru)	13 in.	Visual	Canals and "lakes" at canal junctions
Barnard	1892	Lick	36 in.	Visual	Some canals but not fine lines
Lowell/Pickering/ Douglass	1894 1895	Lowell	12 in. 18 in.	Visual	Canals artificial
Antoniadi	1894/1896	Juvisy (France)	9.6 in.	Visual	42 canals; 1 double doubling is illusion
Cerulli	1896	Teramo (Italy)	15.5 in.	Visual	Illusion—optical origin
Lampland	1905	Lowell	24 in.	Photographic	Canals photographed
Todd/E. Slipher	1907	Chile	18 in.	Photographic	Canals photographed
Antoniadi	1909	Meudon	33 in.	Visual	Canals resolved
Hale	1909	Mount Wilson	60-in. reflector	Visual	Much detail, no canals
Trumpler	1924	Lick	36 in.	Visual/photo-graphic	Strips of vegetation

Sources of data M. J. Crowe, *The Extraterrestrial Life Debate* (Cambridge, 1986); W. G. Hoyt, *Lowell and Mars* (Tuscon Ariz., 1976), and primary sources. Instruments are refractors unless specified otherwise
Source Dick (1998), 28 (Table 2.1); used with permission. (Observational highlights of Martian canals, 1877–1924)

subjects frequently drew lines. One of Lowell's staff, C. O. Lampland, reported attaining photographs showing dozens of canals, but the images prepared by Lampland, although impressive, were only three-eighths of an inch in diameter and proved unconvincing. In addition, although new spectroscopic observations

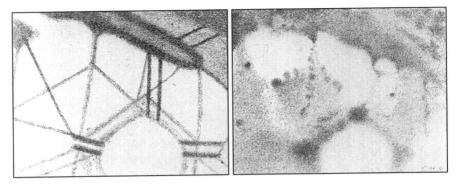

Fig. 1.5 Drawings of one region of Mars by Schiaparelli (*left*) and Antoniadi (*right*)

made at Lowell Observatory by Vesto Slipher supported the existence of water vapor in the Martian atmosphere, W. W. Campbell's new observations again gave a negative result. Maunder and Antoniadi published numerous papers arguing the illusion theory of the canals, winning over astronomers as influential as Simon Newcomb. A key development occurred when it became clear that Campbell's spectroscopic results fit very well with the claims of the illusion theorists. Another key development consisted of a drawing of a specific region of Mars made by Antoniadi using the high quality telescope at Meudon Observatory, which drawing appears alongside a drawing of the same area by Schiaparelli (see Fig. 1.5). Whereas Schiaparelli's drawing showed numerous canals, Antoniadi's showed diffuse detail, but no canals.

In short, by about 1912 Maunder and Antoniadi, in conjunction with contributions from W. W. Campbell, Vincenzo Cerulli, Simon Newcomb, and others, had built a case against the claims of Schiaparelli, Lowell, Flammarion, and their allies, a case that convinced the astronomical community that the canal sightings are illusory, which left Mars bereft of evidence for higher forms of life. As the Spanish astronomer J. Comas Sola put it in 1910: "The marvelous legend of the canals of Mars has disappeared with this opposition" (Crowe 2008, 509).

1.5.8 The Extraterrestrial Life Debate at the End of the Nineteenth Century

This survey of the last half of the nineteenth century cannot do justice to the large number of intellectuals who became involved with ideas of extraterrestrials. Literary and philosophical figures from America as prominent as Whitman, Twain, and Peirce, from Britain as important as Tennyson, Hardy, and Newman, from Germany as well known as Engels and Strauss, from France as prestigious as Balzac and Hugo, and from Russia the great novelist Dostoevsky entered the debate.

On the substantive level, we suggest that the most significant change that occurred during the nineteenth century (completed by 1910 or so) was the exit of the extraterrestrials from our solar system. On the methodological level this led to an attenuated Copernican Principle and a weakened Principle of Plenitude. Moreover, it probably called into question the belief that the discovery of extraterrestrial intelligent life was only a few years distant. Influenced by the increased naturalism that went with the successes of evolutionary theories in biological and physical sciences, including geology, religious and metaphysical claims became less evident in the debate. On the other hand, immense improvements in the instrumentation of astronomy, including spectroscopy and photography, as well as major advances in physical science, including an understanding of energy, of atomism including both chemical and physical aspects, and of stellar regions, as well as a far deeper understanding of biological forms all gave indications that the debate about extraterrestrials might be coming closer to resolution.

1.6 Overview of Part I

In this chapter, we have provided a historical overview of the extraterrestrial life debate from antiquity through the start of the twentieth century. In broad outline, we have seen waxing and waning of belief in extraterrestrial life. Two schools of thought in antiquity, Aristotelian and Atomistic, reached differing conclusions about its existence. By the Middle Ages, the Aristotelian view, which denied its existence, was predominant in intellectual circles, though this view was occasionally challenged. The Copernican revolution, however, eventually led to a view of the universe as expansive, in which the Earth was merely one planet among many, and the Sun the center of merely one solar system. By the end of the seventeenth century, belief in extraterrestrials was winning many converts, laying the groundwork for a vast corpus from various fields of inquiry—scientific, philosophical, theological, literary—discussing extraterrestrials over the next two centuries. At the same time, and with the benefit of hindsight, we can see that much of this discussion was highly speculative, and included a great number of religious and metaphysical assumptions. In the middle of the nineteenth century, William Whewell wrote a powerful critique of the assumed position, arguing that many factors mitigate against the existence of extraterrestrial intelligent beings. A newfound skepticism about at least the commonality of extraterrestrials, and certainly about their existence in our own solar system, developed in the latter half of the century, with Richard Proctor playing a major role and eventually culminating in the controversy over the canals of Mars. As we entered the twentieth century, then, our confidence in the existence of extraterrestrials had ebbed from its heights a century before.

The remainder of the first part of the present volume consists of five chapters that examine various individuals and concepts at greater length than has been done in this overview chapter. In addition, Michael Crowe, one of the co-authors of this chapter, has added a short addendum to this chapter that contains two brief historical analyses that supplement his previous (and extensive) work on the historical extraterrestrial life debate.

The following chapters, as well as the present chapter, demonstrate that many of the issues still facing us today have precedents in earlier years of the debate. The questions that arise now, even if our scientific information has expanded, are not always greatly different from those that our intellectual predecessors confronted. The dearth of precise information about other solar systems or about other terrestrial planets, for example, leads us to reason about those places, and on their suitability for life, on the basis of analogy. The implications of extraterrestrial life, too, have been the subject of much analysis, and will continue later in this volume.

The first of the five further chapters on the historical debate is by Dennis Danielson. To lead off that chapter, he discusses how the division of the world into sublunary and superlunary realms, the typical Aristotelian view of the cosmos, was dissolved in the sixteenth and seventeenth centuries, focusing especially on the contributions of Thomas Digges (1546–1595). Galileo's telescopic observations further reinforced the analogy between the Earth and other environs of the universe, leading some authors to consider the existence of extraterrestrials on other planets in our solar system. Danielson then draws our attention to imaginative fictions from the mid-seventeenth century that involve the idea of travelling away from the bounds of the Earth to the Moon, and even beyond. Finally, he concludes the chapter with discussion of seventeenth-century examples of "reflexive telescopics," or the notion of considering the Earth from a point of view outside of the Earth itself, as an extraterrestrial would see it. The implications of Earth's similarity to other bodies of the universe were thus grasped and sparked imaginations immediately after the Copernican cosmological shift changed our view of Earth's place in the cosmos. In addition, Danielson's exposition nicely illustrates how analysis of literary works can reveal changes in thought that are less accessible in other sources.

Woodruff T. Sullivan, III, discusses some of the principles—metaphysical, religious, methodological—that supported astrobiological considerations in the eighteenth and nineteenth centuries, focusing especially on William Herschel. Sullivan notes that these principles, and not least the use of analogy, are problematic, this being evidenced by the fact that astronomers believed that the Sun and stars are themselves inhabited. One can also note Herschel's tendency to see planets, moons, stars, and the Earth as having important similarities and hence analogous. Sullivan, following John Stuart Mill's analysis, notes that the principle of analogy can be fruitful not as a method of proof, but as a guide to discovery and formulation of new theories. Sullivan's contribution is especially interesting because Sullivan himself has long been active in contemporary astrobiology, including radio astronomy, but also writes as an author fully sensitive to and practiced in historical research. He also very appropriately stresses the importance of what he labels "the $N = 1$ Problem," that is, the fact many astrobiological inferences are limited by the fact that in searching for other earths we have only Earth as our sample.

The focus of the chapter by Joseph T. Ross is the philosophy of the prominent German philosopher Georg Friedrich Wilhelm Hegel (1770–1831), especially Hegel's views on the trustworthiness of analogical arguments and on the credibility of claims for extraterrestrial life. Ross begins by presenting information on how philosophers from the ancient Greeks to the early nineteenth century (and especially Immanuel Kant) viewed analogical reasoning. Hegel himself, Ross shows, tended to be skeptical of analogical arguments. Moreover, Ross shows that Hegel, far more than most of his contemporaries, expressed reservations about claims for extraterrestrial life. He was especially skeptical of claims for life on the Sun and on moons, believing that only planets are in any significant degree analogous to the Earth. A striking aspect of this essay emerges if Hegel is compared with various astronomers, for example, William Herschel, who tended to view stars, moons, planets, comets, and Earth as five fundamentally similar entities, whereas Hegel viewed them as significantly different. In this, Hegel was closer to the astronomy of the twentieth century than his scientific contemporaries. In important ways, this chapter contrasts with a number of others, not only in regard to Hegel's reservations about analogical inferences but also in regard to the degree of primacy that Hegel assigns the Earth.

Stéphane Tirard examines evolutionary ideas, including the origins of life, from the nineteenth century, and how those ideas were applied to discussions of life on other worlds. Both Charles Darwin (1809–1882) and Herbert Spencer (1820–1903) proposed materialistic theories for the origins of life, thereby suggesting that life could arise elsewhere in analogous fashion; an alternate theory of the nineteenth century, panspermia, pushed away the question of the origins of life, but in fact insisted that life must exist elsewhere in the universe. Tirard then discusses two writers of the late nineteenth and early twentieth century, Camille Flammarion and Edmond Perrier (1844–1921). Both considered the question of how life would evolve on other planets, offering evolutionary paths for life elsewhere based on analogy with the progression of evolution on Earth, though modified to fit the environment of other planets. Interestingly, Perrier restricted extraterrestrial life in our solar system to Venus and Mars, the worlds most like our own Earth.

Florence Raulin Cerceau's chapter focuses on the issue of finding a proper definition of "habitability," a concern that has intensified with the discovery of numerous exoplanets. Her essay opens with a brief survey of ideas regarding extraterrestrial life in the period before 1700, in which she stresses the importance of Copernicanism and also notes that Christiaan Huygens in his *Cosmotheoros* anthropomorphized his extraterrestrials. She then turns to the latter half of the nineteenth century, focusing on the ideas of Richard Proctor, Jules Janssen, and Camille Flammarion. She presents Proctor as taking more seriously than his predecessors the need to discuss habitability in terms of specific features of a planet or moon. She stresses the importance of the new spectroscopic techniques for studying planetary habitability by noting Jules Janssen's emphasis on spectroscopy, including his report of his detection of water vapor in the Martian atmosphere. Concerning Camille Flammarion, she stresses his readiness to argue that extraterrestrials could be dissimilar to humans. Finally, she turns to the efforts around 1950 of Hubertus Strughold, a physiologist, to stress the need to think of planetary life in terms of such biological categories as ecology.

1.7 An Addendum by Michael J. Crowe: Updating My *Extraterrestrial Life Debate* (1986)

I have added this addendum in order to supplement this survey by two historical analyses that I developed after the publication of my *Extraterrestrial Life Debate, 1750–1915: The Idea of a Plurality of World from Kant to Lowell*, and I have also included some broad remarks on research and teaching on the history of the extraterrestrial life debate (hereafter the ETD).

1.7.1 Blaise Pascal, Copernicanism, and Parallax

A number of authors have noted that we lack solid evidence that the brilliant scientist and religious author Blaise Pascal (1623–1662) was a Copernican. One of these authors, Arthur Lovejoy, reported finding in Pascal "the curious combination of a refusal to accept the Copernican hypothesis with the unequivocal assertion of the Brunonian" (Crowe 1986, 14). It is true that Pascal nowhere in his published writings directly affirmed the Copernican position. What Lovejoy was no doubt thinking of in making his claim about Pascal is various statements from Pascal's *Pensées*, such as:

> The whole visible world is only an imperceptible atom in the ample bosom of nature … .
> … let man consider what he is in comparison with all existence; let him regard himself as lost in this remote corner of nature; and from the little cell in which he finds himself lodged, I mean the universe, let him estimate at their true value the earth, kingdoms, cities, and himself. What is a man in the Infinite? (Pascal 1938, pensée #72).

In my 1986 volume, I accepted Lovejoy's first claim but argued against his second (Crowe 1986, 14–16). I shall not repeat those arguments. Rather what I now shall claim is that Pascal was indeed a Copernican. Because a central theme of the *Pensées* is the infinitization of the universe, because no author before 1660 had written more effectively about the vastness of the universe, and because this vastness is a feature that results directly from the heliocentric and not from the geocentric system, it seems certain that Pascal was Copernican. I suspect that Pascal writing on religious matters in his *Pensées* and adopting positions that upset many of his orthodox contemporaries did not wish directly to assert his Copernican convictions, but his genius and the geometry of his universe are definitely Copernican.

1.7.2 *Immanuel Kant and "The Starry Skies Above Me"*

Inscribed on the tombstone of Immanuel Kant (1724–1804) is arguably the most famous line from his many writings: "the starry heavens above me and the moral law within me." This is from his *Critique of Practical Reason*, where he wrote:

Two things fill the mind with ever new and increasing admiration and awe, the oftener and more steadily they are reflected on: the starry heavens above me and the moral law within me.... The former ... broadens the connection in which I stand into an unbounded magnitude of worlds beyond worlds and systems of systems.... The former view of a countless multitude of worlds annihilates, as it were, my importance as an animal creature, which must give back to the planet (a mere speck in the universe) the matter from which it came (Kant 1949, 258–259).

This leads to two questions, one of which I sought to answer in my 1986 volume, and other of which I shall address now. The first question is: What was Kant's conception of the "starry heavens," and the other is: was he correct in his view? In my 1986 volume, where I analyzed about eight of Kant's writings discussing extraterrestrials, I attempted to show that Kant was not only referring to the three thousand or so stars visible on a clear night, but that he was thinking of those stars as typically encircled by inhabited planets, as worlds filled with life, indeed intelligent life. He was not envisioning stars as merely big and blazing balls of fire. This is what filled his "mind with ever new and increasing admiration and awe" and so deeply moved his heart (Crowe 1986, 47–55). It seems unnecessary to repeat all the evidence for this conclusion in the present paper. Rather I shall now turn to the second question, which I believe can now be definitively answered. That question is whether the numerous stars that Kant saw and we see on a dark night are typically seats of life. Twentieth-century studies of the nature, lifetimes, and locations of stars make it possible to answer this question.

Two factors are above all important in determining whether a star is visible to us: (1) its distance and (2) the luminosity of the star—how much light it produces. The importance of distance is evident from the fact that given two stars A and B with equal luminosity, if star A is twice as far from us, it will appear one quarter as bright. The overall rule is brightness is inversely proportional to distance squared. The luminosity of a star is an even more important factor in determining whether we can see it. We now know that stars differ very greatly in luminosity. One star may be millions of times more luminous than another. This fact is understated in the classification of stars as dwarfs and giants or even supergiants. The net effect of this is that an examination of the three thousand or so stars visible to us shows that the great majority of them are giant or supergiant stars.

A passage from Joel Achenbach's book *Captured by Aliens* supports this overall point about giant stars:

Most of the stars we see with the naked eye are extremely hot, bright supergiants that are several times the size of the Sun and, more important, much younger. The supergiants reach their demise after only a matter of some millions of years, not billions. That's too short a period, probably, for the evolution of intelligent life. Everything we know about life on Earth tells us that it requires a tremendous span of time to evolve into anything like a thinking organism. We don't know for sure, but life probably needs billions of years, not millions, to reach the level of worms and plankton, never mind intelligence. The next time you look at a star filled night sky and wonder who might be out there, slap yourself upside the head and remember that *those* stars for the most part, are not friendly to life. Those are just the showy stars, the flamboyant stars, the most extravagant examples of nuclear physics (Achenbach 1999, 281).

We now need to examine what characteristics determine whether a star—assuming that it has both a habitable zone and a planet moving in that zone—will exist for a sufficiently long period that higher forms of life will develop. Our benchmark for estimating how long in takes for higher forms of life to develop is what has happened in our solar system, where we now know that it has taken near 4.5 billion years to form and develop conditions adequate for us. It turns out that although no simple linear relationship applies, we can say that for the most part, the lifetime of a star varies inversely with its mass. This entails that the supergiants and giants do not have lifetimes long enough for higher forms of life to develop on planets in their habitable zones. Thus it is clear that among the stars visible to Kant on a clear night nearly all must be barren of higher forms of life.

It should be stressed that this analysis applies only to the relatively small number of stars visible to humans. It leaves out of consideration stars of mass significantly lower than that of our Sun because they lack naked eye visibility. Such stars do face some problems, for example, their planets (if they have such) tend to become tidally locked (as is our Moon) so that they always keep the same side to their suns. Some recent work indicates that such stars may turn out to be reasonably good candidates for habitability.

1.7.3 On Investigating the History of the Extraterrestrial Life Debate

Persons interested in extending their knowledge of the ETD may wish to know that my long historical treatment (Crowe 1986) now has a companion in the form of a source book (Crowe 2008), which was specifically designed for classroom use; in fact, Dr. Dowd and myself have for a number of years used it as a text for the first half of a University of Notre Dame course called The Extraterrestrial Life Debate: A Historical Perspective. Professor Peter Ramberg of Truman State University has also made extensive use of the source book. This 2008 volume incorporates a useful bibliography of publications on the ETD.

Scholars wishing to take up any topics treated in my two volume may wish to know that my main research files (running about thirty linear feet) developed while writing these volumes and teaching in this area were donated in 2011 to the Adler Planetarium and Astronomical Museum in Chicago, where they have been carefully referenced and cataloged and will shortly become available to researchers.

References

Achenbach, Joel. 1999. *Captured by Aliens: The Search for Life and Truth in a Very Large Universe*. New York: Simon and Schuster.

Brooke, John H. 1977. "Natural Theology and the Plurality of Worlds: Observations on the Brewster-Whewell Debate." *Annals of Science* 34: 221–286.

Clerke, Agnes. 1887. *Popular History of Astronomy in the Nineteenth Century*, 2nd ed. Edinburgh: A. and C. Black.

Comte, Auguste. 1968. *Cours de philosophie positive*, vol. 2 of *Oeuvres de Auguste Comte*. Paris: Éditions Anthropos.

Crowe, Michael J. 1986. *The Extraterrestrial Life Debate 1750–1900: The Idea of a Plurality of Worlds from Kant to Lowell*. Cambridge: Cambridge Univ. Press. Repr. 1999, as *The Extraterrestrial Life Debate 1750–1900*. Mineola, NY: Dover.

Crowe, Michael J. 2006. *Mechanics from Aristotle to Einstein*. Santa Fe: Green Lion.

Crowe, Michael J., ed. 2008. *The Extraterrestrial Life Debate, Antiquity to 1900: A Source Book*. Notre Dame: University of Notre Dame Press.

Crowe, Michael J. 2011. "The Surprising History of Claims for Life on the Sun." *Journal of Astronomical History and Heritage* 14:169–179.

Dales, Richard C. 1990. *Medieval Discussions of the Eternity of the World*. New York: E. J. Brill.

Dick, Steven J. 1982. *Plurality of Worlds: The Origins of the Extraterrestrial Life Debate from Democritus to Kant*. Cambridge: Cambridge University Press.

Dick, Steven J. 1998. *Life on Other Worlds: The Twentieth-Century Extraterrestrial Life Debate*. Cambridge: Cambridge University Press.

Evans, Ifor. 1954. *Literature and Science*. London: Allen and Unwin.

Finocchiaro, Maurice A. 2002. "Philosophy versus Religion and Science versus Religion: The Trials of Bruno and Galileo." In *Giordano Bruno: Philosopher of the Renaissance*, ed. Hilary Gatti, 51–96. Aldershot: Ashgate.

Goodman, Matthew. 2008. *The Sun and the Moon: The Remarkable True Account of Hoaxers, Showmen, Dueling Journalists, and Lunar Man-Bats in Nineteenth-Century New York*. New York: Basic Books.

Gregory, Brad S. 2012. "Giordano Bruno Superstar." Review of Rowland. 2008. *Books and Culture* 18, no. 2 (March/April):19–21. Also available through booksandculture.com.

Hennessey, Roger. 1999. *Worlds without End: The Historic Search for Extraterrestrial Life*. Charleston: Tempus Publishing.

Herschel, John. 1833. *Treatise on Astronomy*. London: Longman.

Herschel, William. 1912. *The Scientific Papers of Sir William Herschel*. London: Royal Society and Royal Astronomical Society.

Kant, Immanuel. 1949. *Critique of Practical Reason and Other Writings in Moral Philosophy*. Translated by Lewis White Beck. Chicago: University of Chicago Press.

Kozhamthadam, Job. 1994. *The Discovery of Kepler's Laws: The Interaction of Science, Philosophy, and Religion*. Notre Dame: University of Notre Dame Press.

Mach, Ernst. n.d. *The Principles of Physical Optics: An Historical and Philosophical Treatment*. New York: Dover.

Manning, R.J. 1993. "John Elliot and the Inhabited Sun." *Annals of Science* 50: 349–364.

Meadows, A.J. 1970. *The Early Solar Physics*. Oxford:Pergamon Press.

Pascal, Blaise. 1938. *Thoughts*. Translated by W.F. Trotter. Harvard Classics edition. New York: P. F. Collier and Son.

Rowland, Ingrid D. 2008. *Giordano Bruno: Philosopher/Heretic*. New York: Farrar, Straus and Giroux.

Sheehan, William, and Thomas Dobbins. 2001. *Epic Moon: A History of Lunar Exploration in the Age of the Telescope*. Richmond: Willmann-Bell.

Singer, Dorothea. 1950. *Giordano Bruno: His Life and Thought, with Annotated Translation of His Work "On the Infinite Universe and Worlds."* New York: Henry Schuman.

Westman, Robert. 1980. "The Astronomer's Role in the Sixteenth Century: A Preliminary Study." *History of Science* 18: 105–147.

Whewell, William. 2001. "A Dialogue on the Plurality of Worlds." In *Of the Plurality of Worlds, Facsimile*, ed. Michael Ruse. Chicago: University of Chicago Press.

Yates, Frances. 1964. *Giordano Bruno and the Hermetic Tradition*. London: Routledge and Kegan Paul.

Further Readings:

Basalla, George. 2005. *Civilized Life in the Universe: Scientists on Intelligent Extraterrestrials*. New York: Oxford Univ. Press.

Boss, Valentin. 1972. *Newton and Russia: the Early Influences, 1698–1796*. Cambridge: Harvard Univ. Press.

Crowe, Michael J. 1997. "Extraterrestrial Intelligence." In *History of Astronomy: An Encyclopedia*, ed. John Lankford, 207–209. New York: Garland.

Danielson, Dennis. ed. 2000. *The Book of the Cosmos: Imagining the Universe from Heraclitus to Hawking*. Cambridge: Perseus Publishing.

Goldsmith, Donald. ed. 1980. *The Quest for Extraterrestrial Life: A Book of Readings*. Mill Valley CA: University Science Books.

Guthke, Karl S. 1990. *The Last Frontier: Imagining Other Worlds from the Copernican Revolution to Modern Science Fiction*. Translated by Helen Atkins. Ithaca: Cornell University Press.

Jaki, Stanley. 1978. *Planets and Planetarians*. Edinburgh: Scottish Academic Press.

Lovejoy, Arthur. 1960. *The Great Chain of Being*. New York: Harper and Row.

Ross, Joseph. 1991. "Kant's and Hegel's Assessment of Analogical Arguments for Extraterrestrial Life." M.A. thesis, University of Notre Dame.

Tipler, Frank. 1981. "A Brief History of the Extraterrestrial Intelligence Concept." *Quarterly Journal of the Royal Astronomical Society* 22: 133–145.

Westman, Robert. 1975. "The Melanchthon Circle, Rheticus, and the Wittenberg Interpretation of the Copernican Theory." *Isis* 66: 165–193.

Whewell, William. 1859. *Of the Plurality of Worlds: An Essay*, 5th ed. London: John W. Parker and Son.

Chapter 2
Early Modern ET, Reflexive Telescopics, and Their Relevance Today

Dennis Danielson

Abstract The period from the discovery of Tycho's New Star in 1572 to Galileo's "geometrization of astronomical space" in 1610 (and the years following) saw the disintegration of the boundary between the sublunary and superlunary spheres—between the "lower storey" and "upper storey" of the Aristotelian Universe. This establishment of a strong physical affinity between the universe "up there" and the earthly realm "down here" was also complemented by the rise of Copernicanism: for once the Earth was seen as a planet, the other planets could readily be imagined as other Earths. This analogy suggested not only physical but also biological affinities and supported the plausibility of humans' capacity to travel to the Moon and beyond. Robert Burton—given the demise of Aristotle's physics—declared in 1621 that "If the heavens be penetrable ... it were not amiss in this aerial progress to make wings and fly up." John Wilkins and Francis Godwin in the 1630s actively imagined creatures in the Moon and human journeys thither. The epic poet John Milton in 1667 hinted that "every star [is] perhaps a world / Of destined habitation." Moreover, space travel was no one-way street: Thomas Traherne in the 1670s imagined a dweller among the stars visiting Earth and remarking on what must be the condition of its inhabitants. In these and other ways, seventeenth-century writers offered serious and impressive speculation about extraterrestrial life and its possible perceptions of Earth. Such speculations remain pertinent to astrobiological theory today. What Hans Blumenberg in the 1970s called "reflexive telescopics"—the examination of Earth from an imagined extraterrestrial viewpoint—is an important counterpart to the search for life "out there." It serves as a reminder of the obvious but profound premise that Earth is part of the cosmos. At a popular level we often continue to speak of "outer space" as if the old "two-storey" picture of the universe still had some residual legitimacy. However, if Galileo, Wilkins, and other devotees of the New Astronomy were right about Earth's being a full participant in "the dance of the stars," then "outer" is a merely relative and parochial term, not a

D. Danielson (✉)
Department of English, University of British Columbia, Vancouver, BC, Canada
e-mail: danielso@mail.ubc.ca

D. A. Vakoch (ed.), *Astrobiology, History, and Society*, Advances in Astrobiology and Biogeophysics, DOI: 10.1007/978-3-642-35983-5_2,
© Springer-Verlag Berlin Heidelberg 2013

scientific or qualitative one. And it is no trivial claim to assert that the search for intelligent life in the universe has already identified its first specimens.

2.1 Introduction

As the now-classic studies of Crowe (1986) and Dick (1982) have indicated, ET-related discussions began in antiquity and were not completely neglected in the European Middle Ages. Nonetheless, a series of landmark developments in mathematical and observational astronomy drove that discussion forward during the early modern period, and our capacity to conceive of space travel and ultimately of astrobiology was powerfully shaped by two particular sixteenth-century innovations in astronomy. To grasp their significance, we do well to review some of the assumptions of the old astronomy that eventually the theories and observations of astronomers such as Nicolaus Copernicus and Tycho Brahe utterly undermined.

2.2 "Two Storeys" of the Universe

Astronomical teaching in the European universities, resting chiefly on the *Physics* of Aristotle and the *Almagest* of Ptolemy,[1] involved (generally speaking) the following premises (see Fig. 2.1, the Ptolemaic cosmos from Apian 1550). The Universe is immense in size but finite, consisting of ten spheres, the tenth or outer one being the "Prime Mover" and the ninth being the crystalline sphere. Below these is the eighth, the sphere of the fixed stars, followed by the spheres of the seven planets, or wandering stars: Saturn, Jupiter, Mars, the Sun, Venus, Mercury, and the Moon. Beneath the sphere of the Moon is located the (unnumbered) "elementary" sphere, comprising the domains of the four Aristotelian elements, in descending order: fire, air, water, earth. The domains above the Moon and below it—also known respectively as the superlunary sphere and the sublunary sphere—are qualitatively different from each other as to their physics and substance. Motion is governed by different laws peculiar to the two realms, and things in the superlunary sphere are composed not of earth, water, air, or fire, but of a fifth element, or quintessence. There was thus in the Aristotelian cosmos no known physical or scientific basis for presuming any analogy between what Arthur Koestler has dubbed the "two storeys" of the Universe (Koestler 1959, 59–62). The most radical practical difference between the superlunary and sublunary realms is that "up there" nothing ever changes, whereas "down here" *everything* changes: all things come to be and pass away. Indeed, the very processes of what we now call biology—conception, birth, growth, decline, death, and the manifold mutability that

---

[1] See Aristotle (1930) and Ptolemy (1984), but also excerpts from these works in Danielson (2000), Chaps. 6 and 11.

Fig. 2.1 The Ptolemaic universe (from Apian 1550). Courtesy of the Linda Hall Library of Science and Technology

characterizes them all—were held to take place uniquely in the sublunary realm. With the new astronomy and physics of the early modern period, however, those assumptions would gradually undergo a complete revision.

2.3 "The Shadow of Heaven"

The English astronomer Thomas Digges neatly embodies Copernicus's and Tycho's twin contributions to the dissolution of the Aristotelian model of the universe and to subsequent discussions of ET. Like Tycho, Digges studied the Supernova of 1572 to determine whether that new phenomenon was sublunary or superlunary: whether it was truly stellar. And, like Tycho, he concluded on observational and trigonometric grounds that it genuinely was superlunary.[2] In early

---

[2] For the gripping account of the Danish astronomer's first observations of what came to be known as Tycho's Supernova, see Brahe (1929).

1573 Digges published (in Latin) *Alae seu scalae mathematicae, quibus visibilium remotissima coelorum theatra conscendi—Mathematical wings or ladders whereby we may ascend the highest theater of the visible heavens.* Even though this important work on parallax was not specifically about the new star of 1572, its prefatory dedication to William Cecil unambiguously declared the new phenomenon to be "far beyond the sphere of the moon" (Digges 1573, sig. A.iii.^{v}). A demonstration that mutability exists in *both* domains made it possible for humans to begin imagining what else the two domains might have in common. The observation of the new star's coming to be and passing away thus inserted into human thinking the thin edge of a powerful analogy. Even allowing for great differences, it created grounds for adumbrating similitudes, and it raised the question, as Milton would write almost a century later:

> What if Earth
> Be but the shadow of heaven, and things therein
> Each to other like, more than on earth is thought? (5.574–576).

In short, Tycho's and Digges's observations of change taking place in the realm of the stars began to erase the boundary between the lower and upper storeys of the Universe, so that these no longer needed to be thought of as radically distinct or characterized by utterly dissimilar physics, substances, and beings.

Then in 1576 Digges published *A perfit description of the Caelestiall Orbes*, in which he included the main cosmological parts of book I of Copernicus's *De revolutionibus*—the very first translation of that work from Latin into any vernacular language. Digges also included an influential graphic of the Copernican system that continues to appear regularly in books on the history of astronomy (see Fig. 2.2). Digges's foreword "To the Reader" begins by mentioning the older model—"according to the doctrine of Ptolemy, whereunto all universities … have consented"—and continues:

> But in this our age one rare wit (seeing the continual errors that from time to time more and more have been discovered, besides the infinite absurdities in their theorics, which they have been forced to admit that would not confess any mobility in the ball of the Earth) hath by long study, painful practice, and rare invention delivered a new theoric or model of the world, showing that the Earth resteth not in the center of the whole world, but only in the center of this our mortal world or globe of elements which environed and enclosed in the Moon's orb, and together with the whole globe of mortality is carried yearly round about the Sun, which like a king in the midst of all reigneth and giveth laws of motion to the rest, spherically dispersing his glorious beams of light through all this sacred celestial temple. And the Earth itself to be *one of the planets* having his peculiar and straying courses turning every 24 hours round upon his own center whereby the Sun and great globe of fixed stars seem to sway about and turn, albeit indeed they remain fixed (Digges 1576, sig. M.1^r; italics added).

While reveling in both the science and the poetic flavor of Copernicus, Digges also noticeably retained, indeed reinserted, much vocabulary inherited from the system that Copernicanism would displace—"mortal world," "globe of elements," etc.—language that Copernicus's original text does not in fact employ. Moreover, Digges retained these pictorially as well as textually. The main thing presented in his famous graphic is the Copernican planetary system, with Mercury, Venus,

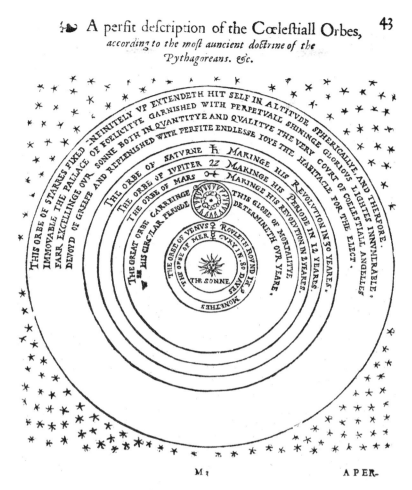

Fig. 2.2 Thomas Digges's version of the Copernican Universe (from Digges 1576). Courtesy of Owen Gingerich

Earth, Mars, Jupiter, and Saturn circling the central Sun. But if we look closely, we notice that the terrestrial system—the Earth with the Moon circling it—is simply a shrunken, simplified version of the old Ptolemaic sublunary sphere enclosing the familiar Aristotelian elements (compare Fig. 2.1). It is impossible to determine Digges's motivations for offering this particular presentation of the universe. However, deliberately or not, his graphic would have permitted a sixteenth-century audience to contemplate the encompassing Copernican cosmology without having to jettison the "local arrangements" in which they had been taught to feel so at home. And crucially, the picture is modular in one further important respect. Just as Digges neatly cut-and-pasted the familiar sublunary core of the Ptolemaic universe into that of Copernicus, so in turn he quietly but dramatically inserted

Copernicus's finite cosmos—what we now know simply as the solar system—into an infinite super-cosmos of stars that "infinitely up extendeth itself in altitude spherically" (see Fig. 2.2). In short, accepting the Copernican cosmos as Digges presented it would have permitted one to continue believing in Ptolemy's sublunary sphere and yet at the same time to venture forth imaginatively into an infinite universe inherited from Atomists such a Lucretius.

2.4 "Each Star an Island"

Such hybridity, or syncretism—the tendency to combine and harmonize disparate elements from different philosophical traditions—is one of the most prominent and occasionally endearing or befuddling elements of Renaissance-humanist thought. A further relevant example of it in Digges's time is offered by a long Latin poem by the Italian Marcello Palingenius (ca. 1500–1543) called the *Zodiacus vitae*, which Digges quotes approvingly a number of times in *A perfit description*'s foreword. Although in many ways a philosophical hodgepodge, this poem must have appealed to sixteenth-century readers, for it was printed in England seven times in Latin between 1569 and 1599 (four of these by Digges's own printer), and three times in Barnaby Googe's English translation (1565, 1576, 1588). For our purposes, it is notable not only for continuing to offer the familiar gloomy estimation of life in the sublunary sphere ["all that nature framed beneath the Moon, is nought, and ill" (Palingenius 1565, sig. GG.vii.^{r})], but also for proposing a number of times the likely existence of extraterrestrials. Palingenius argues this claim from the immense largeness of the universe relative to the Earth ("the seas and earth … are[,] compared to the skies[,] as nothing") and from the quasi-religious belief that God's creativity tends to fill all the places he creates.[3] Of the Earth, accordingly, Palingenius writes:

> Shall then so small and vile a place so many fish contain [,]
> Such store of men, of beasts and fowls and th'other void remain?
> Shall skies and air their dwellers lack? He dotes that thinketh so.
> (Palingenius 1565, sig. X.v.^{r})

Thus he concludes that "Each star an island shall be thought," and "doubtless heaven, stars, and air inhabitants enjoys" (Palingenius 1565, sig. X.vi.^{r}).

Palingenius was neither an astronomer nor a Copernican, yet his writings are evidence that talk of ET and of inhabited stars in an immense created universe—atomism without atheism, as it were—had already appeared prominently on the scene in the sixteenth century some decades before the writings of Giordano Bruno, whom general accounts often give quite disproportionate credit for introducing such ideas.[4] Digges, whose work also preceded both Bruno's and Kepler's, was

---

[3] On this "Principle of Plenitude," see the previous chapter, Crowe and Dowd (2013).

[4] On, for example, Bruno's tenuous grasp of Copernicanism, see McMullin (1987).

indeed an astronomer and a Copernican (England's first thoroughgoing one). He lucidly recognized, moreover, that Copernicus's cosmology was proposed as no mere mathematical model: "Copernicus meant not as some have fondly excused him to deliver the grounds of the Earth's mobility only as mathematical principles, fained and not as philosophical truly averred" (sig. M.1^r). Digges was thus clear—as many interpreters even into the seventeenth century were *not* clear—that the Copernican proposal offered no mere saving of the appearances, no purely instrumentalist model. Yet, as already indicated, Digges deftly fused a realistic Copernicanism with familiar vocabulary inherited from the Aristotelian/Ptolemaic model, repeatedly referring to Earth, for example, as a "dark star" (sig. M.2^r).

As far as ET is concerned, however, the crucial word here is simply "star." I have argued elsewhere that one of the great original impediments to the acceptance of Copernicanism was not, as many have assumed, its *demotion* of the central Earth but rather its demotion of the Sun from its planetary status and *exaltation* of Earth into the heavens—not a suitable place for something that in Aristotelian physics was supposed to be the sump of the Universe, or in Giovanni Pico's choice words the "excrementary and filthy parts of the lower world" (Pico 1948, 224; see also Danielson 2001, and 2006, 75–78). In the old cosmology, the planets (as their names even today still suggest) were identified with divinity—something to which the upstart Copernican Earth now cheekily seemed to aspire. Most decisively for the present discussion, the reconception of Earth as a planet, a wandering *star*, established a firm analogy between it and the other planets.

2.5 "The Light of the Earth"

The period from the discovery of Tycho's New Star in 1572 to Galileo's pursuit of telescopic observations of the heavens in 1609 and the years following accordingly saw further disintegration of the imagined boundary between the sublunary and superlunary spheres—between the lower and upper storeys of the Aristotelian Universe. This cosmological "homogenization"—the establishment of a strong physical uniformity between the Universe "up there" and the earthly realm "down here"—was complemented by the rise of Copernicanism. For the analogy just mentioned was a two-way street: not only was Earth reconceived as a planet, but also the other planets could now readily be imagined as other Earths.

Galileo's telescopic observations of the Moon powerfully reinforced the same analogy, particularly as complemented by his demonstration that geometrical dimensions could be calculated in the heavens. We so take this application for granted that we may miss how radical it was at the time. By contrast with Aristotelian tenets concerning qualitatively different sublunary and superlunary spaces, in Galileo's Universe geometry (literally "Earth measure") applies up there as well as down here. This is what Samuel Edgerton refers to as the "geometrization of astronomical space" (Edgerton 1991; Danielson 2000, Chap. 25). As soon as Galileo saw the similarity between mountains on Earth and mountains on the Moon, he set out by means of measuring shadows and angles to compute the elevations of the lunar mountains. He noticed that some peaks appeared illuminated even though they stood on the dark side of the "terminator," the line on the Moon dividing light and dark. And, using basic trigonometry, he calculated that those mountains are higher than any in the Alps. As Copernicus's student Rheticus had written some years earlier, deliberately echoing the Lord's Prayer, thus we behold "God's geometry in heaven and on earth" (quoted in Danielson 2006, 82). As Johannes Kepler so poetically expressed it in his response to Galileo's *Sidereus nuncius*, "Geometry ... shines in the mind of God" (Kepler 1610, 43). Little wonder, then, that in such an age Milton should imagine the Creator himself as exercising his geometry by means of eternal drafting tools—his golden compasses—to shape and "circumscribe / This universe" (7.226–227).

A further highly significant recognition arising from Galileo's lunar observations—again, which we might simply take for granted but which had radical implications—concerns the behavior of light. Consider first that in Aristotle's Universe the natural tendency of all things is to fall or flow downward, or inward, toward the center. Earth was therefore generally thought to receive, but not to emit, light, just as it was the recipient literally of astronomical or astrological influences. Thus, from a cosmic and extraterrestrial perspective, Earth was inevitably dark, as implied by the residual Aristotelian/Ptolemaic vocabulary even of a genuine Copernican such as Digges ("dark star" etc.).

Yet once Copernicanism had fully grasped Earth's star status, that idea concerning terrestrial darkness could not long endure. Hence the importance of Galileo's empirical confirmation of Earth's brightness as reported in *Sidereus nuncius*. Conducting the first telescopic examination of the dark side of the Moon, Galileo discerned that it is in fact bathed in gentle light reflected from the Earth, just as Earth reciprocally receives light from the Moon:

> ... In its cycle each month the Moon gives us alternations of brighter and fainter illumination. But the benefit of her light to the Earth is balanced and repaid by the benefit of the light of the Earth to her. ... This is the law observed between these two orbs: whenever the Earth is most brightly enlightened by the Moon, that is when the Moon is least enlightened by the Earth, and vice versa.
>
> That is all I need say for now on this subject, which I will consider more fully in my *System of the Universe*, where many arguments and experimental proofs will be provided to demonstrate a very strong reflection of the Sun's light from the Earth—this for the benefit of those who assert, principally on the grounds that it has neither motion nor light, that the Earth must be excluded from the dance of the stars (Danielson 2000, 149–150).

Even today, scientists and the general public often only dimly grasp the extent to which Copernicanism thus raised, not lowered, the cosmic status of the Earth; and it did so in part by theorizing Earth as a light-bearer and light-sharer (along with the Moon and other planets). It did so, in other words, by making Earth part of a dynamic cosmic community—no longer, as Galileo's words indicate, "excluded from the dance of the stars." Indeed, from the demonstration of the Moon's and the Earth's "grateful exchange" of light, John Wilkins, three decades later in England, would go on to extrapolate and emphasize what we now assume to be the homogeneity or uniformity of space. Within this space, he declared, these twin planets inhabit "but one region" and enjoy a mutually beneficial light-sharing relationship "as loving friends" (Wilkins 1638, 153). The Aristotelian/Ptolemaic dark, isolated, sump-like Earth could hardly have undergone a more radical imaginative transformation, nor one of greater consequence for humankind's capacity to imagine—and perhaps form relationships with—beings elsewhere in the Universe.

2.6 "To Make Wings, and Fly Up"

Johannes Kepler instantly recognized these implications. Upon reading Galileo's *Sidereus nuncius* in 1610, Kepler immediately asserted the probability "that there are inhabitants not only on the moon but on Jupiter too," and went on to speculate that the "Jovians" may enjoy four moons (unlike us, who have only one) as consolation for the fact that they are less ideally located in the universe than we Earthlings. Kepler also boldly prophesied the day when we might launch our own lunar and planetary expeditions. For surely "settlers from our species ... will not be lacking" and "given ships or sails adapted to the breezes of heaven, there will be those who will not shrink from even that vast expanse" (Kepler 1610, 39–41).[5]

Not only such imaginative journeys but also the science that supported them—the robust physical analogy between the upper and lower storeys of the Universe, complemented by a planetary Earth and changes proven to be taking place in the heavens—received ever greater acknowledgment in the seventeenth century. In his encyclopedic *Anatomy of Melancholy* in 1621, Robert Burton, recognizing the demise of Aristotelian physics and the abolition of crystalline spheres, endorsed the *sine qua non* of space travel: that "If the heavens be penetrable ... it were not amiss in this aerial progress to make wings, and fly up" (Burton 1621, 325). Moreover, the achievements of Copernican astronomers such as Galileo and Kepler led to what historian David Cressy has called "England's lunar moment" (Cressy 2006, 967). In 1638, two influential works appeared that helped awaken more thoughts of space travel than ever before. The first of these, published posthumously, was Francis Godwin's imaginative fiction *The Man in the Moon: or a Discourse of a Voyage Thither*. Some elements of Godwin's narrative, such as the tethered flock of geese that conveys the main character to the Moon, are indeed fanciful. But the journey offers a vivid, non-Aristotelian account of physical features such as gravitation as well as the daily rotation of the Earth "according to the late opinion of Copernicus." What Godwin's fiction perhaps most movingly conveys, however—something actualized powerfully and photographically in

---

[5] Kepler extended his brilliant exploration of possible lunar travel, environment, and perspectives much further in his posthumous *Somnium* (Kepler 1967). For more on how Kepler and other early moderns prepared the way for eventual spacetravel, see Danielson (2011).

late 1968 by the Apollo 8 mission—is a vision of our own planet as a "new star" masked "with a kind of brightness like another moon" (Godwin 1638, 90, 92).

In a second English work appearing in 1638, *The Discovery of a World in the Moon*, John Wilkins extrapolated from his enthusiastic presentation of Copernican astronomy the idea of an inhabited Moon. Like Godwin, Wilkins not only vividly described conditions on the Moon but also imagined the shining appearance of our native globe from space. With some relish, he supported the scientific credibility of this imagined vision by citing the authority of two contemporary anti-Copernican Continental philosophers:

> Thus also Carolus Malapertius, whose words are these … "If we were placed in the Moon, and from thence beheld this our Earth, it would appear unto us very bright, like one of the nobler planets." Unto these doth Fromondus assent, when he says … "I believe that this globe of Earth and water would appear like some great star to any one who should look upon it from the Moon" (Wilkins 1638, 149–150).

Admitting the difficulties of a lunar voyage but building, like others, on the recent success of journeys to earthly places such as America (another kind of "New World"), Wilkins concluded by eloquently reprising the prophetic strains of Kepler. He could not, he admitted, conjecture how one might sail to the Moon. "We have not now any Drake or Columbus to undertake this voyage, or any Daedalus to invent a conveyance through the air. However, I doubt not but that time who is still the father of new truths … will also manifest to our posterity that which we now desire but cannot know" (Wilkins 1638, 107).[6]

2.7 Reflexive Telescopics

In the epic *Paradise Lost* (1667 and 1674)—which John Tanner has called "perhaps the greatest description of space travel in high-brow fiction" (Tanner 1989, 268)—John Milton hinted that "every star [is] perhaps a world / Of destined habitation" (7.621–622) and presented his anti-hero Satan as an astronaut (literally a sailor among the stars). In the second-to-last stop on his journey to tempt humankind, the Adversary alights on the Sun to asks directions of its resident angel, who offers him a thoroughly Galilean prospect of the Earth, which appears as a globe that "shines" and so *can be seen*, just like the other wandering stars, in particular like its "neighboring Moon / (So call that opposite fair star)" (3.727–728). In the earlier words of Robert Burton, the Earth "shines to them in the Moon, and to the other planetary inhabitants, as the Moon and they do to us" (Burton 1621, 326–327). Repeatedly in the seventeenth century, therefore, both scientists and poets not only looked outward into a newly conceived Universe but also exercised "reflexive telescopics," a phrase coined by Hans Blumenberg in the 1970s—shorthand for the imagined examination of Earth from an extraterrestrial viewpoint,

---

[6] The other prominent mid-seventeenth account of a lunar voyage was that of Cyrano de Bergerac (posthumously published in 1657).

complementing terrestrial observation of and speculation concerning what is "out there." According to Blumenberg, no sooner had Galileo trained his telescope upon the Moon than the question arose, How would the Earth appear through a telescope? (Blumenberg 1987, 675). At the end of this chapter I shall return to one of the most detailed and beautiful instances, written by Thomas Traherne in the 1670s, of such a scenario: that of a dweller among the stars approaching our Earth and remarking on what must be the condition of its inhabitants.[7]

First, however, two further late seventeenth-century writers deserve mention. Bernard le Bouvier de Fontenelle, popularizer of a Cartesian version of Copernicanism, was introduced in the previous chapter and is known for colorfully pondering the implications of an infinite universe, in which for example the Milky Way comprises an "ant-hill of stars, ... seeds of worlds." Yet it should be noted how carefully Fontenelle strove, within this expanded model of the universe, to deny that the cosmic immensities negated the value of the small and the local. In his dialogue, the philosopher accordingly assures the beautiful marquise: "The infinite multitude of other worlds may render this [world] little in your esteem, but they do not spoil fine eyes, a pretty mouth, or make the charms of wit ever the less: These will still have their true value ... in spite of all the worlds in the Universe." Our gaze outward into the cosmos must be complemented by a due regard for the undoubted value of things within our own world, even granted that we are "but one little family of the Universe" (Fontenelle 1688, 141, 136, 94).

The language of kinship was extended extraterrestrially by the Dutch scientist Christiaan Huygens in his late-seventeenth-century re-articulation the familiar analogy between our planet Earth and other planets upon which the ET hypothesis was chiefly founded. In his *Celestial Worlds Discovered* (1698), Huygens summarized key assumptions whose foundations had been a-building for more than a century:

> A man that is of Copernicus's opinion, that this Earth of ours is a planet, carried round and enlightened by the Sun, like the rest of them, cannot but sometimes have a fancy, that it's not improbable that the rest of the planets have their dress and furniture, nay and their inhabitants too as well as this Earth of ours: especially if he considers the later discoveries made since Copernicus's time of the attendants of Jupiter and Saturn, and the champaign and hilly countries in the Moon, which are an argument of a relation and kin between our Earth and them (Huygens 1698, 1–2; see also Aït-Touati 2011, 95–129).

The kinship of which Huygens writes is precisely what forms the foundation too of reflexive telescopics, an intellectual exercise still highly relevant to wider considerations of astrobiology today.

I conclude this chapter, then, with an application of reflexive telescopics, beginning with a brief interpretation of the frontispiece (Fig. 2.3) that appeared in John Wilkins's re-publication in a single volume in 1684 of two works he had published in 1638 and 1640. The first of the works mentioned on this title page (discussed earlier under its original title, *The Discovery of a World in the Moone*) is the

---

[7] In the context of this volume it might be remarked that if Ted Peters can design a questionnaire that asks Earthlings their opinions concerning extraterrestrials (Peters 2013), it is certainly a reasonable exercise to ponder what extraterrestrials might think about us.

Fig. 2.3 Frontispiece from
Wilkins (1684). Courtesy
of the Linda Hall Library of
Science and Technology

Discourse concerning a New World, Wilkins's extrapolation from the work of both
Galileo and Kepler regarding the nature of the Moon as an earthlike planet with,
by analogy, earthlike inhabitants. As already acknowledged, every analogy has two
sides, and the "discovery" of an earthlike Moon was thus appropriately followed
by an examination of what Wilkins called "another planet"—by which of course
he meant the planet Earth, still apparently a novel-sounding idea even a 141 years
after the publication of Copernicus's *De revolutionibus*.

And as for Copernicus, there he is in Wilkins's frontispiece posing the hypothesis that in real life he actually *never* posed merely as a hypothesis: "What if
it be thus?" But on the other side of the title stand Wilkins's other two heroes,
Galileo and Kepler, who represent the twin pillars of astronomy then as now: (1)

Fig. 2.4 Detail from
Fig. 2.3, frontispiece from
Wilkins (1684). Courtesy
of the Linda Hall Library of
Science and Technology

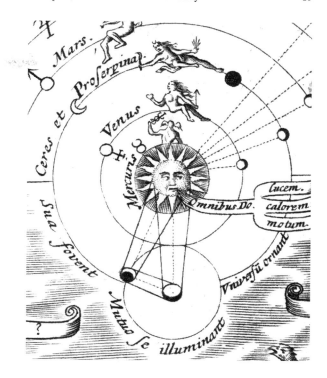

Observation (Galileo with his telescope, "Here be his eyes"); and (2) Mathematics (Kepler: "Yes, and his wings"—a metaphor reaching back through Digges to Rheticus and even to Plato; see Danielson 2006, 25–26). The other thing to notice in the top third of Wilkins's frontispiece is that the Copernican cosmos offered here is the *English* Copernican cosmos inherited from Digges, with the stars not in a nice neat band or stellar sphere, but seeming to spill off over the edge of the page and so suggesting a possibly infinite Universe.

Nonetheless, even within that immensity there is a coherent family of Sun-and-planets (see Fig. 2.4) who in Wilkins's charming but serious cartoon literally eye each other. The Sun "gives light, warmth, and motion" to all of them, while the Moon and Earth, in accordance with their reciprocal luminosity as discovered by Galileo, "enlighten each other." No wonder that Kepler, Wilkins, and their inheritors, having postulated beings on the Moon and having arrived at the basic but decisive Galilean realization that Earth is visible from "out there," began imagining how Earth might appear to ET. In a manuscript discovered only in the 1990s but dating from the 1670s, Thomas Traherne offered just such a scenario, that of a "Celestial Stranger" discovering our planet for the first time:

Had a man been always, in one of the stars, or confined to the body of the flaming Sun, or surrounded with nothing but pure ether, at vast and prodigious distances from the Earth, acquainted with nothing but the azure sky, and face of heaven, little could he dream of any treasures hidden in that azure veil afar off. ... Should he be let down on a sudden, and see the sea, and the effects of those influences he never dreamed of; such strange kind of creatures; such mysteries and varieties; ... such never heard of colors; such a new and lively green in the meadows; such odoriferous and fragrant flowers; such reviving and refreshing winds, ... it would make him cry out How blessed are thy holy people, how divine, how highly exalted! Heaven it self is under their feet! ... The Earth seems to swell with pride, that it bears them all; all its treasure[s] laugh and sing to serve them. ... Verily this star is a nest of angels! ... This little star so wide and so full of mysteries! So capacious, and so full of territories, containing innumerable repositories of delight, when we draw near! Who would have expected, who could have hoped for such enjoyments? (Traherne 2002, 112–114).

Traherne's reflexive-telescopic thought experiment remains significant today for a number of reasons, and I end with these more personal reflections. First, I worry that at some popular level the search for exoplanets and for ET may potentially dilute the profound sense of responsibility and admiration we ought to have for our own local, precious, precarious planet. Vividly imagining how an extraterrestrial being might view our home and native star—and indeed exclaim concerning its glories—may mitigate any too cavalier attitude toward the availability (at least to us) of alternative habitats for life in the Universe. Second, a related point: If reflexive telescopics indeed offer a legitimate exploration of the possibilities of interplanetary or interstellar consciousness, then Earthlings must not be excluded or bracketed off from scientific theorizing about life in the Universe. Is it not misguided, we may reasonably inquire, to worry too much about purging our search for ET of "anthropocentrism"? Yes, of course we ought to remain open-minded about what forms other life or intelligence might manifest, but surely it is arbitrary and artificial not to pay special attention to the single sample we actually have of the very category of thing we are searching for.[8]

The editor of this volume has commented on the need to beware how prior assumptions impeded the acceptance of new discoveries that were later widely endorsed. So let me also suggest, against the backdrop of the achievements of the early modern period, that we continue the process of purging remnants of latent Aristotelianism that in the twenty-first century might still cloud our thinking about life in the Universe. For example, we still often carelessly speak of "outer space," as if the old "two-storey" picture of the Universe retained some residual legitimacy. However, if Galileo, Wilkins, and other devotees of the New Astronomy were right about Earth's being a full participant in the dance of the stars, then "outer" is a merely relative and parochial term, not a scientific or qualitative one. And it is a fact with truly cosmological implications that the search for life in the Universe has already identified its first specimens—whom, and whose astonishing home planet, we have compelling reasons to cherish and seek to preserve.

---

[8] See Sullivan's (2013) comments about the "N = 1" problem in Chap. 3.

References

Aït-Touati, Frédérique. 2011. *Fictions of the Cosmos: Science and Literature in the Seventeenth Century*. Translated by Susan Emanuel. Chicago: University of Chicago Press.

Apian, Peter. 1550. *Cosmographia*. Antwerp.

Aristotle. 1930. "Physics" and "On the Heavens." In *The Works of Aristotle*, ed. W. D. Ross, vol. 2. Oxford: Clarendon Press.

Blumenberg, Hans. 1987. *The Genesis of the Copernican World*. Translated by Robert M. Wallace. Cambridge, MA: MIT Press.

Burton, Robert. 1621. *The Anatomy of Melancholy*. Oxford.

Cressy, David. 2006. "Early Modern Space Travel and the English Man in the Moon." *American Historical Review* 111(October): 961–982.

Crowe, Michael J. 1986. *The Extraterrestrial Life Debate, 1750–1900: The Idea of a Plurality of Worlds from Kant to Lowell*. Cambridge: Cambridge Univ. Press.

Crowe, Michael J., and Matthew F. Dowd. 2013. "The Extraterrestrial Life Debate from Antiquity to 1900." In *Astrobiology, History, and Society: Life Beyond Earth and the Impact of Discovery*, ed. Douglas A. Vakoch. Heidelberg: Springer.

Danielson, Dennis, ed. 2000. *The Book of the Cosmos: Imagining the Universe from Heraclitus to Hawking*. Cambridge: Perseus Publishing.

Danielson, Dennis. 2001. "The Great Copernican Cliché." *American Journal of Physics* 69(10): 1029–1035.

Danielson, Dennis. 2006. *The First Copernican: Georg Joachim Rheticus and the Rise of the Copernican Revolution*. New York: Walker.

Danielson, Dennis. 2011. "Ancestors of Apollo." *American Scientist* 99 (March-April):136–143.

de Bergerac, Cyrano. 1657. *L'Autre Monde: ou les États et Empires de la Lune*. Paris.

Dick, Steven J. 1982. *Plurality of Worlds: The Origins of the Extraterrestrial Life Debate from Democritus to Kant*. Cambridge: Cambridge Univ. Press.

Digges, Thomas. 1573. *Alae seu scalae mathematicae, quibus visibilium remotissima coelorum theatra conscendi*. London.

Digges, Thomas. 1576. *A Perfit Description of the Caelestiall Orbes*. London.

Edgerton, Samuel, Jr. 1991. "Geometrization of Astronomical Space." *The Heritage of Giotto's Geometry: Art and Science on the Eve of the Scientific Revolution*, 223–253. Ithaca: Cornell Univ. Press.

Fontenelle, Bernard le Bouvier de. 1688. *A Discovery of New Worlds*. Translated by Aphra Behn. London. (Original title: *Entretiens sur la Pluralité des Mondes.*).

Galileo Galilei. 1610. *Sidereus nuncius*. A good modern edition is *Sidereus Nuncius: or The Sidereal Messenger*. Translated by Albert van Helden. Chicago: University of Chicago Press, 1989. Quotations in the present chapter are from the version in Danielson (2000), Chap. 24.

Godwin, Francis. 1638. *The Man in the Moon: or a Discourse of a Voyage Thither*. London. Reprinted in ed. William Poole. Peterborough, Ontario: Broadview, 2009. References are to this edition.

Huygens, Christiaan. 1698. *Celestial Worlds Discovered, or, Conjectures Concerning the Inhabitants, Plants and Productions of the Worlds in the Planets* (original title: *Cosmotheoros*). London.

Kepler, Johannes. 1610. *Dissertatio cum Nuncio sidero*. Translated by Edward Rosen as *Kepler's Conversation with Galileo's Sidereal Messenger*. New York and London: Johnson Reprint Corporation, 1965.

Kepler, Johannes. 1967. *Somnium: The Dream*. Translated by Edward Rosen. Madison: University of Wisconsin Press.

Koestler, Arthur. 1959. *The Sleepwalkers: A history of man's changing view of the Universe*. London: Hutchinson.

McMullin, Ernan. 1987. "Bruno and Copernicus." *Isis* 78(1): 55–74.

Milton, John. 1667, 1674. *Paradise Lost*. London. The edition cited in this chapter is *Paradise Lost*, ed. John Leonard. London: Penguin, 2000.

Palingenius, Marcellus. 1565. *The Zodiake of Life*. Translated by Barnaby Googe. London.

Peters, Ted. 2013. "Would the Discovery of ETI Provoke a Religious Crisis?" In *Astrobiology, History, and Society: Life Beyond Earth and the Impact of Discovery*, ed. Douglas A. Vakoch. Heidelberg: Springer.

Pico, Giovanni. 1948. "Oration on the Dignity of Man." In *The Renaissance Philosophy of Man*, ed. Ernst Cassirer, et al., 223–254. Chicago: University of Chicago Press.

Ptolemy, Claudius. 1984. *Almagest*. Translated by G. J. Toomer. New York: Springer.

Sullivan, Woodruff T., III. 2013. "Extraterrestrial Life as the Great Analogy, Two Centuries Ago and in Modern Astrobiology." In *Astrobiology, History, and Society: Life Beyond Earth and the Impact of Discovery*, ed. Douglas A. Vakoch. Heidelberg: Springer.

Tanner, John. 1989. "'And Every Star Perhaps a World of Destined Habitation': Milton and Moonmen." *Extrapolation* 30(3): 267–279.

Traherne, Thomas. 2002. *Poetry and Prose*, ed. Denise Inge. London: SPCK.

Tycho Brahe. 1929. *De Nova Stella*. Translated by John H. Walden. In *A Source Book in Astronomy*, ed. Harlow Shapley and Helen E. Howarth. New York: McGraw-Hill.

Wilkins, John. 1638. *The Discovery of a World in the Moone*. London.

Wilkins, John. 1684. *A discovery of a new world; or, a discourse tending to prove that 'tis probable there may be another habitable world in the moon. … Unto which is added, A discourse concerning a new planet, tending to prove, that 'tis probable our earth is one of the planets.* London. (This work's frontispiece is dated 1683 and shows a variant version of the title).

A Note on References

For ease of reading I have modernized spelling and capitalization in all quotations from early modern publications; titles, however, have been left in their original spelling. For works lacking page numbers, in-text citations indicate instead the signature by letter, number, and whether recto or verso (e.g.: sig. X.vi.^{r}). In-text citations of Milton's *Paradise Lost* indicate book and line numbers (e.g. 7.226–227).

Chapter 3
Extraterrestrial Life as the Great Analogy, Two Centuries Ago and in Modern Astrobiology

Woodruff T. Sullivan III

Abstract Mainstream ideas on the existence of extraterrestrial life in the late 18th and early 19th centuries are examined, with a focus on William Herschel, one of the greatest astronomers of all time. Herschel viewed all of the planets and moons of our solar system as inhabited, and gave logical arguments that even the Sun, and by extension all of the stars, was a giant planet fit for habitation by intelligent beings. The importance for astrobiology both two centuries ago and now of the type of inductive reasoning called "analogy" is emphasized. Analogy is an imperfect tool, but given that we have only one known case of life and of a life-bearing planet, it is very difficult to make progress in astrobiology without resorting to analogy, in particular between known life and possible other life. We cannot overcome the "N = 1 Problem" without resorting to this "Great Analogy" to guide our research.

3.1 Introduction

The core questions of astrobiology are not new. They have always been asked and are central to Western intellectual history: How did life begin? How has it changed? What is the relation of *Homo sapiens* to other species? Does life exist elsewhere? If so, where might it be and what might it be like? For over 2000 years astrobiological ideas have been woven through the realms of natural philosophy, theology, and science—supported by evidence ranging from pure metaphysics to empirical science. Over that same period our perceived place in the cosmos has oscillated within and between the extremes of either (a) being the special, unique product of all creation, or (b) the *plurality of worlds*, in which every star is a Sun with peopled planets.

W. T. Sullivan III (✉)
Department of Astronomy, University of Washington, Seattle, WA, USA
e-mail: woody@astro.washington.edu

D. A. Vakoch (ed.), *Astrobiology, History, and Society*, Advances in Astrobiology and Biogeophysics, DOI: 10.1007/978-3-642-35983-5_3,
© Springer-Verlag Berlin Heidelberg 2013

In this chapter I will examine mainstream ideas on the existence of extraterrestrial life[1] in the late 18th and early 19th centuries, focusing on William Herschel (1738–1822), one of the greatest astronomers of all time. I will emphasize the importance of the type of inductive reasoning called "analogy" both then and now. For Herschel's context I rely heavily on the comprehensive histories of ideas on extraterrestrial life by Dick (1982) and Crowe (1986). The introductory chapter in this volume by Crowe and Dowd (2013) also provides excellent background.

Herschel asserted in 1795 that the Sun had a cool region below its surface and was in fact inhabited, and ever since this has been looked upon as an unfortunate misguided belief by the great astronomer. He in fact took *all* the planets (*and* our Moon) to be inhabited with intelligent beings, as were the myriad unseen planets that he presumed circled other stars. How did he come to such seemingly outlandish conclusions? In this chapter I will argue that this idea of Herschel's was no less rational than others of his that we readily applaud. We must look carefully at not only the science, but also its enveloping context. Whether two centuries ago or today, the scientific enterprise has always been shaped by metaphysics, doctrines, and predilections as received from philosophy, religion, and society. Both William Herschel *and* we today can do no more than tackle questions with the best tools available, apply the cleverest insight we can muster at the time, and struggle to fashion a consensus as to the nature of the world. For the case of extraterrestrial life, where we have only one known example to guide us, I argue that out of necessity the best one can do, in the past as well as today, is to follow as a guide the "Great Analogy," namely that of Earth and its life to other extraterrestrial locales and their possible life (Sullivan and Baross 2007, 5).

3.2 Analogy

Discussion of the concept of analogy and analogues can be found as early as Aristotle (Hesse 1966). Most modern philosophers discuss analogy with respect to the process of creating and verifying scientific models (say, of atomic structure), but a treatment more relevant to the present discussion was published in 1843 by John Stuart Mill in *A System of Logic* (Crowe 1986, 231–232; Crowe and Dowd 2013). In Chap. 20 of Book III Mills analyzes analogy as a form of induction that is incomplete in the sense that one has *not*, as in induction, experimentally ruled out all possible hypotheses except one (which must therefore be correct). Thus in analogical reasoning the conclusion can only be probabilistic, and establishing the correct degree of confidence in the validity of any given conclusion is very

---

[1] Until the 20th century, discussions of extraterrestrial life always centered on *intelligent* beings, with very limited interest in the possible existence of simple organisms.

difficult. Mill even uses the question of possible life on the Moon or planets as an example, deducing that it seems unlikely on the Moon, but somewhat likely on some planets.

Here is an example of Mill's principles of analogy relevant to today's astrobiology:

1. *Two objects A and B are known to be similar in two properties.*
   Earth and exoplanet KOI 77 have (1) similar radii and (2) similar distances from the Sun and a Sunlike star, respectively.
2. *Object A has a further third property, but this is not known for B.*
   Earth also has [an atmosphere, plate tectonics, oceans, a ~300 K temperature, microbial life, complex life....].
3. *Can we deduce that B probably has that third property also?*
   KOI 77 probably also has [an atmosphere, plate tectonics, oceans, a ~300 K temperature, microbial life, complex life....].

The strength of an argument from analogy such as this depends on:

(a) the relevance of the known shared properties to the unknown property
(b) the number and variety of *known* cases [only one (A) in the example]
(c) for the known shared properties [only two in this example]: their variety, their number, their fraction of all relevant properties, and the relative weights to assign each property.

Examining the above three criteria, we see that they are all rather subjective. Regarding (a), who is to say exactly how relevant or not a planet's radius and distance from its star is to whether or not it has an ocean or the presence of microbial life? Regarding (b), we immediately run into astrobiology's "N = 1 Problem," namely that we have only *one* known example of (i) a life-bearing planetary system, (ii) a life-bearing planet, (iii) a form of life, and (iv) an evolutionary history of life (Sullivan and Baross 2007, 5).[2] Finally, regarding (c), myriad complexities and conundrums arise: How do we decide how to weight the various possible similarities and their presence or lack of same–is planetary size more important than distance from its star? How many similarities must we have before we consider an argument by analogy to be convincing enough to, for example, hand a Nobel Prize to someone for first finding "another Earth"?!

After making all of these philosophical and logical points, Mill asks whether analogical argument has any use in science, and then answers in the affirmative, namely that analogy's highest scientific value is as a guide-post, pointing out the direction for more rigorous investigations. His answer is very apt for astrobiology today, for although we frustratingly cannot make airtight inferences, analogies do nevertheless importantly guide useful further research.

---

[2] The N = 1 Problem leads to heroic mathematical and logical efforts to overcome it. For example, see the recent insightful Bayesian analysis by Spiegel and Turner (2012), who try to reach conclusions about planets in general based on the fact that life emerged relatively quickly for the one known case of early Earth.

In the following section I highlight salient ideas on extraterrestrial life leading up to Herschel's time, showing in particular the heavy role of analogy in the prevalent reasoning.

3.3 Eighteenth Century Views on Extraterrestrial Life

As detailed earlier in this volume by Crowe and Dowd (2013), a Principle of Plenitude was central in European natural philosophy and theology from the Middle Ages onward: a perfect and glorious God must bring to be all that is possible, out of the fullness and goodness of his Divine Power. Furthermore, nature exhibits a basic uniformity and unity indicative of God's perfection. When these principles were applied to the question of extraterrestrial life, most concluded that it must be ubiquitous. For example, in the late 17th century the Parisian Bernard le Bovier de Fontenelle wrote a slight book that had enormous influence on the reading public in Europe. *Conversations on the Plurality of Worlds* (*Entretiens sur la pluralité des mondes*) was published in 1686 and over a century ran through almost a hundred editions in many languages.[3] Under the guise of a series of moonlit conversations with a charming but unschooled marquise, Fontenelle lays out the latest in astronomical knowledge and employs the Principles of Plenitude and of the uniformity of nature to assert the existence of inhabitants not only on the planets we know, but also on presumed planetary retinues accompanying every star in the sky. Just as our world exhibits a profound fecundity and diversity, by analogy it would certainly be wasteful of Nature to possess all these other locales without populating them. There is a purpose for each and every element of Creation.

As the mechanical Universe of Isaac Newton took hold in the 18th century, another important way of thinking called *natural theology* took its place alongside scriptural or revealed theology based on the Bible. Natural theology sought to reconcile religious beliefs and findings from natural philosophy by studying the "Book of Nature" to learn of God (and even prove his existence). It had an important influence on mainstream science throughout Europe and America as late as the mid 19th century,[4] with particular persistence and strength in Britain. For example, the Scottish minister Thomas Chalmers wrote an influential treatise entitled *A Series of Discourses on the Christian Revelation, Viewed in Connection with the Modern Astronomy* (1817). One argument of particular interest to today's astrobiology concerned microscopic realms, which, Chalmers said, revealed worlds and "tribes of animals" every bit as unknown and vast and fascinating as those seen in telescopes. Infinity in one direction was balanced by infinity in the

---

[3] A modern English translation with excellent commentary was published in 1990 by the University of California Press.

[4] Even today, there are those who very much work in this tradition by promoting the necessity, based on scientific findings, for so-called Intelligent Design. For a remarkable astrobiological example, see *The Privileged Planet* by Gonzalez and Richards (2004).

other. Since God's beneficence had applied to these realms even before we were aware of them, so God cared for humans even though we might be insignificant on a cosmic scale.

Three writers in the tradition of natural theology who were important for setting the stage for William Herschel were Thomas Wright of Durham, Immanuel Kant, and James Ferguson (Crowe and Dowd 2013). Wright, an Anglican cleric in the north of England, published *Original Theory or New Hypothesis of the Universe* in 1750. He thought it a ridiculous notion that all the stars and their presumed planets could have been made for "this diminutive World, our little trifling Earth." And in turn all those planets were populated because neither Nature nor God does anything in vain. Just as there are fish in every river, by analogy there were beings on every planet around every star.

Immanuel Kant, the great philosopher of Königsberg, early in his career wrote *Universal Natural History and Theory of the Heavens* (1755). He interpreted the Principle of Plenitude as implying that the entire Universe is full of an infinite number of forming and decaying worlds:

> Analogy does not leave us to doubt that these systems have been formed and produced in the same way as the one in which we find ourselves, namely, out of the smallest particles of elementary matter that filled empty space—that infinite receptacle of the Divine Presence.

Finally, William Herschel was directly influenced by James Ferguson, an English astronomer and lecturer, who in 1756 published his highly successful textbook *Astronomy, Explained upon Sir Isaac Newton's Principles*. In 1774, when Herschel was learning astronomy (as a fulltime musician), he bought the latest edition of Ferguson and learned not only about the planets, but also their "rational inhabitants," who were "creatures endowed with capacities of knowing and adoring their beneficent Creator." The stars must have a purpose and that purpose must be to illuminate nearby planets. Ferguson spoke of "a general analogy running through (all of Creation) and connecting all the parts into one scheme, one design, one whole." All the planets and their moons "are much of the same nature with our Earth, and defined for the like purposes."

3.4 William Herschel

Herschel was born in Hanover (now in Germany), but spent most of his life in England. After a successful career as a musician (and amateur astronomer) until his early forties, upon his spectacular discovery in 1781 of Uranus, the first non-telescopic planet, he became a fulltime astronomer, aided by the patronage of King George III. Over the next few decades he built unprecedentedly large and accurate reflector telescopes and became an indefatigable and astute observer of the heavens. Assisted by his sister Caroline, he singlehandedly revolutionized astronomy, then dominated by a concern with our solar system and accurate positions of stars and planets (Hoskin 2011). Herschel instead took his contemporaries outside the solar system by surveying the entire northern sky for faint nebulae, star clusters, and double stars. He also produced the first quantitative map of our stellar system (a process he called the "construction of the heavens"). To understand the nature of his thousands of newly catalogued objects, he brought into astronomy the novel idea of deducing their slow evolution by taking their present properties to be exemplars of various stages of maturity during the lifetime of any individual object. Furthermore, he demonstrated for the first time that some stars were in orbit about each other ("binary stars" was his neologism) and thus that gravity applied to the sidereal universe beyond the solar system.

But Herschel did not neglect the solar system—40% of his publications dealt with observations of the Sun, planets, moons, comets, and the new *asteroids* (again his new term). He discovered four moons of Saturn and Uranus, and investigated in scrupulous detail many properties of the visible features of planets, including the Martian polar caps and the Saturnian rings. The Sun too was of great interest and, almost as a byproduct, he discovered (in 1800) infrared radiation from the Sun. It is also with the Sun that we encounter what is often put forward as one of the few serious mistakes of Herschel's career, namely his assertion in 1795 that the Sun had a cool region below its surface and was in fact inhabited. He in fact argued that *all* the planets (*and* our Moon) were inhabited with intelligent beings, as were the myriad unseen planets that he presumed circled other stars.

While still a fulltime musician in the fashionable resort town of Bath, Herschel became quite serious about his astronomy and was active in the Bath Philosophical Society, a group of gentlemen interested in all aspects of the natural world. He fabricated his first metal mirrors and began telescopic observations of everything from stars and nebulae to planets and the Moon. After a few years of experience, he grew bold enough in 1780 to submit his first paper to the prestigious Royal Society in London. But, as Crowe (1986, 62–66) has shown [also summarized in this volume in Crowe and Dowd (2013)], the originally submitted manuscript contained many claims concerning intelligent life on the Moon, backed up by his observations of, for instance, towns and a huge forest on the Moon. The Astronomer Royal (Neville Maskelyne), acting as referee, insisted that Herschel remove almost all of his enthusiastic arguments for lunarians. Herschel, anxious to break into the elite world of the Royal Society, obliged and focused on his detailed determination of the heights of lunar mountains (from measurements of their shadow lengths). But he also managed to slip in the statement: "[study of the Moon leads to] the great probability, not to say almost absolute certainty, of her being inhabited" (Herschel 1780, 508). In the paper as finally published, this was the lone sentence referring to lunarians.

In subsequent publications over 40 years he similarly severely limited his mentions of extraterrestrial life, while nevertheless maintaining his conviction in the ubiquity of alien beings. Note that unlike most of his predecessors, Herschel always backed his arguments for extraterrestrial life with empirical evidence: detailed observations of planetary surfaces, clouds, rings, and satellites (Schaffer

1980, 100–102). Following are some of his published statements from the seventy papers he published in the *Philosophical Transactions of the Royal Society*. (All bolding of phrases is mine.)

From Herschel (1784, 260 and 273) on detailed observations of Mars:

> The analogy between Mars and the earth is, perhaps, by far the greatest in the whole solar system.
> [Mars] has a considerable but moderate atmosphere, so that **its inhabitants** probably enjoy a situation in many respects similar to ours.

From Herschel (1785, 258) on the "construction of the heavens" (structure of the Milky Way system):

> As we are used to call the appearance of the heavens, where it is surrounded with a bright zone, the Milky-Way, it may not be amiss to point out some other very remarkable Nebulæ which cannot well be less, but are probably much larger than our own system; and, being also extended, **the inhabitants of the planets that attend the stars** which compose them must likewise perceive the same phænomena. For which reason they may also be called milky-ways by way of distinction.

From Herschel (1792, 5) on detailed observations of Saturn's rings:

> This opening in the ring must be **of considerable service to the planet**, in reducing the space that is eclipsed by the shadow of the ring to a much smaller compass.

Explanation: The "opening in the ring" refers to what is today called Cassini's Division, and Herschel presented detailed evidence that it was truly a gap, not just dark material. "Service to the planet" is an obtuse way to say "service to the Saturnians," who, because of the gap, are not shaded as badly by the ring(s) during the various Saturnian seasons, and who need every bit of solar warmth that they can muster at 10 AU from the Sun![5]

From Herschel (1805, 272) on observations of Saturn:

> There is not perhaps another object in the heavens that presents us with such a variety of extraordinary phenomena as the planet Saturn: a magnificent globe, encompassed by a stupendous double ring: attended by seven satellites:...all the parts of the system of Saturn occasionally reflecting light to each other: **the rings and moons illuminating the nights of the Saturnian**....

From Herschel (1814, 263) on the evolution of stars and nebulae:

> **Stars**, although surrounded by a luminous atmosphere, may be looked upon as so many **opaque, habitable, planetary globes**; differing, from what we know of our own planets, only in their size, and by their intrinsically luminous appearance.

In addition to the above short interjections, on one occasion Herschel included several pages of arguments for extraterrestrial beings. This, however, was not for any planet or moon, but for the *Sun*, and, by extension, all other stars. Based on detailed observations and plausible physics for his time, Herschel argued first that the Sun was in essence a giant planet, and secondly that it was probably inhabited by beings adapted to its decidedly unearthlike conditions.

---

[5] The usefulness of the Cassini Division to the Saturnians was a point that Herschel undoubtedly took from Ferguson's book, although he does not acknowledge this.

3.5 Herschel and the Inhabited Sun

It is extremely difficult for today's reader not to think that the Sun has always obviously been considered an extremely hot ball of gas. It has of course long been realized that the Sun was the source of all heat and light for the Earth, but it does not necessarily follow that (a) the surface of the Sun is extremely hot, or (b) that the interior of the Sun must be even hotter and in the gaseous state. In fact it was not until the early 20th century that physics and astronomy established that the Sun's entire interior was probably extremely hot and gaseous, not liquid or solid.

Most natural philosophers in Herschel's time took the Sun to be solid. Herschel himself (1795, 62) argued that "by calculation from the power [the Sun] exerts upon the planets we know [it] to be of great solidity." He does not give any further details, but he must be referring to the fact that Newtonian theory allowed one to calculate the mass of the Sun, which when combined with its volume yielded a mean density greater than that of water, implying that it was solid and planet-like.

Secondly, through years of carefully observing sunspots, Herschel confirmed that sunspots were actually depressions relative to the bright solar surface. He then hypothesized that the darkness of sunspots indicated that one was peering through "openings" in the bright clouds above the actual dark surface of the Sun below, not unlike an observer on the Moon looking at the surface of the Earth through gaps in the cloud cover. He then concluded (Herschel 1795, 63):

> The Sun, viewed in this light, appears to be nothing else than **a very eminent, large, and lucid planet**.... Its similarity to the other globes of the solar system with regard to its solidity, its atmosphere, and its diversified surface; the rotation upon its axis, and the fall of heavy bodies, leads us on to suppose **that it is most probably also inhabited, like the rest of the planets, by beings whose organs are adapted to the peculiar circumstances of that vast globe**.

Continuing, he emphasized that this was not some wild speculation (as in the past), but rather an eminently scientific conclusion, based on detailed observations and plausible deductions:

> Whatever fanciful poets might say, in making the sun the abode of blessed spirits, or angry moralists devise, in pointing it out as a fit place for the punishment of the wicked, it does not appear that they had any other foundation for their assertions than mere opinion and vague surmise; but now **I think myself authorized,** *upon astronomical principles* [italics in original]**, to propose the sun as an inhabitable world**, and am persuaded that the foregoing observations, with the conclusions I have drawn from them, are fully suffi-cient to answer every objection that may be made against it.[6]

Lastly, Herschel employed "analogical reasonings" concerning the "construc-tion and purposes of the sun." The word *purposes* here is telling, for we see that he is implicitly making a teleological argument regarding the Creator's perfect

---

[6] See Crowe (2011) for a full historical account of the notion of an inhabited Sun, an idea that started long before Herschel and, abetted by his authority and arguments, lasted well past his time.

design, where each element of the design has achieved its highest possible purpose. Combined with the Principle of Plenitude, this purpose would be habitation by intelligent beings. Herschel then makes the case by analogy that the Moon should be inhabited because of all of its other similarities to the Earth: considerable size, mountains and valleys, variety of seasons and day/night cycle, stars and planets rising and setting, heavy bodies falling, and its own "capital satellite" (the Earth!).[7] And just as the lunarians would make a mistake in thinking that the Earth hanging in their sky had no denizens, so would we in dismissing the Sun as having no purpose other than to act as our attractive center and source of heat and light. He then concludes:

> From experience we can affirm, that the performance of the most salutary offices to inferior planets, is not inconsistent with the dignity of superior purposes; and, **in consequence of such analogical reasonings**, assisted by telescopic views, which plainly favor the same opinion, we need not hesitate to admit that **the sun is richly stored with inhabitants**.

Herschel's final leap is to link our solar system and the sidereal universe beyond: "But if stars are suns, and suns are inhabitable, we see at once what an extensive field for animation opens itself to our view." He concludes by pointing out that by analogy

> we may have an idea of **numberless globes that serve for the habitation of living creatures**. But if these suns themselves are primary planets, we may see some thousands of them with our own eyes; and millions by the help of telescopes.

In other words we cannot see the (inhabited) planets associated with each star, but we *can* actually see the inhabited "primary planets," i.e., the stars themselves. The stars/suns thus have a purpose in accord with the Principle of Plenitude (Herschel 1795, 71):

> **Many stars, unless we would make them mere useless brilliant points, may themselves be lucid planets.**

3.6 Conclusion

It is indeed a bold and provocative enterprise that astrobiologists tackle when they test the world and develop views on fundamental topics such as the role of life in the Universe. As we astrobiologists try to extend the Copernican Principle to the *biological* world, the issues become even more profound than for Copernicus, because the uniqueness of us humans, as well as our form of life, is more deeply vested in our psyche than is the uniqueness of any *physical* aspect of our home planet.

---

[7] Herschel (1795, 66) also lists the properties of the Moon that greatly *differ* from those of Earth: no seas, no atmosphere, no dense clouds and thus no rain, very different types of seasons, days, and climates. He then argues, however, that this diversity is no problem and only means that the lunarians will have adapted to these conditions and thus be notably different from us.

William Herschel *and* we today similarly tackle questions with the best tools available, apply the cleverest scientific insight we can muster, and struggle to fashion a consensus as to the nature of the world. To assay the chances of extraterrestrial life, Herschel and we can do no more than apply the best current science and analogical reasoning to overcome the "N = 1 Problem." As Mill has cautioned us, we must tread very carefully when employing astrobiology's Great Analogy—such an analogy can seldom *prove* anything, but it often can and does lead to the next steps in our research program.

Our best efforts in science are surely steadily improving in their usefulness and their verisimilitude to the natural world, but we should not forget the many historical examples that illustrate strong influences from the prevailing cultural milieu. What are these cultural biases and assumptions *today*? For us, as fish ensconced in the stream, they are frustratingly difficult to recognize. Boldly trying to "flop out on the land for just a minute," I offer a provocative first list, hardly exhaustive:

- the uniformity of physical laws throughout the cosmos (which assumption we share with the 18th century)
- the dogma that miracles are not allowed
- the Copernican Principle that our life and our Earth are *not special*
- the degree to which our investigations *and their results* are strongly influenced by society's wishes, e.g., through patronage such as that from NASA and its Astrobiology Roadmap, which even as it guides also excludes many possibilities
- the degree to which our investigations *and their results* are strongly influenced by our own psychological needs, e.g., the quest for *another Earth*, tellingly sometimes called a search for "the Holy Grail"

I do not offer this list to denigrate today's enterprise, but only to acknowledge the essential humanity woven through science, a powerful way to understand the world.

References

Crowe, Michael J. 1986. *The Extraterrestrial Life Debate, 1750–1900: The Idea of a Plurality of Worlds from Kant to Lowell*. Cambridge: Cambridge University Press.

Crowe, Michael J. 2011. "The Surprising History of Claims for Life on the Sun." *Journal of Astronomical History and Heritage* 14: 169–179.

Crowe, Michael J., and Matthew F. Dowd. 2013. "The Extraterrestrial Life Debate from Antiquity to 1900." In *Astrobiology, History, and Society: Life Beyond Earth and the Impact of Discovery*, ed. Douglas A. Vakoch. Heidelberg: Springer.

Dick, Steven J. 1982. *Plurality of Worlds: The Origins of the Extraterrestrial Life Debate from Democritus to Kant*. Cambridge: Cambridge University Press.

Gonzalez, Guillermo, and Jay Richards. 2004. *The Privileged Planet: How Our Place in the Cosmos Is Designed for Discovery*. Washington, DC: Regnery Publishing.

Herschel, William. 1780. "Astronomical Observations Relating to the Mountains of the Moon." *Philosophical Transactions of the Royal Society of London* 70: 507–526.

Herschel, William. 1784. "On the Remarkable Appearances at the Polar Regions of the Planet Mars, the Inclination of Its Axis, the Position of Its Poles, and Its Spheroidical Figure; with a Few Hints Relating to Its Real Diameter and Atmosphere." *Philosophical Transactions of the Royal Society of London* 74: 233–273.

Herschel, William. 1785. "On the Construction of the Heavens." *Philosophical Transactions of the Royal Society of London* 75: 213–266.

Herschel, William. 1792. "On the Ring of Saturn, and the Rotation of the Fifth Satellite Upon Its Axis." *Philosophical Transactions of the Royal Society of London* 82: 1–22.

Herschel, William. 1795. "On the Nature and Construction of the Sun and Fixed Stars." *Philosophical Transactions of the Royal Society of London* 85: 46–72.

Herschel, William. 1805. "Observations on the Singular Figure of the Planet Saturn." *Philosophical Transactions of the Royal Society of London* 95: 272–280.

Herschel, William. 1814. "Astronomical Observations Relating to the Sidereal Part of the Heavens, and Its Connection with the Nebulous Part; Arranged for the Purpose of a Critical Examination." *Philosophical Transactions of the Royal Society of London* 104: 248–284.

Hesse, Mary. 1966. *Models and Analogies in Science*. Notre Dame: University of Notre Dame Press.

Hoskin, Michael. 2011. *Discoverers of the Universe: William and Caroline Herschel*. Princeton: Princeton University Press.

Schaffer, Simon. 1980. "'The Great Laboratories of the Universe': William Herschel on Matter Theory and Planetary Life." *Journal for the History of Astronomy* 11: 81–111.

Spiegel, David S., and Edwin L. Turner. 2012. "Bayesian Analysis of the Astrobiological Implications of Life's Early Emergence on Earth." *Proceedings of the National Academies of Science.* 109: 395–400.

Sullivan, Woodruff T., III, and John A. Baross. 2007. *Planets and Life: The Emerging Science of Astrobiology*. Cambridge: Cambridge University Press.

Chapter 4
Hegel, Analogy, and Extraterrestrial Life

Joseph T. Ross

Abstract Georg Wilhelm Friedrich Hegel rejected the possibility of life outside of the Earth, according to several scholars of extraterrestrial life. Their position is that the solar system and specifically the planet Earth is the unique place in the cosmos where life, intelligence, and rationality can be. The present study offers a very different interpretation of Hegel's statements about the place of life on Earth by suggesting that, although Hegel did not believe that there were other solar systems where rationality is present, he did in fact suggest that planets in general, not the Earth exclusively, have life and possibly also intelligent inhabitants. Analogical syllogisms are superficial, according to Hegel, insofar as they try to conclude that there is life on the Moon even though there is no evidence of water or air on that body. Similar analogical arguments for life on the Sun made by Johann Elert Bode and William Herschel were considered by Hegel to be equally superficial. Analogical arguments were also used by astronomers and philosophers to suggest that life could be found on other planets in our solar system. Hegel offers no critique of analogical arguments for life on other planets, and in fact Hegel believed that life would be found on other planets. Planets, after all, have meteorological processes and therefore are "living" according to his philosophical account, unlike the Moon, Sun, and comets. Whereas William Herschel was already finding great similarities between the Sun and the stars and had extended these similarities to the property of having planets or being themselves inhabitable worlds, Hegel rejected this analogy. The Sun and stars have some properties in common, but for Hegel one cannot conclude from these similarities to the necessity that stars have planets. Hegel's arguments against the presence of life in the solar system were not directed against other planets, but rather against the Sun and Moon, both of which he said have a different nature from Earth and planets. Although he did not explicitly discuss the possibility of life on comets, the fourth type of body in his theory of the solar system, it is clear that he

J. T. Ross (✉)
University of Notre Dame, Notre Dame, IN, USA
e-mail: jross@nd.edu

D. A. Vakoch (ed.), *Astrobiology, History, and Society*, Advances in Astrobiology and Biogeophysics, DOI: 10.1007/978-3-642-35983-5_4,
© Springer-Verlag Berlin Heidelberg 2013

rejected the views of Bode and Johann Heinrich Lambert, who did defend this possibility. Again, Hegel's critique of the use of analogical argument is important here. The Sun, comets, and moons are not analogous to the Earth or to the planets; these are four different bodies with different forms of motion and different physical constitutions. Only planets have completeness according to Hegel because only they have water, air, earth, and light, and completeness in this sense is necessary for life. Hegel discerned a need to make distinctions in nature rather than to consider superficially different realities as fundamentally similar. Celestial bodies should not be considered, according to Hegel, as all of one type or nature, as one kind.

4.1 Introduction

Analogical arguments have long played a role in the extraterrestrial life debate, in spite of their sometimes acknowledged limitations for determining the likely nature and prevalence of life beyond Earth. In the eighteenth century, for example, Immanuel Kant questioned the reliability of analogical arguments and emphasized their empirical and subjective nature, but he himself persisted in using these arguments primarily for the suggestion that life is likely on the Moon. Shortly afterwards, Johann Gottfried Herder made extensive use of analogy in *Ideen zur Philosophie der Geschichte der Menschheit* to suggest the possibility of life on other planets (Herder 1970). Similarly, Johann Elert Bode (1816) and William Herschel (1795) defended by means of analogical reasoning the habitability of celestial bodies including the Sun and stars, which they considered similar to the Earth in their essential features. Unlike Kant, these later thinkers seldom reflected on the nature and validity of analogical reasoning, although they also used these arguments in their writing. They accepted analogy often uncritically as an important scientific tool that allowed them to learn about celestial bodies that are too distant to be observed in detail even with the telescope.

Georg Wilhelm Friedrich Hegel's contribution to the question of extraterrestrial life and his analysis of analogical arguments for life on other celestial bodies have often been misunderstood. Several histories of the extraterrestrial life debate have made reference to Hegel's views on this question. These discussions have asserted that Hegel did not allow for the possibility of life beyond the Earth, although the majority of his predecessors and colleagues in philosophy and natural science did consider such life possible and likely. The consensus of these critiques is that Hegel completely rejected the possibility of life beyond the Earth because of his emphasis on humanity (Zöckler 1866, 1879; Huber 1878; Guthke 1983).

Hegel scholars have not dealt very extensively with the question of extraterrestrial life. Michael John Petry mentioned the topic only very briefly in the notes to his English translation of the *Naturphilosophie* (Hegel 1970a). He apparently believed that Hegel's reference to the topic had to do with the mythological and superstitious beliefs about "the Man in the Moon and the Woman in the Sun" rather than with the contemporary discussions by astronomers and philosophers, although he did concede that "astronomers themselves have not been free from these superstitions" and cites as instances Tycho Brahe and William Herschel (Hegel 1970a, II, 225). Scholarly treatments of extraterrestrial life by Herschel, Bode, and Kant are more likely points of departure for Hegel's remarks here rather than proverbs and mythological expressions.

Hegel's treatment of this topic needs a more thorough examination than it has received up to now. It is necessary to place his statements in the context of the history of discussions of analogy. In this chapter I will argue that Hegel did indeed allow for life beyond the Earth—at least on other planets of our solar system. Hegel, it will be seen, did not enthusiastically endorse analogical reasoning, especially in those forms that were used in his time to suggest that the Moon and Sun as well as the planets other than the Earth are inhabited because they are similar to the Earth.

To preview Hegel's use of analogical arguments for life on other celestial bodies, it will be helpful to present briefly Hegel's theory of astronomy, especially his theory of celestial bodies. Unlike Kant and many astronomers of Hegel's own time, Hegel differentiated celestial bodies into several classes that have distinctive qualities: planets, moons, comets, Sun, and stars. Hegel's critique of analogical arguments for life on other celestial bodies was basically a denial of the assumption that all celestial bodies are similar to the Earth. If celestial bodies are differentiated according to the five classes of entities that Hegel proposed, then analogical arguments can only be valid within a particular class. Because Earth is a planet, an analogical argument from a property of the Earth—the property of having inhabitants, for example—should only be ascribed to other planets, i.e., other "Earths."

4.2 A Brief History of the Multiple Meanings of Analogy

As John Stuart Mill (1973, I, 554) remarked, "There is no word… which is used more loosely, or in a greater variety of senses" than analogy. A distinction can be made between (1) analogy as a proportion and (2) analogy as a comparison of one thing to another. The original meaning of the word was "proportion," and it was used in mathematics to express that four or more terms have the same relation among themselves, such that the first is to the second as the third is to the fourth, etc. A proportion can also be established as an equality or similarity of relationships, not only between numbers, but also entities, properties, and qualities. This is the original sense of *analogia*, used by the Pythagoreans and in Greek mathematics, and many philosophers and scientists have considered, and still consider, the mathematical sense of analogy as proportion to be the only proper one (Hänssler 1927).

More frequently, however, "analogy" is used in the non-mathematical sense of similarity or comparison of properties or qualities where no ratio is intended. This kind of analogy is sometimes considered a form of inductive reasoning or syllogism. Induction reasons from a property, quality, or law of an individual to the presence of the same property, quality, or law in all individuals of a kind or type by examination of all instances. In contrast, analogy argues for the presence of the property, quality, or law in one individual to the presence of the property, quality, or law in a similar individual of the same kind. Sometimes an analogy results in a universal-like induction, but it is based not on all individual instances but only on a significant number of similar instances (imperfect induction). Analogy is sometimes considered inferior to induction because it is based on less evidence (Mill 1973), but others consider it superior because perfect induction is impracticable if not impossible (Stebbing 1930; Breidbach 1987; Eaton 1931).

A third sense of analogy is the ontological assumption that "nature is conformable to itself," i.e., that nature is harmoniously put together or that natural beings are similar to one another. This sometimes takes the form of a principle of cosmological simplicity or ontological economy. These three kinds of analogy are not mutually exclusive, and they often blend into one another.

Plato used the term αναλογια (*analogia*) in the sense of a proportion in the *Republic* (511d6–e4) to show the relationship of intellectual knowledge to understanding, belief, and pictorial apprehension as four parts of a line.[1] The proportional sense of analogy acquired a cosmological application in the dialogue *Timaeus*, where analogy was used as a bond that establishes the best unity possible where the middle term of the proportion has the same ratio to the first and last terms (31b–32a). Aristotle also used αναλογια as a proportion. He defined αναλογια in the *Nicomachean Ethics* as "the equality of ratios and has at least four terms" (1131a31). Aristotle used analogy in the sense of proportion in *Historia Animalium* to show an identity of function: "Some animals have parts neither the same in kind nor in a relation of excess and defect but rather in accordance with analogy... For a feather is in a bird what the scale is in a fish" (486b18–22). The feather and scale, dissimilar in appearance and texture, have a similar function in the bird and fish, respectively. They have a common logos or reason, but they are not in the same kind or class.

Plato did not use the word αναλογια for similarity, but the search for similarity and comparisons is central to the search for forms. Plato questioned, however, whether one could make use of similes and comparisons without first knowing the nature of the things being compared: "Now will such a one, not knowing the truth of each thing be able to discern the small and the great similarity of the unknown thing in other things? It will be impossible" (*Phaedrus* 262a9–11).

Arguments based on similarities also found a place in Aristotle's philosophy, but in the rhetoric and dialectic rather than in the logical and deductive treatises. In *Topica* Aristotle called the finding of similarities one of the four "tools by which we gain possession of deductive and inductive arguments" (105a21–25). He differentiated between arguments based on similarities of proportions and arguments based on similarities of properties (138b23–26).

Aristotle discussed a form of dialectical syllogism, argument by example, in the *Prior Analytics* in relation to induction, which had a profound influence on the

---

[1] All translations from Greek and German sources are by the author. Citations to works by Aristotle and Plato use the Bekker number and Stephanus pagination, respectively.

history of analogical reasoning. A property or quality is said to belong to one thing on the basis of the presence of that property or quality in a thing similar to it:

> It is obvious that example functions not as a part to a whole nor as a whole to a part, but as a part to a part when both fall under the same [class] and one of the two is known. It is distinct from "induction" in that "induction" shows that the extreme is present in the middle by means of all the individual instances and does not join the conclusion with the extreme. "Example," by contrast, both joins the conclusion with extreme and does not demonstrate by means of all the instances (68b38-40, 69a13-19).

That is, analogy applies the property or quality to a particular entity, but induction does not.

Induction and analogy are forms of argumentation that result in probable, not certain, conclusions, and they are often considered complementary to each other. This discussion of analogy is found in the *Prior Analytics* immediately after induction and is contrasted with induction on the basis of the evidence it uses (all instances or some) and on the basis of the type of conclusion it makes (universal or particular). This treatment of arguments by example in *Prior Analytics* forms one of the classic texts for analogical syllogisms, although it does not use the word αναλογια at all. Induction argues from individuals to a universal statement about all the individuals. Analogical argument uses a known property or condition in one term to establish the presence of the property or condition in a second term that is considered similar to the first term, i.e., similar by virtue of both belonging to one kind. The conditions for an argument by example are: the known presence of the property in the first term, the unknown presence of the property in the second term, and the known similarity of the two terms to each other because they fall under the same class or kind. Similarity is a key element in this form of reasoning about qualities or properties in unknown objects or beings on the basis of known objects or beings.

The shift in the meaning of αναλογια from proportion to similarity is apparent in Cicero and Quintilian. Cicero glossed αναλογία in his translation of the *Timaeus* passage discussing the proportion that makes the best cosmic bond: "comparison or proportion" (Cicero 1975, 186–188). The application of the word analogy in the sense of example in Aristotle's discussion of argument by example is found in Quintilian's *Institutio Oratoria* (1963, I, vi., 3–4): "All these things require a keen power of discernment, especially *analogia*, which those who translate it literally from the Greek call *proportio*. Its meaning is that one refers what is doubtful to something else similar that does not need investigation and one demonstrates [proves] uncertain things by those that are certain."

The Ciceronian and Quintilian understanding of analogy came into play in the Renaissance humanistic philosophy. In the Renaissance period, dialectical reasoning was emphasized by humanistic philosophers, who turned to Cicero and Quintilian for their understanding of dialectical reasoning and who sought to bring it into association with demonstrative reasoning. Rather than being opposite in nature and superior to probable opinion, the necessarily true was considered a part of what is probable (Oeing-Hanoff 1972, 182).[2]

---

[2] See also Ernst Cassirer's treatment of Giacomo Zabarella (Cassirer 1911, Bd. 1, 136–140).

The most important treatment of analogy for eighteenth- and nineteenth-century philosophy is that found in Isaac Newton's *Regulae Philosophandi* at the beginning of Book III of *Principia*, entitled "Systema Mundi." In Regula III, one finds a clear ontological use of analogy: "One certainly ought not blindly construct dreams contrary to the evidence of experiments, nor ought one draw back from the analogy of nature since it is accustomed to be simple and harmonious with itself" (Newton 1972, II, 553). Analogy figures also in Regula II although the word is not explicitly used: "And therefore the same causes are to be ascribed to natural effects of the same kind as far as this is possible. As of human respiration and in beasts; of the fall of stones in Europe and in America; of light in kitchen fire and in the Sun; of reflection of light on Earth and on the planets" (Newton 1972, II, 552). If the effects are similar, then the causes ought to be considered similar as well. Newton stressed that these are "effects of the same kind." This rule is used to ascribe causes where the causes are unknown or unknowable. One does not or cannot know the cause of light in the Sun nor the cause of the reflection of light on planets, but they are known on Earth. Because of the mere distance, one cannot investigate the phenomena to determine the cause, but Newton wanted to argue from an analogical syllogism or argument from example that if one can assign the same cause of human respiration to that in beasts and of the fall of stones in America and in Europe, then similarly one can assign causes for the light in the Sun and the reflection of light on other planets. For Newton, because the light caused by the Sun and by a kitchen fire are of the same kind, the cause must be the same. The Sun is thus similar to a kitchen fire. Newton was furthering the analogical arguments of Kepler and Galileo by insisting that the heavenly bodies are similar to the Earth, specifically that the Sun is similar to fiery things on the Earth. Newton formulated these statements as rules of reasoning, not philosophical positions or "hypotheses" they are not proofs but methodological principles.[3]

John Locke considered analogy as a type of judgment in his *Essay concerning Human Understanding*. In book IV, chapter 14 Locke presented a concept of judgment as a determination of the truth of a proposition "without perceiving a demonstrative evidence in the proofs" (Locke 1959, II, 361). Rather than certain knowledge, judgment has only probability, "likeliness to be true" (Locke 1959, II, 365). "The conformity of anything with our own knowledge, observation and experience" and the "testimony of others, vouching their observations and experience" are grounds for probability, according to Locke. There are two types of propositions to which probabilistic judgments are applied: propositions concerning "matters of fact," which are "capable of human testimony," and propositions "concerning things, which, being beyond the discovery of our senses, are not capable of any such testimony" (Locke 1959, II, 374–375). The second type of propositions involves the use of analogy:

> In things which sense cannot discover, analogy is the great rule of probability.... Such are,
> 1. The existence, nature, and operations of finite immaterial beings without us; as spirits,

---

[3] For a discussion of analogy in Newton's thought and its influence on eighteenth-century philosophy and science, see Gilardi (1988).

angels, devil, &c., or the existence of material beings, which, either for their smallness in themselves, or remoteness from us, our senses cannot take notice of; as whether there be any plants, animals and intelligent inhabitants in the planets, and other mansions of the vast universe. 2. Concerning the manner of operation in most parts of the works of nature: wherein, though we see the sensible effects, yet their causes are unknown and we perceive not the ways and manner how they are produced.... Analogy in these matters is the only help we have, and it is from this alone we draw all our grounds of probability.... This sort of probability, which is the best conduct of rational experiments, and the rise of hypothesis, has also its use and influence: and a wary reasoning from analogy leads us often into the discovery of truths and useful productions which would otherwise be concealed (Locke 1959, II, 379–382).

Locke thus established analogy as a necessary tool to find probable truth about things, not perceivable by sense experience, viz. immaterial beings and objects too far away for direct observation and the inner connections of nature, such as the unknowable causes of sensible effects.

Gottfried Wilhelm Leibniz was one of the most outspoken proponents of analogies: "The hypothetical method *a posteriori*, which takes its point of departure from experiments, customarily supports itself on analogies." (Leibniz 1955, 315) Analogies serve the purpose, according to Leibniz, of allowing one to conclude from the particular to the general and, furthermore, help to make predictions about unexperienced objects (Leibniz 1955, 315–316). In an essay from 1674, "Schediasma de Arte Inveniendi Theoremata," Leibniz considered analogy an imperfect induction: "Since we are unable to run through all the instances, one must choose which are to be considered before the rest, and this is now brought back to analogy; and therein consists the entire art of experiments." (Leibniz 1955, 425) Experiment chooses an ideal or crucial phenomenon to test a theory, similar to what Bacon called *instantiae crucis*. The ontological sense of analogy is apparent in Leibniz's assertion that analogy is based on the fact that objects or phenomena which agree or are opposite in many qualities will also agree in other given qualities that are related to them. In nature Leibniz found a "series, an order, a progression, which is the result of many analogies or comparisons" (Leibniz 1955, 355). Scientists have found many analogies in studying plants, insects and comparative anatomy of animals. There is an "analogy of things which can be extended beyond our observations and there is no difference except between the large and small, the sensible and the insensible" (Leibniz 1962, 472).

Kant made extensive use of analogy in his pre-critical writings both in the sense of proportions and similarity. The inductive sense of analogy is emphasized throughout the course of his philosophical and scientific writings as important and valuable when understanding lacks infallible proofs (Kant 1902, I, 315). Kant sometimes asserted that analogy combined with observation has the same degree of certainty as formal proofs (Kant 1902, I, 255). A cosmological underpinning for analogy was offered by Kant's statements that "things are not alien and separate from one another" (Kant 1902, I, 364). Natural entities form together a system where each is related to another, and the affinity of entities is a consequence

of a common cause or source (Kant 1902, I, 364). On the basis of these ontological statements about the analogy of nature, Kant argued for a theory that the
stars are similar to the Sun and the centers of systems of celestial bodies similar
to the solar system (Kant 1902, I, 247–307). The components of those systems
are similar to the Earth, the basic example of celestial body in our own system
(Kant 1902, I, 327–328). Kant's understanding of analogy underwent considerable change from 1755, the publication date of *Allgemeine Naturgeschichte und
Theorie des Himmels*, until the 1770s when he treated analogy in his lectures on
logic. In these lectures, Kant considered analogy and induction as two similar
forms of reasoning, which proceed from the particular to the general (Kant 1902,
XXIV/1, 287). These forms of argumentation are syllogisms of the understanding as opposed to syllogisms of reason (Kant 1902, XXIV/1, 286–297). Analogy
argues from some properties in common between two entities to all properties
in common, while induction argues from some individuals of the same kind
having a particular property to all individuals of that kind having that property.
Analogical arguments are necessary for empirical rather than logical knowledge,
and, if one were to dispense with them, much empirical knowledge where one
must be content with probability rather than certainty would also be discarded.
Using a phrase that will be echoed throughout his lectures on logic even into the
1780s, Kant called analogy and induction the "crutches of the understanding"
(Kant 1902, XXIV/1, 287).

Kant also thought of analogy in terms of proportions or relationships.
The affinity is easy to understand in arguments from similar effects to similar
causes where a proportion can be established among the various terms, but Kant
seems to see a conformity even in cases of arguing on the basis of some properties in common to all properties in common where no proportion is apparent
(Kant 1902, XXIV/1, 478). In the *Kritik der reinen Vernunft*, Kant distinguished
sharply between mathematical and philosophical analogies. Mathematical analogies are "constitutive" and *a priori* (determinative and independent of experience) while philosophical analogies are only "regulative" and *a posteriori*
(heuristic and dependent on experience) (Kant 1902, IV, 123; Kant 1902, III,
160–161). Philosophical analogies, as they are defined in the *Kritik der reinen
Vernunft*, are proportions. The "analogies of experience" are rules for finding an
event or entity that is related to another event or entity according to the relations
of substance to accident, cause to effect, and the relation of reciprocal interaction
of two entities.

Kant described the analogical syllogism as imperfect induction (Kant 1902, III,
514). To argue from the consequences of a statement to its truth is only permissi-
ble, according to Kant, if one examines all possible consequences of the statement
to assess their truth. Because this goes beyond human capabilities, one must have
recourse to the argument that, if all examined consequences are in agreement with
the assumed reason, then all possible unexamined consequences are also in agreement. Kant warned that these proofs must be used cautiously and, furthermore,
that the arguments must be considered hypothetical and should never be assumed
to be a demonstrated truth.

To the discussion of induction and analogy Kant introduced two new principles, universalization and specification, in the lectures on logic that were recorded by Jäsche (Kant 1902, IX, 133):

> Induction concludes from the particular to the universal in accordance with the principle of universalization: Whatever is an attribute of many things of a kind is also an attribute of the rest. Analogy concludes from particular similarity of two things to total similarity in accordance with the principle of specification: Things belonging to a kind that agree with respect to many qualities agree also with respect to the remaining qualities that we can observe in some but not in others.

Related to this characterization of analogy as specification was Kant's statement that things are empirically included in a class by analogical reasoning. Kant also cautioned about transferring specific properties of one entity of a genus to another entity in the same genus. The example Kant gave here was the argument for intelligent beings on the Moon. One cannot conclude to humans on the Moon on the basis of the similarity of the Moon to the Earth, but only to the presence of rational beings because humans are specific to Earth, but the genus is rationality: "We reason by analogy only to rational Moon inhabitants; in analogy only the identity of the ground *par ratio*, is required" (Kant 1902, XVI, 760). This kind of cautionary remark is reminiscent of the proportional sense of analogy between dissimilar entities that have a common relationship. One can argue to the fact that God and an artist have a common nature because there is a common relationship between God and the world and the artist and her work. The relationship is the same but the entities are different and do not even have a common nature or essence. Here, however, Kant was arguing for a similarity but not identity of properties.

4.3 Hegel and Analogy

Hegel presented an account of analogical reasoning in the *Phänomenologie des Geistes* (Hegel 1968, IX, 143) in the section dealing with observation of nature: "One cannot observe that all stones fall to the Earth but only that very many of them do. One concludes then by analogy that with probability the rest of them do as well." Induction is an inference based upon observation of a property or law in all instances while example uses only some instances to conclude that all or most have a property or law. Analogy has sometimes been considered an imperfect induction. Arguments by example rely upon a situation or condition which is known and does not need to be proved and from that basis one concludes to the presence of the state or condition in a similar thing or circumstance. Hegel here did not conclude to the presence of the condition in a particular object or term, but used a small but significant number of instances or terms to justify a conclusion to a universal. Analogy is here an induction, but an imperfect induction. Natural science in fact follows this imperfect form of induction rather than ideal induction. Analogy, however, does not have complete justification. Induction and analogy only give probable knowledge; no matter how great the probability, it is not truth, although reason instinctually takes these conclusions and laws for truth.

Hegel used analogical syllogism in the sense of a comparison of one entity to another in the lecture notes on the doctrine of syllogisms from his period as Dozent in Jena (1800–1806). Induction and analogy are considered forms of "the syllogism of abstraction." The syllogisms present conclusions that relate the singular or individual to a universal. In induction, Hegel said, the individual becomes the universal by abstracting from its individuality and finding a common essence in all individuals (Hegel 1974, 332). In induction, the singular is itself general because it is "all singulars." By contrast, "the analogical syllogism equates two singulars and fuses the properties of the one with the properties of the other" (Hegel 1974, 333). Hegel argued that analogical syllogism here uses the first singular in a two-fold sense, as a singular but also as a universal. His example of analogical syllogism refers to the question of inhabitants on the Moon:

> The Earth has inhabitants, the Moon is an Earth–Earth is not only Earth as such, but rather Earth in general…. The conclusion of the induction (the Moon has inhabitants) has the same form on the face of it—a property is attributed to a subject—as any other syllogism. It has significance, however, only on the basis or the form of mediation: that the subject is attributed this property insofar as a singular is understood in a general sense and in this general sense a property has a definite existence, through its general nature (Hegel 1974, 334).

The 1827 *Enzyklopädie der philosophischen Wissenschaften* also treats induction and analogy, but now under a heading of the reflective syllogism. The middle of an inductive syllogism is the totality of individuals having a property or universal. Insofar, however, as an immediate empirical individual differs from the universal and therefore no completeness can be guaranteed, induction is based on analogy, in which the middle is an individual but in the sense of an essential universal, of its species or essential definiteness (Hegel 1968, XIX, 152). Induction, however, cannot be complete or perfect because the existent singular is subject to "free contingency" and therefore the syllogism is imperfect (Hegel 1970b, IV, §72, 26). The singular term is equated with the universal, as *its* universal and is considered the summation of all singulars. In induction the singular is itself general; it is "all singulars."

The 1816 *Wissenschaft der Logik* uses the same example of the inhabitability of Earth and the Moon for analogical syllogisms (Hegel 1968, XII, 115): "Here an individual is the middle, but according to its general nature; furthermore another individual is the extreme, which has the same general nature as that one, e.g., The Earth has inhabitants / The Moon is an Earth, / Therefore the Moon has inhabitants." Analogical conclusions can be very superficial when the universal, which the two have in common, is a mere quality, or the quality is subjectively understood and the identity of both is a mere similarity.

In the student notes of Hegel's 1831 lecture on logic recorded by Karl Hegel, analogy is again presented under the reflective syllogism. Again, there are three forms of the reflective syllogism: universal, inductive, and analogic syllogisms. In the first form, universality is implied in the premise by a general statement, e.g., "The body that is not supported, falls to the center-point" (Hegel 2001, 196). "All bodies" are meant in this statement because it is talking about body in general, which is body in the abstract sense. This syllogism becomes an induction when the bodies are mentioned as individuals: "This syllogism is based on induction, all individuals, human is now the middle and is specified as all individuals: The induction includes the same specification, but not all in general, but rather as the individuals" (Hegel 2001, 197). Once again, Hegel remarks that induction is inadequate because not all individuals can be examined, and one "takes refuge in analogy, where the middle is again an individual but taken in the sense of a universal." All humans up to this point have died, a very limited quantity of humans, but one concludes on this basis to the generalization that other humans are also mortal. The second example of analogy is the comparison of Moon and Earth: "The Earth has inhabitants, the Moon is an Earth; here the Earth [and the Moon] are included under a universal, being a celestial body, and one concludes that the Moon also has inhabitants." It depends, according to Hegel, on whether "the Earth has inhabitants in accordance with its [the Earth's] quality in itself, and not according to its universality."

Hegel emphasized, as also did Kant, Locke, and Leibniz, that analogy needs to be used with caution or else very superficial arguments are made. In contrast to these other philosophers, Hegel seemed to emphasize the weaknesses and superficiality of analogical arguments more than their benefits for the advance of knowledge. As he had said already in the *Phänomenologie des Geistes*, the success rate of analogical reasoning is such that if one were to use the historical instances of analogical reasoning as the basis of an analogical argument, analogical reasoning would not be found to be successful in discovering truth. As Aristotle had emphasized deductive reasoning as the basis of knowledge rather than dialectical reasoning, Hegel also felt that a better form of argument is available for philosophy: a categorical syllogism, which is based upon definitions. Rather than accidental and superficial qualities or properties, the categorical syllogism uses the general nature or essence of the terms in its premises. A predicate is combined with a subject on the basis of its substance, which is the universal or kind. In his interpretation of the objective universal in Hegel's theory of the categorical syllogism, Georg Sans suggested a connection with Hegel's understanding of Plato's universal. In the lectures on the history of philosophy, Hegel said that for Plato the universal was not merely the "ideal" but the "only real" (Sans 2004, 191). Sans suggested that the terminology of kind had not only logical but also naturalphilosophical connotations: "The middle of the categorical syllogism according to Hegel is the essential nature of the singular, not an arbitrary definiteness or property of the singular."

4.4 Earth as Such and Earth in General

Hegel said that analogical reasoning uses Earth both as a singular individual and as a generality or universal. Properties of the singular entity Earth may not be properties of Earth as a universal or class. The question that needs to be addressed now is Hegel's description of Earth in general, Earth as a class of entities, and what other entities are included in Earth in general.

The comparison of heavenly bodies to the Earth extends back to the beginnings of Greek thinking about the cosmos. While some pre-Socratic thinkers considered the Moon, Sun, and stars to have a fire-like nature, others insisted on the Earth-like nature of the planets and stars (Diels 1965, 356). The most striking accounts that liken heavenly bodies to the Earth are to be found in the descriptions of the Moon by Anaxagoras, Democritus, and the Pythagoreans. Anaxagoras and Democritus asserted that the Moon was a solid like the Earth, with mountains, plains, and clefts. While Plato considered the planets like the Earth to be divine gods, their material nature was a combination of fire and earth.

Aristotle's cosmology made a sharp differentiation between the celestial realm and the terrestrial realm. The celestial realm beyond the Moon is eternal and unchanging while the terrestrial realm is the region of generation and corruption. The four elements of earlier science and philosophy have a place in the terrestrial realm while the celestial realm is the region of ether. All of the terrestrial elements have finite rectilinear motion, but ether has an eternal circular motion. The difference of the two realms is blurred in the case of the Moon, which according to Aristotle is the dwelling place of pure fire. Ether mixes with terrestrial elements in the sphere of the Moon.

The Moon has a certain degree of similarity to the Earth in some medieval philosophical thought as well. While other planets are self-luminous and uniform in appearance, the Moon reflects the light of the Sun rather than generating its own light and shows dark features on its surface.

The dramatic shift in the understanding of the Moon and planets occurred in the Copernican theory of the Sun and planets, but an even more dramatic shift was the result of Galileo's telescopic observations, which showed the degree of similarity between the surface of the Moon and that of the Earth, as well as the similarity of the Jupiter-moon system with the Sun-planet system. The planets and Moon were revealed to have surface features like the Earth and were therefore considered to be composed of the same terrestrial corruptible matter and elements as the Earth.

By the eighteenth century telescopic observation had convinced philosophers and scientists that not only was the Moon similar to the Earth but all the planets were, and some philosophers and scientists even considered the Sun and comets to be composed of a solid terrestrial core with water and an atmosphere or their equivalents. Analogy in the sense of cosmological economy and simplicity figured prominently in these theories of the planets, Moon, comet and Sun.

Kant, Wolff, and Herschel are among those who used analogical arguments to suggest that the Sun and planets are similar to the Earth and that therefore these bodies are inhabited or inhabitable like the Earth. Kant and Wolff both thought the planets and comets are in the same species of celestial bodies, i.e., they are similar to the Earth. Sunspots were considered by Herschel to be dark land masses beneath a fiery atmosphere, which lent support to analogical thought about the similarity of all celestial bodies to the Earth.

The paradigm of Earth for other celestial bodies led scientists to look for evidence of water and air on the surfaces of the Moon, planets, and comets. Even when there was no clear convincing evidence, those who were persuaded by a principle of analogical uniformity and simplicity insisted that an atmosphere and water must be present even if only in small amounts.

Hegel, however, did not so eagerly embrace a homogeneous celestial world, but instead found in the solar system a system or organism having four kinds of celestial bodies: a central Sun, comets, moons, and planets. Structures having four components are common in nature rather than the trinitarian structure of the mental or spiritual world: thesis, antithesis, synthesis. In the world of nature, antithesis has a two-fold aspect. Movement is essential to matter, according to Hegel's conceptual analysis, and he found in the heavens four types of circular motion: a rotatory motion of a stationary central body; a non-rotating motion revolving around a central body; a non-rotating motion revolving around bodies, which themselves revolve around the central body; and finally, a rotatory motion revolving around a central body. The first type of motion is that of the Sun; the second the motion of comets; the third that of moons; and the final type the motion of the planets. For Hegel, the fourth type of motion—the planetary motion—is the most complete and includes the motions of the other three types.

In Hegel's view, comets, the Moon, Sun, and planets are not at all of similar material composition, and he found in them the four Empedoclean elements of Greek physics: air, water, fire and earth. The Sun, comets, and the Moon are represented by the elements air, water, and fire, respectively: "Fire corresponds to the definition of the lunar and water the definition of the cometary body" (Hegel 2002b, 99). Hegel associated light with air, both being invisible and simple or abstract elements. The association of water with the comet is an easy association. Hegel called comets clouds, completely without any solid core. The Moon was associated with fire in Aristotle's cosmology, and that may be the basis for the fiery nature of the Moon in Hegel's cosmology. The primary description of fire here is something "rigid" and flammable. Earth, however, was not a separate element for Hegel. Earth is the composite, concrete and perfect body, which subsumes the other elements (air, water, and fire into itself). Earth is the only complete celestial body: "If the seed of unrest is present in it, then it is a totality. The Earth is this totality. The planet is the living body, the unity" (Hegel 2002b, 68). Hegel does use Earth as a synonym of planet or as a paradigm of planet in general, and he is often erroneously understood to be speaking only of Earth as an individual when he means planet or Earth in general, not as an individual. One of the clearest statements of this is in Boris von Uexkuell's notes of the lecture on natural philosophy in Berlin 1821–1822: "The actualization of this living activity is represented in the [meteorological] process. Without the association with the Sun, without light, a processless body would result. But along with that, the body must in its essence be this universal individuality. That these processes are also present on the various planets is a necessary consequence. The planets that are closer to the Sun appear to be veiled in clouds" (Hegel 2002b, 103). The terminology is very revealing in the light of Sans' description of the categorical syllogism:

"universal individuality" is what separates analogical reflective syllogisms from categorical syllogisms.

Hegel explicitly rejected the possibility of life on the Sun, comets and the Moon because they lack the dynamic living process, which for Hegel is shown by the presence of clouds indicating the presence of air and water, and therefore a living meteorological process: "That humans might be present on the Moon, follows on empty teleological reasoning, which people ascribe to the wisdom of God" (Hegel 2002a, 66). In contrast to the dynamic living Earth, the Moon has no atmosphere and no water: "It is said of the Moon, it has no atmosphere. It has no water, no meteorological process and that is because, it has no water. The Moon is a waterless crystal" (Hegel 2002a, 65–66).

While Hegel explicitly rejected life on the Sun, comets, and the Moon, nowhere did he deny that other planets are living or have life. Planets are "living" according to Hegel, not just Earth. Other cosmic bodies—comets, Moon, and Sun—only have a single element; they are elementary bodies, not composite or concrete or perfect or living bodies, like the Earth and planets, which Hegel explicitly says is the seat of intelligibility.

Hegel also rejected the use of analogy to make stars the centers of systems like the solar system. He did not think that there was enough evidence of rational systems of organization and structure such as are found in the solar system. In this respect, those who have said Hegel was not a pluralist are correct. Hegel did not delight in endless repetitions. He preferred a finite cosmos like Aristotle and like Kepler, who also rejected Bruno's comparison of the stars to suns with similar planetary organizations. In a passage in the 1827 and 1832 *Naturphilosophie*, Hegel denied the possibility of life on the Sun and stars as well as on the Moon:

> Piety would introduce human beings, animals and plants into the Sun and Moon; but only a planet can accomplish this. Natural entities that have turned in on themselves, such concrete forms that preserve themselves against the universal, have not yet developed on the Sun; in the stars and in the Sun only light-matter is present (Hegel 1970b, IX, 114–115).

Hegel denied in this passage the presence of organic material on the Sun and stars as well as on the Moon. The argument against life on the Moon, considered above, is different from that against life on the Sun and stars, but both arguments are motivated by physical reasons.

Hegel's view of the nature of the Sun constitutes a denial of the similarity of the Earth and the Sun. In physical constitution, the Sun is closer to the comet than to the Earth: "Comet and Moon repeat thus in an abstract way the Sun and planet" (Hegel 1970b, IX, §279, 129). Whereas the comet is made of water, the Sun is made of light, but both lack a solid core or nucleus, which is characteristic of the planetary and the lunar natures. The difference between Earth and the Sun consists not only in their physical nature but also in their motion. The Sun and the Earth are very dissimilar entities. The Sun is a body of light rotating on its axis, whereas the Earth is a planet rotating on its axis but also revolving about the Sun as a central body. Because Hegel sided with those who considered the Sun a body of light and not a solid body capable of supporting organic beings—the position of Johann Elert Bode and William Herschel—he considered the Sun uninhabitable.

For Hegel the Sun was a more developed entity than the stars because it has become a central body, whereas the stars are simply points of light, which have not progressed beyond a punctual form and have no relationship with each other or with other bodies (Hegel 2000, 124). Since the Sun and the stars are only composed of light, they are not concrete realities and cannot sustain concrete beings, which have a more complex, complete reality than they themselves are. Human beings, plants, and animals are not to be found on the stars and the Sun, but only on planets, according to Hegel.

4.5 Conclusion

Analogical syllogisms are superficial, according to Hegel, insofar as they try to conclude that there is life on the Moon even though there is no evidence of water or air on that body. Similar analogical arguments for life on the Sun made by Bode and Herschel were considered by Hegel to be equally superficial, although here it is more a matter of which theory of the Sun one chooses to accept. Hegel's own theory that the Sun is a body of light was not commonly accepted, but still the planetary theory of the Sun offered only superficial similarities between the Sun and the Earth. Like the arguments for life on the Moon, they focused on only surface similarities and could not establish the physical conditions necessary for life: water and air. Analogical arguments for life on other planets would have been judged less harshly by Hegel. Planets, after all, have meteorological processes and therefore are "living" according to his philosophical account. Still, they are analogical arguments. It is better to present such theories in terms of philosophical proofs, by explanations and definitions rather than by flawed cognitive methods that can lead to error.

Whereas William Herschel was already finding great similarities between the Sun and the stars and had extended these similarities to the property of having planets or being themselves inhabitable worlds, Hegel rejected this analogy. Hegel expressed reservations about the validity of analogical reasoning, which has a certain usefulness in science but which is ambiguous and lacks necessity. The Sun and stars have some properties in common, but one cannot conclude from these similarities to the necessity that stars have planets. He also questioned the modern emphasis upon infinity. An infinity of worlds and the immensity of distances between them were problematic for Hegel. Like Plato and Aristotle, Hegel preferred finitude and limit to the infinite and unlimited.

Qualifications are, furthermore, necessary in statements about the solar system. It is unclear whether Hegel thought intelligence in the sense of "Geist" developed on other planets of the solar system, but he did allow that life developed at least on Mercury and Venus up to the point of the geological entity. The solar system was not geocentric for Hegel; it was "planetocentric," although his statements are subject to misinterpretation because he used "planet" and "Earth" interchangeably. His arguments against the presence of life in the solar system were not directed against other planets, but rather against the Sun and Moon, both of which he said have a different nature from Earth and planets. Although he did not explicitly discuss the possibility of life on comets, the fourth type of body in his theory of the solar system, it is clear that he rejected the views of Bode and Lambert, who did defend this possibility. Again, his critique of the use of analogical argument is important here. The Sun, comets, and moons are not analogous to the Earth or to the planets; these are four different bodies with different forms of motion and different physical constitutions. Only planets have completeness according to Hegel because only they have water, air, earth, and light, and completeness in this sense is necessary for life. Hegel discerned a need to make distinctions in nature rather than to consider superficially different realities as fundamentally similar. Celestial bodies should not be considered, according to Hegel, as all of one type or nature, as one kind.

References

Bode, Johnn Elert. 1816. *Betrachtung der Gestirne und des Weltgebäudes*. Berlin: Nicolai.

Breidbach, Olaf. 1987. *Der Analogieschluß in den Naturwissenschaften oder die Fiktion des Realen?: Bemerkungen zur Mystik des Induktiven*. Frankfurt am Main: Athenaum.

Cassirer, Ernst. 1911. *Das Erkenntnisproblem in der Philosophic und Wissenschaft der neueren Zeit*, Vol. 1. Berlin: Bruno Cassirer.

Cicero, M. Tullius. 1975. *De divination, De fato, Timaeus*. ed. Remo Giomini. Leipzig: Teubner.

Diels, Hermann. 1965. *Doxographi Graeci*. Berlin: Walter de Gruyter.

Eaton, Ralph M. 1931. *General Logic: An Introductory Survey*. New York: Charles Scribner's Sons.

Gilardi, Roberto. 1988. "Hume, Newton e il 'principio di analogia.'" *Rivista di filosofia neo-scolastica* 80: 63–104.

Guthke, Karl S. 1983. *Der Mythos der Neuzeit: Das Thema der Mehrheit der Welten in der Literatur- und Geistesgeschichte von der kopernikanischen Wende bis zur Science Fiction*. München: Francke.

Hänssler, Ernst Hermann. 1927. *Zur Theorie der Analogie und des sogenannten Analogieschlusses*. Basel: G. Bohm.

Hegel, Georg Wilhelm Friedrich. 1968. *Gesammelte Werke*, herausgegeben im Auftrag der Deutschen Forschungsgemeinschaft. Hamburg: Felix Meiner.

Hegel, Georg Wilhelm Friedrich. 1970a. *Hegel's Philosophy of Nature*, ed. M. J. Petry. Vol. 3. New York: Humanities.

Hegel, Georg Wilhelm Friedrich. 1970b. *Werke in zwanzig Bänden*, ed. Eva Moldenhauer und Karl Markus Michel. Vol. 9. Theorie Werkausgabe. Frankfurt am Main: Suhrkamp.

Hegel, Georg Wilhelm Friedrich. 1974. *Dokumente zu Hegels Entwicklung*, ed. Johannes Hoffmeister. Stuttgart: Frommann-Holzboog.

Hegel, Georg Wilhelm Friedrich. 2000. *Vorlesung über Naturphilosophie Berlin 1823/24. Nachschrift Griesheim*, ed. Gilles Marmasse. Hegeliana, Bd. 12. Frankfurt am Main: Peter Lang.

Hegel, Georg Wilhelm Friedrich. 2001. *Vorlesungen über die Logik: Berlin 1831. NachschriftK. Hegel*, ed. Udo Rameil with Hans-Christian Lucas. Vol.10. Hamburg: Meiner.

Hegel, Georg Wilhelm Friedrich. 2002a. *Vorlesungen über die Philosophie der Natur Berlin 1819/20*, ed. Martin Bondelli and Hoo Nam Seelman. Vol. 10. Hamburg: Meiner.

Hegel, Georg Wilhelm Friedrich. 2002b. *Vorlesungen über Naturphilosophie Berlin 1821/22. Nachschrift Uexküll*, ed. Gilles Marmasse and Thomas Posch. Frankfurt am Main: Peter Lang.

Herder, Johann Gottfried. *Ideen zur Philosophie der Geschichte der Menschheit*. Wiesbaden: R. Löwit. 1970.

Herschel, William. 1795. "On the Nature and Construction of the Sun and Fixed Stars." *Philosophical Transactions of the Royal Society of London for the Year MDCCXCV,* 85: 46–72.

Huber, Johannes. 1878. *Zur Philosophie der Astronomie*. Münich: Theodor Ackermann.

Kant, Immanuel. 1902. *Kant's gesammelte Schriften*. Berlin: Walter de Gruyter.

Leibniz, Gottfried Wilhelm. 1955. *Schöpferische Vernunft: Schriften aus den Jahren 1668–1686,* ed. Wolf von Engelhardt. Münster: Böhlau Verlag.

Leibniz, Gottfried Wilhelm. 1962. *Nouveaux Essais*. Berlin: Akademie Verlag.

Locke, John. 1959. *An Essay Concerning Human Understanding*. ed. Alexander Campbell Fraser. Vol. 2. New York: Dover Publications.

Mill, John Stuart. 1973. *A System of Logic Ratiocinative and Inductive, Being a Connected View of the Principles of Evidence and the Methods of Scientific Investigation,* ed. J. M. Robson. Vol. 2. Toronto: University of Toronto.

Newton, Isaac. 1972. *Philosophiae naturalis principia mathematica,* ed. Alexandre Koyré and I. Bernard Cohen with Anne Whitman. Vol. 2. Cambridge, MA: Harvard University Press.

Oeing-Hanhoff, Ludger. 1972. "Dialektik." In *Historisches Wörterbuch der Philosophie*, ed. Joachim Ritter. Vol. 2. Basel: Schwabe & Co.

Quintillian. 1963. *Institutio Oratorio*. Trans. H. E. Butler. Vols 4. Cambridge, MA: Harvard Univ. Press.

Sans, Georg. 2004. *Die Realisierung des Begriffs: Eine Untersuchung zu Hegels Schlusslehre*. Berlin: Akademie Verlag.

Stebbing, L.S. 1930. *A Modern Introduction to Logic*. New York: Crowell.

Zöckler, Otto. 1866. "Der Streit über die Einheit und Vielheit der Welten." *Der Beweis des Glaubens* 2: 353–376.

Zöckler, Otto. 1879. *Die Geschichte der Beziehungen zwischen Theologie und Naturwissenschaft*. Vol.2. Gütersloh.

Chapter 5
The Relationship Between the Origins of Life on Earth and the Possibility of Life on Other Planets: A Nineteenth-Century Perspective

Stéphane Tirard

Abstract In this chapter we examine how, during the second part of the nineteenth century and the beginning of the twentieth century, assumptions about the origins of life were specifically linked to the development of theories of evolution and how these conceptions influenced assumptions about the possibility of life on other planets. First we present the theories of the origins of life of Charles Darwin (1809–1882) and Herbert Spencer (1820–1903) and underline how they were linked to the knowledge of physical and chemical conditions of environments. These two examples lead us to think about the relationship between the origin of life, evolutionary biology, and geology, particularly the uniformitarian principle. An important point is the extension of the comprehension of terrestrial conditions of emergence and evolution of life to other planets. We claim that there was a sort of extended uniformitarian principle, based not only on time, but also on space. Second, after a brief look at panspermia theory, we compare two examples of assumptions about life on other planets. The French astronomer Camille Flammarion (1842–1925) and the French biologist Edmond Perrier (1844–1921) presented views that consisted in complex analogies between life on Earth and life on other planets. We analyze how they used neo-Lamarckian biological concepts to imagine living beings in other worlds. Each planet is characterized by a particular stage of biological evolution that they deduce from the state of living beings on Earth. The two scientists explained these different states with neo-Lamarckian principles, which were based on environmental constraints on organisms. Therefore these descriptions presented a sort of history of life, including the past and the future. We claim that their assumptions could be some intellectual exercises testing neo-Lamarckian theories. Moreover the description of human beings on other planets, and particularly the *Martian epianthropus* presented by Perrier, were complex utopias, which finally spoke about us and about an ideal future.

S. Tirard (✉)
Centre François Viète d'épistémologie et d'histoire des sciences et des techniques,
Université de Nantes, Nantes, France
e-mail: stephane.tirard@gmail.com

D. A. Vakoch (ed.), *Astrobiology, History, and Society*, Advances in Astrobiology
and Biogeophysics, DOI: 10.1007/978-3-642-35983-5_5,
© Springer-Verlag Berlin Heidelberg 2013

5.1 Introduction

During the second part of the nineteenth century, scientists presented assumptions about life on other planets. These assumptions were not simple speculations, but they were complex epistemological constructions based on several scientific fields: origins of life, theories of evolution, and geology. The links between these three fields occurred in the middle of the nineteenth century, notably in the works of Charles Darwin and Herbert Spencer. It is necessary to analyze them to understand the basis for the elaboration of analogies regarding life on other planets.

In addition, we will focus on two French scientists, the astronomer Camille Flammarion and the biologist Edmond Perrier. In their books on plurality of worlds they presented states of livings beings on other planets. We will analyze their explanations and particularly their modalities for introducing scientific concepts in elaboration of analogies. We want to question the extension of theories of evolution and also geological conceptions to other planets. We will ask several questions. Is it possible to conceive a sort of uniformitarian principle, not limited to time, but extended to space, therefore to other planets? Could we assume that these assumptions about life on other planets have a double status? First, are these assumptions some intellectual tests for theories of evolution? Second, are they utopias, and if so, what sort of utopias?

5.2 Origins of Life, Evolution, and Environmental Conditions

The development of evolutionism led to the conception of the origins of life, which explained how life was able to occur on the Earth that had previously lacked life. The two examples presented here reveal how, during the second part of the nineteenth century, several assumptions were formulated in response to the problem of the origin of life on Earth. They will show also how the problem of the origin of life, as well as the problem of evolution, was connected to the description of the environmental conditions on Earth.

Charles Darwin (1809–1882) himself never expressed his view about origins of life in a developed manner. In 1859, in the last sentence of *The Origin of Species*, he briefly suggested how life began on Earth:

> There is grandeur in this view of life, with its several powers, having been originally breathed into a few forms or into one; and that, whilst this planet has gone cycling on according to the fixed law of gravity, from so simple a beginning endless forms most beautiful and most wonderful have been, and are being, evolved (Darwin 1985, 458–460).

However, in 1871 in a letter to his friend and colleague, the botanist Joseph Dalton Hooker (1817–1911), he described the possible scenario of the emergence of life on Earth. He wrote this famous sentence:

> It is often said that all the conditions for the first production of a living organism are now present, which could have been present. But if (and oh what a big if) we could conceive in

some warm little pond with all sort of ammonia and phosphoric salts,—light, heat, electricity and c. present, that a protein compound was chemically formed, ready to undergo still more complex changes, at the present day such matter would be instantly devoured, or absorbed, which would not have been the case before living creatures were formed (as cited in Calvin 1961, 4–5).

This view of life is completely integrated in a materialistic conception. We have to notice two important points. First, Darwin described the conditions under which life could occur on Earth. Second, he conceived the impossibility of a plurality of origins of life in the context of the Earth. According to him, life on Earth prevents life from emerging again on the Earth. In other words, spontaneous generation cannot exist on Earth because life already exists on Earth. Therefore, this sentence of Darwin's is not only a materialistic description of the emergence of life on Earth, and a point of view about spontaneous generation, but an argument for the historicity of life on Earth. All the steps of the evolution of life on Earth were unique, even the first, which was absolutely unrepeatable. Each step of evolution of life on Earth participated in the definition of the conditions of the evolution of the subsequent step (Tirard 2010).

As a philosopher, Herbert Spencer (1820–1903) attempted to describe all the aspects of the universe. His project was to give all the explicative principles from the formation of the universe to human psychology. In his conception of evolution, Spencer was more Lamarckian than Darwinian. Spencer reduced the role of contingency. According to him there is continuity in evolutionary processes in all aspects of matter (Tirard 2011).

Regarding life, his *Principles of Biology* described the gap between mineral matter and living bodies. Spencer conceived very progressive transformations, and there was not a precise beginning of life. However, the process is comprehensible on the basis of physical and chemical laws. It could be entirely described by laboratory approaches, and Spencer tried to deduce each step of transformation: "In the early world, as in the modern laboratory, inferior types of organic substances, by their mutual actions under fit conditions, evolved the superior types of organic substances, ending in organizable protoplasm" (Spencer 1898, 700).

Therefore Darwin and Spencer expressed two views of the beginning of life on Earth. They described progressive processes of the evolution of matter in conceptual contexts of their own theories of evolution. The emergence of life was considered as a possible step of evolution, in the specific conditions of the primitive Earth. Indeed, the two previous propositions showed how conceptions of the origins of life depended on the knowledge of the matter of life and on the conditions on primitive Earth.

These theories were very connected to the conception of the history of Earth. Indeed the evolution of life and the evolution of the Earth have to be connected in comprehensive explanation of the origins of life on our planet. The theories of Darwin and Spencer are epistemologically strongly different. Darwin introduced a historicity founded on contingency, and for his part Spencer rejected hazard and claimed a sort of chemical and physical determinism. However, neither Darwin nor Spencer believed in spontaneous generation.

In these two assumptions, it appears that these scientists deduced the conditions on the primitive Earth from the current conditions on Earth. This possibility of establishing an analogy between past and present is a crucial dimension in the elaboration of scenarios.

During the last third of the nineteenth century the idea of the possibility of life on other planets was broadly accepted. How did some scientists use the knowledge of life on Earth to describe the life on other worlds? Is there an epistemological obstacle to using knowledge about evolution of life on Earth in other contexts? In other words, how can scientists argue that life exists on other planets, and how can they explain its emergence?

It is very notable that the main scenarios of life on other planets give central reference to life on the Earth. Indeed, conceptions of life on other planets were often arrived at by an analogical method. These thoughts were probably dependent on the fact that during the nineteenth century, conceptions of the history of Earth changed. Particularly, the introduction of the principle of the actual causes by Charles Lyell (1797–1875) induced a new vision of the evolutionary process of changes of the Earth. This principle, which was named uniformitarianism by the English philosopher William Whewell (1794–1866), was presented by Lyell in his *Principles of Geology* in 1830. We can suggest that there was a sort of extension of the uniformitarian principle concerning conditions and phenomena on Earth to other planets. This helps to transpose the connection between geology and the origins of life to the context of other planets.

5.3 Evolution and Panspermia: A Complex Relationship

During the second part of the nineteenth century, the possibility of panspermia was central to discussions about the origin of life. Without describing the debate between theories of an evolutionary emergence of life and panspermia, we would like to highlight how complex the relationship was between evolutionary theories and the possibility of life on other planets. Panspermia presents an interesting case for the ambiguous evolution of its own status.

First, we must recall that this theory was a fixist response to Darwin's theory of evolution. Indeed, in 1871 in his Presidential address at the annual congress of the British Association for the Advancement of Science (BAAS), William Thomson (Lord Kelvin) (1824–1907) advocated his panspermia theory in opposition to Darwin's evolutionism (Thomson 1872, lxxxiv–cv). Fundamentally, Thomson was opposed to evolution. He argued that all evolutionary processes imagined by Darwin would be too long given the age of the Earth, which Thomson estimated himself as a physicist and astronomer (Tirard 2006). According to Thomson, life would have arrived on Earth by means of a meteorite, carrying germs of life, such as spores or seeds. In Thomson's case, it was more than a empirical obstacle that induced a particular theory of panspermia; his opposition to biological evolutionism was quite ideological.

Second, in opposition to this first period, we see with panspermia how terrestrial conceptions can be used to imagine the evolution of life on other planets. It is indeed well known that the Swedish scientist, Svante Arrhenius (1859–1927), claimed life is eternal, spread everywhere in the universe. According to him, microscopic germs of life are distributed to new planets where, life could develop itself:

> Finally, we perceive that, according to this version of the theory of panspermia, all organic beings in the whole universe should be related to one another, and should consist of cells which are built up of carbon, hydrogen, oxygen, and nitrogen. The imagined existence of living beings in other world in whose constitution carbon is supposed to be replaced by silicon or titanium must be relegated to the realm of improbability. Life on other inhabited planets has probably developed along which are closely related to those of our earth, and this implies the conclusion that life must always recommence from its very lowest type, just as every individual, however, highly developed it may be, has by itself passed through all the stages of evolution from the single cell upward (Arrhenius 1908, 228–229).

Therefore panspermia pushed away the question of the origin of life. Life is eternal. Germs of life are presents everywhere and all the time. Life is universal, eternal, and therefore unique. However, the question of the evolution of life still exists. Arrhenius' assumption notably showed having a conception of evolution of life on Earth was important for imagining evolution on other planets. In the case of panspermia, the unity of life was a central notion. Arrhenius applied a simple analogy between all of the planetological contexts, and it appears that the evolutionary process was crucial.

5.4 Emergence and Evolution of Life on Other Planets

Historians of science have conscientiously studied the topic of the plurality of worlds, and this chapter cannot be a review of their works (Dick 1982, 1996; Raulin-Cerceau 2006). We only want to focus on the use of analogy between the Earth and other worlds in the elaboration of assumptions about the existence of life on other planets.

We will study two French examples, an astronomer and scientific writer, Camille Flammarion, and a biologist, Edmond Perrier. They are very interesting because, beyond providing complete descriptions of different planets, using analogical reasoning they also deduce the state of living beings in these other environments.

5.4.1 Astronomical Culture and the Conception of Life on Other Planets: Camille Flammarion

Camille Flammarion (1842–1925) was a central person in the French astronomical community at the end of the nineteenth century. He was not an academic astronomer, however, he owned his observatory, and wrote numerous popular books on astronomy, and most of them were very important.

The plurality of worlds was constantly present in Flammarion's books. He dedicated specifically three books to this topic: *La pluralité des mondes habités* (1862), *Les Terres du ciel* (1884), *La Planète Mars et ses conditions d'habitabilité* (1892). The title *Les Terres du ciel* revealed Flammarion's principle: the other planets are analogous to the Earth. The book is divided into eleven parts, in which all the planets are systematically treated: Mars, Venus, Mercury, the Earth, the Moon, Jupiter, the system of Saturn, Uranus, Neptune and the last chapter devoted to "life in infinity." He studied the possibility of life on each of these bodies. According to Flammarion the plurality of the worlds is obvious. In this book he is interested by the varied conditions in which life exists on the other planets.

Flammarion wrote a synthesis of previous descriptions of life on other planets. It is important to underline that Ernst Haeckel particularly influenced his description of the evolution of life on Earth. For example, he used Haeckel's famous picture about the development of embryos that showed how the continuity of the steps of transformation during embryological development recapitulated the steps of evolution. Flammarion generalized Haeckel's theory of evolution to other planetological contexts, letting him imagine the stage of biological evolution on each planet. For example, he claimed that on Mars, beings must be very close to humans. They come from transformations analogous to terrestrial ones, in the specific context of this planet.

Life in Flammarion's work is completely included in his comprehension of the universe by the mean of astronomy, the most fundamental science. This last discipline describes the condition on planets. When linked to his conception of evolution—the neo-Lamarckian theory formulated by Haeckel—it provided a means to claim that life exists on other planets and to describe the ways of its transformations.

5.4.2 Lamarckian Biological Culture and the Conception of Life on Other Planets: Edmond Perrier

When he wrote his little book, *La vie dans les planètes*, Edmond Perrier (1844–1921) was an important and very respected zoologist (Perrier 1911). He was recognized as a leader of the French community of neo-Lamarckian biologists.[1] He was notably well-known for his works on animal colonies. Therefore Perrier was not a specialist of astronomy, however, he gave his own descriptions of life on other planets.

His book, which deserves analysis, begins by a question asked to Perrier by his colleague the physicist Charles Edouard Guillaume, "Did you never mind to the form that animals on Jupiter must have?" (Perrier 1911, 6). Perrier recognized that he never thought of that, and immediately he claimed the possibility of an

---

[1] On this community, see Loison (2011).

analogy. The Earth is a planet like the others. Everything that it produced was the result of immutable properties of matter and of forces acting on it. What was been produced was not able to be *not* produced. The knowledge of all the properties on Earth authorizes one to conceive of the conditions of other planets. Perrier admitted that experimental proof would be lacking, however, he said "It is not forbidden to dream" (Perrier 1911, 8).

Perrier wrote that some of his main references came from Flammarion's *La Pluralité des Mondes, Terres du Ciel* and *La Planète Mars et ses conditions d'habitabilité*. As we noted, these books were very important in the diffusion of astronomical knowledge, notably in France. For Flammarion astronomy provided the elements of a description of the conditions on other planets, and these conditions could be used for the comprehension of the ways of evolution, understood in neo-Lamarckian terms.

Therefore, for Perrier, studying life on other planets is not simply a description of obvious facts as for Flammarion. When he imagined life on other planets, Perrier used his own conception of Lamarckian transformism. On one hand, this exercise is possible because of Lamarckian transformism. On the other hand, it offers a thought experiment testing his Lamarckian theory.

This little book showed how for Perrier transformation of life depends of characteristics of the environment. He thought that there is no chance in living processes and that biology could become an exact science. Without developing his arguments, we have to note that at first he strongly criticized the panspermian theory; therefore, according to him, the problem of life on other planets begins with the problem of the emergence of life. Later he dedicated an entire chapter to the question of the origin of life on Earth. His view clearly relied on chemical facts. He claimed that complex synthesis was able to occur. Carbon, hydrogen, oxygen, and nitrogen reacted to produce complex molecules able to resist destruction and even to rebuild themselves. According to Perrier, at this moment, life was created. Perrier insisted on the universality of this phenomenon: "If such clusters of atoms and molecules were able to occur at the origin, in some point of the universe, we can say that they occur everywhere such atoms were present" (Perrier 1911, 33). This sentence indicates very clearly how knowledge about Earth can be transposed to other contexts.

Before examining the specific cases of Venus and Mars, Perrier dedicated a chapter to study the notion of habitability, regarding planets of the solar system. He deduced that Jupiter and the planets situated in the external part of the solar system are uninhabited. All the planets situated inside the asteroid belt share important common characteristics: having a density close to that of Earth, turning on their axis in 24 h, being solid, and having an atmosphere. Finally, except for Mercury, they all have oceans. Therefore, Perrier claimed that there are only three inhabitable planets: Earth, Mars, and Venus.

According to him the same laws act on Mars and Venus, and they produce the same effects. We can expect to find the same organisms on these planets: algae, mushrooms, trees, much vegetation (e.g., savannas, woods), sponges, and microorganisms. In the water there are fishes, and amphibians are preparing their conquest of continents. Some details, which can be explained, distinguish these living beings from terrestrial ones.

On Venus, because of the high temperature, plants and animals have the same dimensions as in our tropical regions, and they are concentrated around the polar regions. Living beings are in comparable conditions as on Earth during the secondary period, or perhaps the primary period. Perrier described life on Earth during these periods and, as Venus is younger than the Earth, he assumed that this description corresponds to the current stage of the life on this planet.

His reasoning is completely founded on a neo-Lamarckian conception of transformation of living beings. Indeed, he argued that transformations need the action of environment during a very long time. However, under the effect of the same laws, they followed the same path on Earth and on Venus. He wrote that if all the motions were measured, we could, after observing the forms of terrestrial living beings, accurately calculate the forms of living beings on the other planets (Perrier 1911, 108). According to Perrier, Venus displays an earlier stage of the history of Earth. On this planet mammals are poorly developed, and human beings are not present.

On Mars the situation is very different, and Perrier dedicates a chapter to its description. An important difference between the Earth and Mars is the gravity. He reminds his reader that this constraint heavily influenced evolution on Earth. For this reason, martian reptiles have long limbs, perfect for running and jumping. Mammals are very light, like antelopes. The bears took the proportions of rapid greyhounds, and among big cats, tigers and lions have long legs, as do our cheetahs. Moreover, on Mars the higher temperature accelerates the process of evolution and of development. Trees and vegetables are bigger than on the Earth, and the development of insects is also more active: "Therefore Mars is the planet of flowers and butterflies" (Perrier 1911, 80). The low boiling temperature of water led to another problem regarding the drying of organisms, and on Mars animals have developed a thickened skin.

Perrier gave an accurate description of the "*Martian epianthropus*." The etymology of this name is not so obvious, however, it seems that Perrier wanted to describe a "surhomme," a superman. Martian humanoids are very tall because of the low gravity. They are blond because of the light, and they look like our Scandinavians. Among other differences, their eyes are bigger and have a better accommodation faculty than ours.

According to Perrier, in *Martian epianthropus*, there is only one race. This homogeneity is the result of many hybridizations. The consequence is a perfect unity, which, according to Perrier, is the goal of humanity. Therefore Perrier described a utopian world with a perfect political harmony and particularly with no wars.

We have to note that even if there was no dogmatically religious view in his text, Perrier referred to God as the origin of causes. It seems the particular non-academic status of this text authorized him to extend his thought in this way. However, even if Perrier was probably deistic, he only explained evolution on the basis of scientific explanation, and this remark is more philosophical than religious.

5.5 From Analogy to Utopia

At the end of the nineteenth century, the thought of life on other worlds also provided a way to imagine the future of life on Earth. With Lamarckian conceptions, analogous conditions lead more easily to analogous phenomena. In Perrier's case, we can ask ourselves if the main goal of his text is really the question of life on the other planets, or if it is an exercise about neo-Lamarckian principles.

Two points are very important in the two previous conceptions: the specific conditions of each different planet, and secondly, the process of biological evolution used for the description of the process. They are fundamental for the production of the analogy with Earth. Indeed in this epistemological context it is possible to say that there is a sort of implicit uniformitarian principle extended to the other planets. It permits us to conceive the steps of evolution of life on other planets. On Mars evolution is delayed, and life on Mars is described based on Earth's past; on Venus, evolution is earlier, and it indicates the future of Earth.

Describing a Martian superhuman, Perrier prolonged the consequences of the planetological analogy in a very complex utopia. Indeed, if we agree with the definition that a utopia is an imaginary country, where an ideal government leads happy people, we can say that Perrier's view is utopian. We have seen how human life is ideal on Mars. However, perhaps we could suggest that there are two fitted utopias. Indeed, before the social and political utopia there is another utopia, more metaphorical. Is it possible to say that all his conception of a very directed evolution would be a sort of biological utopia? In this case the ideal government would be the Lamarckian causes of transformism.

5.6 Conclusion

During the second part of the nineteenth century, the central concepts of evolution were applied to the other planets. A complex analogy was built between the history of the Earth and the assumed history of other worlds. This analogy was based on geological, astronomical, and biological knowledge.

Indeed, the elaboration of assumptions about the possibility of life on other planets seemed to require two main components. On one hand, the history of the Earth and history of life on Earth are associated. The assumption about the origin of life on Earth depended on knowledge about primitive conditions, which were known on the basis of the uniformitarian principle. On the other hand, this first relation is transposed into a relationship between the history of every planet and the possible life on it. In this case we suggest speaking about a sort of extended uniformitarian principle, based not only on time, but also on space.

These assumptions about life on other planets corresponded to different motivations. For Flammarion life on the planets was the ultimate prolongation of the dream that he wanted to offer to his readers. His presentation of astronomy was a serious scientific discourse and a meditation on the other worlds and the universe. Astronomy was, for him, the queen of the sciences because it explores the universe. Flammarion anticipated the discovery of life on other planets, however, he presented it as obvious. The fascinating discovery of the universe leads to the discovery of other worlds and probably other intelligent beings, therefore astronomy leads to the discovery of other intelligent beings.

Perrier followed a comparative way, however, his goal was not exactly the same. As a biologist, the description of the evolution of life on the other planets was for him a sort of experiment testing his Lamarckian theory.[2] In this exercise, he was able to describe the past and the future of life. In the end, the description of the *Martian epianthropus* appeared as the description of the perfect human in a perfect world. Is that our future?

Moreover, in a more philosophical way, this thought about life on other planets is an ideal topic to imagine utopias, with ethical or political views, and perhaps also in biological terms. We have to remember that the first meaning of the word utopia, invented by Thomas Moore, was "nowhere." Planets, so far from our world, are probably the ideal place for a perfect imaginary world.

To finish this conclusion we can highlight how thin is the limit between these sorts of assumptions and fiction. We notice that in his references Perrier did not neglect the novelist H.G. Wells (1868–1946). In the exercise of assumption, the methods of scientists and of certain novelists are probably very close. The facts of the current and present nature are the foundation of analogies elaborated by scientists regarding a near future. In 1946, Aldous Huxley, in the preface of the second edition of his famous novel, *Brave New World*, claimed that a book on future is interesting only if the described things can be conceived. In other words, we could say that some scientific analogies and also good literature have the capacity to open our mind and to enlarge our world.

References

Arrhenius, Svante. 1908. *Worlds in the Making: The Evolution of the Universe*. London: New York: Harper and Brothers Publishers.
Calvin, Melvin. 1961. *Chemical Evolution*. Oxford: Clarendon Press.
Darwin, Charles. 1985. *The Origin of Species by Means of Natural Selection or the Preservation of Favoured Races in the Struggle for Life*. London: Penguin Books.
Dick, Steven J. 1982. *Plurality of Worlds: The Origins of the Extraterrestrial Life Debate from Descartes to Kant*. Cambridge: Cambridge University Press.
Dick, Steven J. 1996. *The Biological Universe*. Cambridge: Cambridge University Press.
Flammarion, Camille. 1862. *La pluralité des mondes habités*. Paris: Gauthier-Villars.
Flammarion, Camille. 1884. *Les Terres du ciel*. Paris: C. Marpon et E. Flammarion.
Flammarion, Camille. 1892. *La Planète Mars et ses conditions d'habitabilité*. Paris: Gauthier-Villars.

---

[2] He notably tested his own theory of instinct.

Loison, Laurent. 2011. *Qu'est-ce que le néolamarckisme ?: Les biologistes français et la question de l'évolution des espèces*. Paris: Vuibert.

Lyell, Charles. 1830. *Principles of Geology*. London: John Murray.

Perrier, Edmond. 1911. *La vie dans les planètes*. Paris: Edition de la Revue.

Raulin-Cerceau, Florence. 2006. *A l'écoute des planètes*. Paris: Ellipses.

Spencer, Herbert. 1898. *The Principles of Biology*. London: Williams and Norgate.

Thomson, William. 1872. "Address of Sir William Thomson (President)." *British Association for the Advancement of Science, Edinburgh, Report—1871*. London: John Murray.

Tirard, Stéphane. 2006. "William Thomson (Kelvin), Histoire physique de la Terre et histoire de la vie." In *Pour comprendre le XIXe siècle Histoire et philosophie des sciences à la fin du siècle*, ed. Jean-Claude Pont, 297–306. Geneva: Olski.

Tirard, Stéphane. 2010. "Origin of Life and Definition of Life, from Buffon to Oparin." *Origins of Life and Evolution of Biospheres* 40(2): 215–220.

Tirard, Stéphane. 2011. "Spencer et les origines de la vie. La double induction comme méthode." In *Penser Spencer*, ed. Daniel Becquemont and Dominique Ottavi, 81–95. Paris: Presses Universitaire de Vincennes.

Chapter 6
Pioneering Concepts of Planetary Habitability

Florence Raulin Cerceau

Abstract Famous astronomers such as Richard A. Proctor (1837–1888), Jules Janssen (1824–1907), and Camille Flammarion (1842–1925) studied the concept of planetary habitability a century before this concept was updated in the context of the recent discoveries of exoplanets and the development of planetary exploration in the solar system. They independently studied the conditions required for other planets to be inhabited, and these considerations led them to specify the term "habitability." Naturally, the planet Mars was at the heart of the discussion. Our neighboring planet, regarded as a sister planet of Earth, looked like a remarkable abode for life. During the second part of the nineteenth century, the possibility of Martian intelligent life was intensively debated, and hopes were still ardent to identify a kind of vegetation specific to the red planet. In such a context, the question of Mars' habitability seemed to be very valuable, especially when studying hypothetical Martian vegetation. At the dawn of the Space Age, German-born physician and pioneer of space medicine Hubertus Strughold (1898–1987) proposed in the book *The Green and Red Planet*: *A Physiological Study of the Possibility of Life on Mars* (1954) to examine the planets of the solar system through a "planetary ecology." This innovative notion, which led to a fresh view of the concept of habitability, was supposed to designate a new field involving biology: "the science of planets as an environment for life" (Strughold 1954). This notion was very close to the concept of habitability earlier designated by our nineteenth-century pioneers. Strughold also coined the term "ecosphere" to name the region surrounding a star where conditions allowed life-bearing planets to exist. We highlight in this chapter the historical aspects of the emergence of the (modern) concept of habitability. We will consider the different formulations proposed by the pioneers, and we will see in what way it can be similar to our contemporary notion of planetary habitability. This study also shows the convergence of the methodological aspects used to examine the concept of habitability, mainly based on analogy.

F. Raulin Cerceau (✉)
Centre Alexandre Koyré, Muséum national d'Histoire naturelle, Paris, France
e-mail: raulin@mnhn.fr

D. A. Vakoch (ed.), *Astrobiology, History, and Society*, Advances in Astrobiology and Biogeophysics, DOI: 10.1007/978-3-642-35983-5_6,
© Springer-Verlag Berlin Heidelberg 2013

6.1 Introduction

The question of planetary habitability is nowadays a topical subject thanks to the continuous discoveries of exoplanets. Exo-Earths, in particular, are the center of interest of astrobiologists. Exoplanets' habitability is estimated according to the specific characteristics of each exoplanet and the precise habitable zone of each newly detected planetary system. As underlined by Impey in this volume (Impey 2013), it remains however unclear if our solar system is "typical" or not. The detection of more and more numerous multiple planetary systems and the presumption that in the Galaxy every star could have at least one planet, have enlarged our understanding of planetary systems. We have nowadays to contemplate studying an "exoplanet zoo" (Impey 2013). Perhaps are we surrounded with millions of habitable Earth-like planets.

However, habitability is not a new concept. It has been defined in scientific terms and widely discussed among the astronomical community during the second part of the nineteenth century—even if this notion was already present during previous centuries (e.g., Fontenelle 1686; Huygens 1698). Some personalities of astronomy of that time have examined in detail what could be, in our solar system, habitability for every planet. In this chapter, we will successively present the pioneering viewpoints of Richard A. Proctor, Jules Janssen, and Camille Flammarion. The planet Mars will be one of the main objects of this study, within a context devoted to the investigation of the puzzling Martian surface.

Besides this, the second part of the nineteenth century was rich in new methods and theories. Biological evolution and spectroscopy represented, respectively, breakthroughs in theory and technique (Dick 1996). On one hand, spectroscopy confirmed the unity of nature by observational methods, leading to the detection of similar molecules in different planetary or stellar environments. This new science strengthened the idea that the buildings blocks of life were common in the universe. On the other hand, Darwin's theory of evolution provided a scientific background in which physical evolution of the universe became conceivable, along with mechanisms of evolution from inorganic matter to life (Dick 1996).

At the dawn of the Space Age, nearly one century later, the question of habitability reappeared in a completely different context. While the first programs for the launch of artificial satellites were starting, the problem of life in space (human life in outer space) and the question of other life elsewhere, began to be examined in concrete terms. We will present in this chapter Hubertus Strughold's viewpoint—as a pioneer in space medicine—about planetary ecology, a concept similar to habitability.

Finally, this chapter will highlight how close the early concepts of habitability could be to the contemporary ones, when one considers the recent discoveries about exoplanets and the exploration of satellites of giant planets in our solar system.

6.2 The Question of the Plurality of Worlds Through the Ages

The historical question of the plurality of worlds has been studied in detail in reference books, such as those of Michael J. Crowe (1986) and Steven J. Dick (1982, 1996). An overview of this debate in the Western intellectual context prior to 1900 is presented in this volume by Michael J. Crowe and Matthew F. Dowd (Crowe and Dowd 2013). This question is entirely linked to the cosmological models that were accepted during each historical period. It is particularly clear that the heliocentric view of the universe has offered new possibilities leading to the admission that other worlds could be inhabited in the universe. Of course, this last idea is older than that. The question of the plurality of worlds goes back to antiquity and was supported by the atomistic philosophers, such as Leucippus, Democritus, Epicurus, and Lucretius. According to their philosophy, there are innumerable worlds that follow one another in an infinite universe. However, these assumptions remained essentially a philosophical school of thought, without calling into question the central place of the Earth in the universe.

The problem of the plurality of worlds reappeared once the Copernican theory (stated by Nicholas Copernicus in *De revolutionibus orbium coelestium*, 1543) had dethroned the central place of the Earth in the universe, in spite of many difficulties this new paradigm has had to face. In this model of the universe, every planet of the solar system turned round the Sun. Hence, the Earth became no more than one planet among others, and one of the main conclusions was that our planet was no more the center of the universe. It turned out to be a planet "like the other ones" in the solar system. It became therefore quite conceivable that other planets could be inhabited. The ideological consequences of that new paradigm were significant.

One of the most famous authors of that time to have defended the idea of the plurality of worlds was Giordano Bruno (1548–1600), an Italian Dominican friar. He published in 1584 a work of great consequence enclosed in his "Italian Dialogues," entitled *The Infinity, the Universe and Its Worlds*. In this writing Bruno defended the idea of infinite inhabited worlds going round innumerable suns located in an infinite universe. His system of thought could be considered as a materialistic pantheism in which God and the world were one. This idea was very disturbing for the Catholic Church, considering that it left no room for a greater infinite conception named God. Bruno was burned at the stake at Rome in 1600 after the Roman Inquisition had accused him of heresy. The question of the plurality of worlds was just one accusation among many pronounced against Bruno (Raulin Cerceau and Bilodeau 2011).

At the same time, astronomical observations strengthened the new paradigm of heliocentrism. In 1610, Galileo Galilei (1564–1642) discovered the four largest moons of Jupiter with his astronomical telescope. If the planet Jupiter was surrounded by moons, it became difficult to maintain that the Earth was the center of the universe. In spite of strong confrontations with the Catholic Church, Galileo largely contributed to promote the heliocentric view of the universe. In the meantime, heliocentrism was demonstrated by Johannes Kepler's works about the planetary motions in the solar system. However, during the first part of the seventeenth century, the idea that other worlds similar to ours could exist was still supported by very few people. Kepler himself was interested in Moon's habitability (*Somnium*, 1634), but he perceived that strong difficulties remained to assert that our planet was like any other bodies of the solar system.

This idea was more openly tackled at the end of the seventeenth century, as curiosity in planetary studies was increasing. Bernard le Bovier de Fontenelle (1657–1757), a French philosopher and writer, published in 1686 his *Entretiens sur la Pluralité des Mondes* (*Conversations on the Plurality of Worlds*). This influential piece of scientific popularization was presented in the form of a pleasant and elegant dialogue between a philosopher and a Marquise (*la Marquise de la Mésengère*). It expounded the Copernican world system and speculated about the inhabitants of other planets in the solar system.

During the same century, Christiaan Huygens (1629–1695), a Dutch astronomer and mathematician, wrote a treatise entitled *Cosmotheoros: or, Conjectures Concerning The Inhabitants of The Planets*, posthumously published in 1698 by his brother (Huygens 1698). This book presented Huygens' speculations on the construction of the universe and on the question of planetary habitability, as deduced from his own astronomical observations and those of other astronomers. However Huygens' viewpoint could be considered as anthropocentric, since he proposed that "men" (and animals too) living on other planets were very similar to the terrestrial ones (same mind, same body, same senses).

Our study will be focused on the second part of the nineteenth century, marked by many developments in astronomy and an explosion of interest in the plurality of worlds, especially through specific attention paid to the planet Mars.

6.3 The Nineteenth-Century Pioneers of Planetary Habitability

During the second part of the nineteenth century, astronomical research was stimulated by the increasing observations of the planet Mars. Distinguished astronomers attempted to penetrate the secrets of its surface. The canals controversy, introduced in 1877 by Giovanni Schiaparelli (1835–1910) and considerably developed by Percival Lowell (1855–1916) from the very end of that century to the beginning of the twentieth century (see Lowell 1909) intensified the importance attached to the study of the red planet. In such a context of high hypotheses, a few assumptions were however commonly accepted:

- Mars has great similarities with our planet.
- The red planet could present seas, continents, and seasons, like the Earth.
- Mars would be a planet older than ours.
- This planet could be then inhabited by living beings superior to humans.
- The Martian world seemed to be very exciting and represented a fantastic ground for speculative studies about the possibility of life elsewhere.

Could Mars be an abode for life? Lowell's book title about Mars (*Mars, as The Abode of Life*, 1909) is representative of the speculations of that time on Martian life. Lowell, convinced of the presence of complex life on the red planet, formulated pertinent comparisons between Mars and the Earth. He examined what could be a "Martian ecology," and as such, was perhaps the first "exoecologist," as assumed by Markley (2005). Lowell imagined, indeed, a kind of model of "sustainable development."

If the possibility of another intelligent life was a great subject of discussion among the astronomical community, especially during the canals controversy, the hypothesis of Martian vegetation was largely accepted. Since the middle of the nineteenth century, variations of color were observed on the Martian surface. They were interpreted as evidence of seasons correlated with the presence of some kind of vegetation. It should be noted that at the end of the nineteenth century, while the Martian canals controversy was fading (except, perhaps, in Lowell's works), the vegetation hypothesis was at its height.

It became then significant to determine the parameters required *at a minimum* to allow the presence of some simple forms of life on the Martian surface. Planetary habitability began to be a scientific field worth studying in detail, especially thanks to pioneers, namely the astronomers Richard A. Proctor, Jules Janssen, and Camille Flammarion.

6.3.1 Richard A. Proctor (1837–1888)

The British astronomer Richard Anthony Proctor, famous for his first detailed map of the planet Mars (1867) and his talent in astronomy popularization, stated that habitability was a determining factor to answer the question of a potential existence of other life forms in the universe.

In his famous book *Other Worlds Than Ours* (1870) he examined systematically the planets of our solar system. He studied their criteria of habitability, depending on physical and environmental parameters such as climate, seasons, atmosphere, geology, and gravity. According to Proctor, defining planetary habitability was a very difficult task, but this difficulty could be overcome in considering *analogy* with our planet. Proctor's methodology was based on comparison between the terrestrial environmental parameters and all the environmental parameters characterizing every planet. However, a planet could not be necessarily inhabited, at any time. The example of the Moon showed that the question of habitability was not valid for any celestial body. Basically, in Proctor's opinion, the existence and diversity of life forms should depend on the *specific* conditions prevailing on the surface of each planet.

One important point concerning Proctor's study of habitability is that he took into account Darwin's theory (*The Origin of Species*, 1859). The question of adaptation is tackled all throughout the book, especially in the chapter entitled "What Our Earth Teaches Us." This point—considered here by Proctor while evolutionary ideas were developing—has been re-discussed after the reformulation of

Darwin's theory by synthetic evolutionists in the early 1960s, as established by Vakoch (2013). These ideas, indeed, influenced those who speculated on the possibility of extraterrestrial life.

According to Proctor, Darwin's works have demonstrated that a correlation existed between the environmental changes (along with their rhythm and intensity) occurring in a specific habitat and the survival (or not) of the living species in this habitat (Proctor 1870). One conclusion of this observation is that specific conditions of environment could be appropriate only to specific species. Considering analogies with the terrestrial model, Proctor thought that if many other worlds could exist, they should be very different from ours (the title of his book is very explicit). Creatures on their surfaces could be very unusual, and could delight in being in environments inhospitable for terrestrial living beings. To summarize, according to Proctor, these other worlds shelter life in other ways (Proctor 1870).

Proctor examined the celestial bodies of our solar system: Mercury, Venus, Mars, Jupiter, Saturn, Uranus and Neptune, the Moon and other satellites, meteors, and comets. A special attention was given to the planet Mars, "the miniature of our Earth," in Proctor's terms (Proctor 1870, 90). Many physical analogies with our planet could be found on Mars: continents, seas, straits, water, which would be largely present on the surface. The atmosphere would contain water vapor with a water cycle equivalent to the terrestrial one. The Martian world described by Proctor would allow any form of life, from the simplest forms of vegetation to life forms much more complex.

Even Mercury and Venus, which would present very different conditions than ours, could have life on their surfaces. However, since the environmental parameters would be quite dissimilar from those known on our planet, these planets could shelter unfamiliar forms of life—some *microscopic creatures* on Mercury, for instance. Proctor thought that the other planets of the solar system offered their own conditions of habitability.

The same argument was applied to the giant outer planets, especially Jupiter. Proctor assumed that Jupiter was not at present a fit abode for living creatures. However, he suggested that one day Jupiter would be a living world that must be very differently constituted from those we are familiar with. The living creatures, if any existed, would be built on a much smaller scale than the inhabitants of the Earth. According to Proctor, Jupiter could probably be inhabited by "the most favored races existing throughout the whole range of the solar system" (Proctor 1870, 115). However, Proctor expressed some doubts about intelligent life in the solar system. In his book *Our Place among Infinities* published in 1875 (in a chapter entitled "A New Theory of Life in Other Worlds"), Proctor withdrew intelligent extraterrestrials not only from most planets of our solar system but also from other stellar systems (Crowe and Dowd 2013).

6.3.2 Jules Janssen (1824–1907)

Jules Janssen was a French astronomer who contributed to founding the scientific field of planetary spectroscopy during the second part of the nineteenth century. Janssen and Sir William Huggins (1824–1910) were spectroscopic pioneers who carried out the first spectroscopic observations in the hope of detecting oxygen and water in the Martian atmosphere. Janssen strongly supported this new scientific method, which made it possible to question planetary atmospheres and search for water vapor, one of the first conditions required for terrestrial life development (Janssen 1929). Janssen assumed that the detection of water vapor in a planetary atmosphere was a crucial condition to expect the presence of life on its surface. And then, the new methods of physical astronomy (corresponding to the birth of astrophysics including spectroscopy) could perhaps lead to solving the problem of extraterrestrial life. In 1867, Janssen announced to have discovered the presence of water vapor in the Martian atmosphere (Launay 2008). (It was in fact terrestrial signatures.)

According to Janssen, the question of habitability was one of the most interesting queries given to human intelligence (Janssen 1929). Spectroscopy, when giving the chemical constitution of planetary atmospheres, could help to determine very important parameters defining the possible conditions for life. Janssen understood that a strong link existed between the planetary environmental conditions (especially the presence of water vapor in the atmosphere and then, liquid water on the surface) and the possibilities for life to appear and to subsist on a planet.

In the meantime, spectroscopy also gave some limits to the possibility of life on other planets. In identifying the presumed components of planetary atmospheres, it eliminated the planets whose atmospheres did not contain water vapor. However, great doubts subsisted about the detection of water vapor in planetary atmospheres. Therefore, considering the beginnings of planetary spectroscopy, no strong conclusion could have been formulated about the hypothetical habitability of the planets of the solar system. In particular, the problem of the chemical constitution (qualitatively and quantitatively) of the Martian atmosphere remained partly unsolved until the 1940s. However that may be, planetary spectroscopy demonstrated, as stated by Janssen, the material unity of the universe, since molecules analogous to the terrestrial ones were detected elsewhere in the universe (Janssen 1929).

6.3.3 Camille Flammarion (1842–1925)

Camille Flammarion, a very well-known French astronomer, published his first book in 1862 when he was only twenty years old, entitled *La pluralité des mondes habités, étude où l'on expose les conditions d'habitabilité des terres célestes discutées au point de vue de l'astronomie, de la physiologie et de la philosophie naturelle* (*The Plurality of Inhabited Worlds*). In this book, which quickly became famous for its support of the doctrine of the plurality of worlds, Flammarion specified some facts related to the problem of habitability (Flammarion 1862):

- The Earth, as a planet, has nothing remarkable.
- The other planets of the solar system are likely to present other conditions of habitability leading to various life forms, probably very different from the terrestrial ones.

- Living beings present on each world are correlated with the "physiological" state of the planet.
- The degree of habitability could be defined considering the analogies and differences existing between each world.

All these points characterize Flammarion's early ideas about planetary habitability, a concept undoubtedly present in his first book. However, these ideas can also be found in most of his subsequent writings. In *Les Terres du Ciel* (*The Lands of the Sky*) published in 1884, Flammarion advocated the diversity of life forms and various possible adaptations on Earth directly connected to changing environments (as proposed by Proctor, according to Darwin's theory). On Earth, different habitats have led to different forms of life. This observation could be extended to other planets in our solar system and even in the (presumed) numerous inhabited worlds of the galaxy. In *Les Terres du Ciel*, Flammarion stressed the various conditions in which life should exist in each world and the large diversity likely to exist in the universe (Flammarion 1884).

Above all, Flammarion fervently developed the topic of habitability in his two volumes wholly devoted to the planet Mars, entitled *La planète Mars et ses conditions d'habitabilité—Synthèse générale de toutes les observations* (*The Planet Mars and Its Conditions of Habitability—General Synthesis of the Whole Observations about Mars*) (volume 1, 1892—volume 2, 1909). Between the first Flammarion's publication (1862) and that one (1892–1909), the Martian canals controversy has strengthened interest in Martian habitability, even if Flammarion considered the canals above all as natural structures. In these two volumes about Mars, Flammarion offered a synthesis of the Martian observations carried out until then, mainly concerning surface structures, atmosphere, and climate.

The methodological aspects used to study habitability are clearly expressed in Flammarion's books, particularly in *The Plurality of Inhabited Worlds*. As well as Proctor, Flammarion assumed that reasoning by analogy was necessary to carry studies about habitability through to a successful conclusion. According to Flammarion, the method of analogy was inescapable to proceed from the "known" to the "unknown." Then, he successively considered the planets of the solar system in order to examine the similarities and differences existing between all these worlds. Considering the planet Mars, the conditions of habitability on this planet and on Earth could be very similar. Climatic environment, physical features, and atmospheric conditions would be analogous enough to establish a parallel between each planet. Following such an assumption, the inhabitants of Mars would present many similarities with those of the Earth (Flammarion 1862).

As many of his contemporaries, Flammarion asserted that analogy was a suitable method to extrapolate the life conditions characteristic of each world. However, if Flammarion used the principle of analogy to study habitability on other planets, he did not support the principle of similarity, a quite different principle. According to Flammarion, we would make a big mistake if we would take our world for the unconditional model of the universe. We can't determine the biological organization of other living beings in the universe depending solely on the similarities with our planet (Flammarion 1862). Flammarion admitted that the question of habitability remained then very enigmatic at the present state of knowledge (at the end of the nineteenth century). It mainly consisted in formulating plausible conjectures and even this fact remained a challenge! Eventually, he came to the conclusion that analogy, even if sure and fruitful, presented limits. More specifically, this method could not be applied to the search for the specific characteristics inherent in each world (Flammarion 1862).

One remarkable point is Flammarion's contest with the principle of anthropomorphism, which was, according to him, too much present in many minds. He highlighted that most of the authors who have attempted to define the nature of the inhabitants of other worlds, have represented creatures similar to humans. Flammarion distinguished "habitability" from "habitation," in such a way that (Flammarion 1862):

- *Habitability* concerned the correlations between the presumed physical and environmental conditions of the planets and their physiological conditions (allowing the presence of living forms),
- *Habitation* concerned the mental and physical state of each "mankind" supposed to be present on other planets.

According to Flammarion, the universe would be filled with various "mankinds" in harmony with the characteristics of their planet. It should be noted that this viewpoint contrasts with Huygens's one formulated two centuries earlier (Huygens 1698) and mentioned in this paper. Through his numerous writings, Flammarion exercised considerable influence over the debate on the habitability and plurality of worlds.

6.4 Concept of Planetary Ecology

In the 1950s, the German physiologist Hubertus Strughold (1898–1987) proposed a notion close to the concept of habitability defined by Proctor, Janssen, and Flammarion. Strughold was one of the pioneers of space medicine, while the Space Age was beginning.[1] He coined the term "planetary ecology" to name the study of the planetary conditions necessary for life. He developed his theory in his book entitled *The Green and Red Planet: A Physiological Study of the Possibility of Life on Mars* published in 1954. He provided a fresh view on the topic of habitability in confronting physical planetary data with physiological data coming from what was known at that time about terrestrial living beings. His viewpoint was inspired by Percival Lowell's book *Mars as the Abode of Life* published in 1909.

---

[1] However, whereas Strughold was considered as "The Father of Space Medicine," he was also unfortunately taken over by Nazis. He emigrated the United States after World War II.

Strughold described Lowell's book as "the most impressive, most original" (Strughold 1954, 6) book about the possibility of life on other planets.

As a physiologist, he defined methods for a biological study of the planets (Strughold 1954). It seemed to him, indeed, that it was necessary to raise the question of life on other planets to the biological plane. It should be underlined that, until then, only astronomers had considered habitability.

In his book, Strughold made a survey of the physiological foundations of *life-as-we-know-it* on Earth. Then, he justified his principles of planetary ecology with some well-established principles of ecology and physiology. Strughold defined planetary ecology "as the science which studies all the planets, including the earth, with regard to their comparative fitness as a biological environment" (Strughold 1954, 2). Strughold's originality was to combine physical or environmental parameters, on one hand, and biological parameters, on the other hand. As did his predecessors, Strughold proceeded by analogy to start his study. Comparisons were made between every biological parameter and between every planetary parameter. Of course, since the first concepts had been formulated by our pioneers at the end of the nineteenth century, many advances have been obtained in planetology and biology. However, in the 1950s, astrobiology [a term coined by L. J. Lafleur in 1941 (Briot et al. 2004)] was a very new field in which biology was still not included.

Strughold's arguments were based on two definitions (Strughold 1954, 2):

- The definition of *physical ecology*: "Ecology is that science which treats of the physical environment of a place or region, with regard to its fitness as a site for the existence and development of living things."
- The definition of *physiological ecology*: "Ecology deals also with the adaptive reactions or responses of living things to their environment, in order to make their existence easier wherever they might be."

According to Strughold, the astronomical discoveries made during the first part of the twentieth century in the field of planetary atmospheres, provided a lot of data, which can be used by biologists. In that way, frontiers between astronomy and biology could be removed, allowing biologists to enter into the discussion about life elsewhere in the universe. One of his arguments was to delineate the limits accepted by living organisms. His study showed that on Earth only specific organisms could survive in extreme environments, in particular characterized by extreme temperatures. This conclusion could be extrapolated to other planets and led to specify the parameters dealing with planetary ecology. He applied this principle to the bodies of the solar system and came to the conclusion, considering temperature parameters, that most of the planets should be excluded except Mars and Venus:

> From the standpoint of temperature alone Mars and perhaps Venus are the only planets, aside from the Earth, which at present possess the prerequisites for living matter as we know it. All the other planets are excluded, for their temperatures lie far outside the range of active life (Strughold 1954, 31).

As did many of his contemporaries and predecessors, he conferred special attention to the planet Mars, which he considered as *a biological environment*. However, in Strughold's opinion, since molecular oxygen (O_2) had still not been detected in its atmosphere, the habitability of this planet could be very restricted. According to Strughold, the presence of molecular oxygen was crucial for the subsistence of living organisms. In spite of this, he considered that the absence of molecular oxygen did not exclude some possible forms of primitive life, like lichens or bacterial life (Strughold 1954).

It must be pointed out that Strughold underlined that the distance to the Sun was a decisive factor to determine the possibilities of life on the planets of the solar system, through the study of the solar constants of each planet. He proposed the pioneering concept of "thermal ecosphere of the sun," including planets capable of supporting life similar to ours (Strughold 1954, 36). This definition given by Strughold, which has been used again by Dole (1964—see following section), was comparable to that of habitable zone defined more than two decades later by Hart (1979).

6.5 Dole's Habitability

In the 1960s, Stephen H. Dole examined the concept of habitability in his book *Habitable Planets for Men* (Dole 1964):

> The use of the term "habitable planet" is meant to imply a planet with surface conditions naturally suitable for human beings, that is, one that does not require extensive feats of engineering to remodel its atmosphere or its surface so that people in large numbers can live there (Dole 1964, 4).

In this case, habitability concerned above all the planetary conditions suitable for human life, even if these conditions could be also convenient to other forms of (terrestrial) life. Dole attempted to delineate the astronomical circumstances (i.e., mass of the planet, period of rotation, age, axial inclination, level of illumination, orbital eccentricity, mass of the star) that produce these requisite environmental conditions. Then he made an estimate of the probabilities of finding these conditions elsewhere in the galaxy. From the probabilities of occurrence of those habitable planets in the galaxy, he deduced the number of habitable planets in the galaxy, which can be expressed as the following product (Dole 1964, 82):

$$N_{HP} = N_s P_p P_i P_D P_M P_e P_B P_R P_A P_L$$

with:

N_s, prevalence of stars in the suitable mass range, 0.35–1.43 solar masses;
P_p, probability that a given star has planets in orbit about it;
P_i, probability that the inclination of the planet's equator is correct for its orbital distance;
P_D, probability that at least one planet orbits within an ecosphere;
P_M, probability that the planet has a suitable mass, 0.4–2.35 Earth masses;

P_e, probability that the planet's orbital eccentricity is sufficiently low;

P_B, probability that the presence of a second star has not rendered the planet uninhabitable;

P_R, probability that the planet's rate of rotation is neither too fast nor too slow;

P_A, probability that the planet is of the proper age;

P_L, probability that, all astronomical conditions being suitable, life has developed on the planet.

From Dole's probability theory and considering the product quoted above, the estimation of the number of habitable planets *for humans* in the galaxy proposed by Dole was 645 million (Dole 1964, 104). Dole himself underlined that any number of the equation was bound to be highly imprecise, since not all factors were known with accuracy. The result of the equation was then to be considered as merely an attempt to formulate an estimation and not a final assessment.

Dole's concept of habitability—for humans—has been proposed while he was carrying out studies on the physical and physiological requirements of human beings in the spacecraft environment. Dole underlined many problems of astronomical interest that are revived today with the study of exoplanets, such as the definition of an "ecosphere" around a star, a definition comparable to the habitable zone (see Dole 1964, chapter entitled "Properties of the Primary"). Considering the question of habitability for humans, Dole redefined the term "ecosphere" (previously proposed by Strughold):

> Ecosphere will be used to mean a region in space, in the vicinity of a star, in which suitable planets can have surface conditions compatible with the origin, evolution to complex forms, and continuous existence of land life and surface conditions suitable for human beings, along with the ecological complex on which they depend. The ecosphere lies between two spherical shells centered on the star. Inside the inner shell, illuminance levels are too high; outside the outer shell, they are two low (Dole 1964, 64).

However, the concept of habitability introduced by Dole was not exactly analogous to the one previously proposed by the pioneers, and the one studied today. Nowadays, the concept of habitability is closer to the proposals coming from the nineteenth-century pioneers than to the belated definition provided by Dole.

6.6 Conclusion

The concept of planetary habitability is today in the heart of discussions dealing with the search for life elsewhere in the universe, especially when considering the increasing detection of exoplanets. Most of the astronomers adhere to a conventional and conservative definition of habitability which corresponds to the zone around a star within which water can be in stable liquid form on the surface of a rocky planet (Impey 2013). Born—with scientific arguments—at the end of the nineteenth century in the astronomical community, while the studies of the Martian surface were intensifying, the notion of habitability has been supported by important personalities of astronomy. It is noteworthy that the principles formulated about habitability by these pioneers are so close to our current concept.

One century ago, habitability dealt with the physical and environmental conditions necessary to make life possible on other planets (Proctor, Janssen, Flammarion). As today, the definition was globally the aptitude of a planet to develop life, from its origin to its diversification. As today the study of habitability required various parameters of astronomical, geophysical, and geochemical interest, which had connections with the nature of the planetary surface, and the atmosphere. Nowadays, criteria of habitability also integrate biological parameters, as it was attempted by Strughold. Above all, the biological parameters are supposed to identify limits in which the (terrestrial) living forms could persist. Extreme environments on Earth are in that case very significant, since they could provide information that could be extrapolated to other celestial bodies.

Habitability is strongly correlated with the presence of permanent liquid water on a planetary surface. This fact was already clear in the minds of our pioneers, at the end of the nineteenth century. However, the exploration of the solar system has recently shown that the satellites of giant planets could also be relevant targets for the search for life elsewhere. If these celestial bodies do not present liquid water on the surface, they could however contain liquid water ocean under their surface (Titan, Europe, Callisto and Ganymede). This example was not conceivable at the time of our nineteenth-century pioneers. It remains today a problematic case because it is not in accord with the usual definition of the habitable zone. This concept is nowadays questioned again within the astronomical community, following the continual discoveries of planetary systems in the galaxy. For instance recently, Barnes et al. (2010) have suggested that the concept of habitable zone should be modified to include the effects of tides. If planets form around low-mass stars, then the terrestrial ones, which are in the circumstellar habitable zone, will be close enough to their host stars to experience strong tidal forces. According to such models, a *tidal habitable zone* can be delimited. For example, if heating rates on an exoplanet are near or greater than that on Io[2] and produce similar surface conditions, then the development of life seems unlikely. On the other hand, if the tidal heating rate is less than the minimum to initiate plate tectonics, then CO_2 may not be recycled through subduction, leading to a runaway greenhouse that sterilizes the planet. These two cases represent potential boundaries to habitability (Barnes et al. 2009). This could change the usual definition of habitable zone and subsequently planetary habitability, if we assume that orbital evolution due to tides has to be considered for any potentially habitable world.

Eventually, it must be pointed out that the methodological choice of analogy, chosen in order to study habitability, is a convergent approach used by the scientists mentioned in this chapter throughout the centuries. The method of analogy has been used for a long time in the debate about pluralism: the logic of the argument "The Earth is inhabited; therefore the planets are," has been widely discussed (see Crowe 1986), and also criticized. In the case of habitability, some questions

---

[2] Where tides drive volcanism that resurfaces the planet at least every million years; tidal heating can drive plate tectonics, including subduction.

could be raised about the parallel established between our planet and the other planetary environments supposed to be suitable for life. Habitability depends on complex criteria, such as those required *at a minimum* for the presence of life, and on the definition of life itself. Then, analogy—as noted by Flammarion in 1862— has limits, even if it remains (up to now) the unique and tangible way to estimate the possibilities of life on other planets.

Acknowledgments   Many thanks to Danielle Briot for her helpful comments, particularly those about the recent definition of the habitable zone as a result of the newly discovered exoplanets.

References

Barnes, Rory, Brian Jackson, Richard Greenberg, and Sean N. Raymond. 2009. "Tidal Limits to Planetary Habitability." *The Astrophysical Journal Letters* 700 (1): L30–L33.

Barnes, Rory, Brian Jackson, Richard Greenberg, Sean N. Raymond, and Rene Heller. 2010. "Tidal Constraints on Planetary Habitability." In *Pathways Towards Habitable Planets, ASP Conference Series* 430: 133–138.

Briot, Danielle, Jean Schneider, and Luc Arnold. 2004. "G.A. Tikhov, and the Beginnings of Astrobiology." In *Extrasolar Planets: Today and Tomorrow*, *ASP Conference Series* 321: 219–220, ed. Jean-Philippe Beaulieu, Alain Lecavelier des Etangs, and Caroline Terquem.

Crowe, Michael J. 1986. *The Extraterrestrial Life Debate 1750-1900: The Idea of a Plurality of Worlds from Kant to Lowell*. Cambridge: Cambridge University Press.

Crowe, Michael J., and Matthew F. Dowd. 2013. "The Extraterrestrial Life Debate from Antiquity to 1900." In *Astrobiology, History, and Society: Life Beyond Earth and the Impact of Discovery*, ed. Douglas A. Vakoch. Heidelberg: Springer.

Darwin, Charles M.A. 1859. *On the Origin of Species by Means of Natural Selection*. London: John Murray.

Dick, Steven J. 1982. *Plurality of Worlds: The Origins of the Extraterrestrial Life Debate from Democritus to Kant*. Cambridge: Cambridge University Press.

Dick, Steven J. 1996. *The Biological Universe: The Twentieth-century Extraterrestrial Life Debate and the Limits of Science*. Cambridge: Cambridge University Press.

Dole, Stephen. 1964. *Habitable Planets for Men*. New York: The Rand Corporation, Blaisdell Publishing Company.

Flammarion, Camille. 1862. *La pluralité des mondes habités; étude où l'on expose les conditions d'habitabilité des terres célestes, discutées au point de vue de l'astronomie, de la physiologie et de la philosophie naturelle*. Paris: Mallet-Bachelier.

Flammarion, Camille. 1884. *Les Terres du ciel; Voyage astronomique sur les autres mondes et description des conditions actuelles de la vie sur les diverses planètes du système solaire*. Paris: Marpon & Flammarion.

Flammarion, Camille. 1892. *La planète Mars et ses conditions d'habitabilité—Synthèse générale de toutes les observations*, vol. 1. Paris: Gauthiers-Villars.

Flammarion, Camille. 1909. *La planète Mars et ses conditions d'habitabilité—Synthèse générale de toutes les observations*, vol. 2. Paris: Gauthiers-Villars.

Fontenelle, Bernard (le Bovier de). 1686. *Entretiens sur la pluralité des mondes*. http://abu.cnam. fr/BIB/auteurs/fontenelleb.html.

Hart, Michael H. 1979. "Habitable Zones about Main Sequence Stars." *Icarus* 37 (1): 351–357.

Huygens, Christiaan. 1698. *Cosmotheoros (The Celestial Worlds Discovered: or, Conjectures Concerning the Inhabitants, Plants and Productions of the Worlds in the Planets)*. London: Timothy Childe.

Impey, Chris. 2013. "The First Thousand Exoplanets: Twenty Years of Excitement and Discovery." In *Astrobiology, History, and Society: Life Beyond Earth and the Impact of Discovery*, ed. Douglas A. Vakoch. Heidelberg: Springer.

Janssen, Jules. 1929. *Œuvres Scientifiques, recueillies et publiées par Henri Dehérain*, vol. 1. Paris: Editions de la Société d'éditions géographiques, maritimes et coloniales.

Kepler, Johannes. 1634 (original publication). 2003. *Kepler's Somnium, The Dream or Posthumous Work on Lunar Astronomy*. Translated by Edward Rosen, New York: Dover Publications.

Launay, Françoise. 2008. *Un Globe-Trotteur de la Physique Céleste, L'astronome Jules Janssen*. Paris: Vuibert and L'Observatoire de Paris.

Lowell, Percival. 1909. *Mars as the Abode of Life*. New York: The MacMillan Company.

Markley, Robert. 2005. *Dying Planet: Mars in Science and the Imagination*. Durham & London: Duke University Press.

Proctor, A. Richard. 1870. *Other Worlds Than Ours: The Plurality of Worlds Studied Under the Light of Recent Scientific Researches*. New York: AL. Burt Publisher.

Raulin Cerceau, Florence, and Bénédicte Bilodeau. 2011. *D'autres planètes habitées dans l'univers?* Paris: Ellipses.

Strughold, Hubertus. 1954. *The Green and Red Planet: A Physiological Study of the Possibility of Life on Mars*. London: Sidgwick and Jackson.

Vakoch, Douglas A. 2013. "Life Beyond Earth and the Evolutionary Synthesis." In *Astrobiology, History, and Society: Life Beyond Earth and the Impact of Discovery*, ed. Douglas A. Vakoch. Heidelberg: Springer.

Part II
The Modern Extraterrestrial Life Debate

Chapter 7
The Twentieth Century History of the Extraterrestrial Life Debate: Major Themes and Lessons Learned

Steven J. Dick

Abstract In this chapter we provide an overview of the extraterrestrial life debate since 1900, drawing largely on the major histories of the subject during this period, *The Biological Universe* (Dick 1996), *Life on Other Worlds* (Dick 1998), and *The Living Universe* (Dick and Strick 2004), as well as other published work. We outline the major components of the debate, including (1) the role of planetary science, (2) the search for planets beyond the solar system, (3) research on the origins of life, and (4) the Search for Extraterrestrial Intelligence (SETI). We emphasize the discovery of cosmic evolution as the proper context for the debate, reserving the cultural implications of astrobiology for part III of this volume. We conclude with possible lessons learned from this history, especially in the domains of the problematic nature of evidence, inference, and metaphysical preconceptions; the checkered role of theory; and an analysis of how representative general current arguments have fared in the past.

7.1 Major Themes of the Debate

When the twentieth century began, the idea of a universe filled with life was widely accepted, completely unproven, and heavily burdened with a long and checkered history that finally held the promise of more successful scientific scrutiny. The challenge was to bring new data to bear on an age-old controversy. The infamous episode of Percival Lowell and the canals of Mars, resolved to the satisfaction of most astronomers by 1912 (Crowe and Dowd 2013), demonstrated just how difficult that challenge could be. Difficulties notwithstanding, the search for life would continue

S. J. Dick (✉)
National Air and Space Museum, Washington, DC, USA
e-mail: stevedick1@comcast.net

D. A. Vakoch (ed.), *Astrobiology, History, and Society*, Advances in Astrobiology and Biogeophysics, DOI: 10.1007/978-3-642-35983-5_7,
© Springer-Verlag Berlin Heidelberg 2013

not only in our solar system with tools ranging from ground-based telescopes to *in situ* observations on Mars, but also in the realm of the stars with the search for extrasolar planets, in laboratories and environments on Earth performing research bearing on the origins of life, and with the radio search for signals from extraterrestrial intelligence. We now examine the major themes of each of these areas in turn.

7.1.1 Planetary Science

In the wake of the demise of the canals of Mars, the red planet remained a focus for the search for life in the solar system. After Lowell's death in 1916, with the close approach of Mars in 1924 attention focused on the possibility of Martian vegetation rather than intelligence. In one particularly important case this was still tied to the old visual method and canals; using the 36-inch Lick Observatory refractor astronomer Robert J. Trumpler concluded that the canals were the result of natural topography but that vegetation caused the dark Martian areas and made the canals visible (Trumpler 1927). But the mid-1920s mark a new era in Martian studies: physical methods of spectroscopy and infrared astronomy now came into widespread use in the attempt to determine temperature and atmospheric conditions. Respected scientists like W. W. Coblentz of the National Bureau of Standards, C. O. Lampland of Lowell Observatory, and Edison Petit and Seth Nicholson at Mt. Wilson Observatory, pioneering in the field of infrared astronomy, determined that the temperature conditions on Mars were adequate for some form of Martian vegetation (Coblentz and Lampland 1924; Petit and Nicholson 1924).

The belief in a harsher, but Earth-like Mars with vegetation was still very much alive at mid-century. At that time astronomers believed Mars had an atmospheric pressure of about 85 millibars at its surface, ten times thinner than Earth's. In 1949 the Dutch-American astronomer Gerard Kuiper had used early near-infrared techniques to discover carbon dioxide, one of the principle gases in the process of photosynthesis (Kuiper 1949). Seasonal vegetation across parts of Mars was commonly accepted, based on visual and photographic observations showing unmistakable seasonal changes on the surface as the polar caps melted, spreading a wave of darkening (Slipher 1927; Barabashev 1952; Kuiper 1955). The second edition of the standard astronomy textbook of the time was pessimistic about the existence of even primitive animal life, but asserted that the existence of vegetation was "more likely than not" (Russell, Dugan, and Stewart 1945, 344). Meanwhile, in the Soviet Union the astronomer Gavriil Adrianovich Tikhov assumed the mantle of the Russian Lowell, with a passion for Martian vegetation rather than Martian canals. In a career spanning many decades Tikhov used reflection spectra to study the optical properties of terrestrial vegetation in harsh climates and applied the results to Martian observations, claiming a new science of "astrobotany" (Tikhov 1955; Tikhov 1960). Tikhov's work, like Lowell's, provoked great criticism in his own country as well as abroad.

Using spectroscopic techniques, others found evidence of oxygen and water vapor in the Martian atmosphere, but in increasingly minute amounts, now known to be largely spurious (Spinrad et al. 1963). Despite the desert conditions revealed by the new physical methods, by 1957 and the dawn of the Space Age the existence of hardy, perhaps lichen-like Martian vegetation was widely accepted, especially in the wake of William Sinton's claims in that year to have discovered infrared bands in the Martian spectrum that were unique to vegetation (Sinton 1957; Sinton 1959).

These hopes were partially dashed in the early 1960s when the Sinton bands were found to be caused by deuterated water in the Earth's own atmosphere, and the water content of the Martian atmosphere was lowered almost to the vanishing point. But hopes were completely dashed two decades into the Space Age when the Viking orbiters and landers in 1976 seemed to demonstrate not only the lack of vegetation on Mars, but also the complete absence of any organic molecules at the two landing sites (Dick 1996, 153). And they showed an average atmospheric surface pressure of only 6 millibars. As we shall see in the next section, the Viking results on organic molecules—the *sine qua non* for life—have been questioned, and in the decades since that time other spacecraft have shown evidence of abundant water flow on Mars in the past. The Mars Global Surveyor and Mars Odyssey missions have both indicated that water ice still exists in plentiful amounts just below the surface, and the Mars Exploration Rovers have found strong evidence for plentiful liquid water below and on the surface in the past.

Nonetheless, evidence for life itself has not been found on Mars. The tantalizing seasonal changes were shown not to be due to vegetation, but to seasonal wind-blown sand. With the discovery at mid-century that Venus was a victim of the greenhouse effect, with temperatures consequently at the 800 °F level, it appeared that the solar system was bereft of life beyond Earth. Hope of microbial life in the solar system has not totally disappeared, due especially to the possibility that organics exist on some of the moons of the outer gas giants, notably Europa, Ganymede, Callisto and Titan. But because Mars had been viewed as a test case for life in the universe, the apparent absence of life there was a correspondingly great blow to the concept of a universe filled with life.

7.1.2 Planetary Systems

Long before the Viking results were in hand, attention had turned beyond the solar system to the possibility of the existence of other planetary systems—a prerequisite for life in the realm of the stars. Since they could not be directly observed, belief in such systems was greatly affected for most of the century by theories of their origin (Dick 1996). The nebular hypothesis of Laplace, whereby planetary systems were theorized to originate from the same rotating gas clouds that formed the stars themselves, indicated that planets were a natural by-product of star formation and, therefore, very abundant (Brush 1996). At the turn of the century, however, this theory was under heavy attack. In its place the geologist T. C. Chamberlin and the astronomer F. R. Moulton, both at the University of Chicago, proposed that solar systems originated by the close encounters of stars, which resulted in the tidal ejection of matter, which then cooled to form small planetesimals, which in turn accreted to form planets (Chamberlin and Moulton 1900). This "planetesimal hypothesis," elaborated and modified by the British astronomer James Jeans from 1916 almost until his death three decades later (Jeans 1917), implied that solar systems were extremely rare, since stellar collisions in the vastness of space were extremely rare. For this reason, during the 1920s and 1930s belief in extraterrestrial life was at a low point; it was difficult to conceive of life without planets.

But the 15 years from 1943 to 1958 saw once again a complete turnabout in opinion (Table 7.1). In 1943 two astronomers independently claimed they had observed the gravitational effects of planets orbiting the stars 61 Cygni and 70 Ophiuchi (Reuyl and Holmberg 1943). Although these observations were proven spurious decades later, they filled a need at the time. Doubts expressed in 1935 about Jeans's stellar encounter hypothesis by the dean of American astronomers, Henry Norris Russell, had grown to a crisis point by the early 1940s. Carl Friedrich von Weizsäcker began the revival of a modified nebular hypothesis in 1944, and the theoretical basis was once again laid for abundant planetary systems. The turnabout involved not only possible planetary companions and the revived nebular hypothesis, but also arguments from binary star statistics and stellar rotation rates. Helping matters along was Russell, whose *Scientific American* article "Anthropocentrism's demise" enthusiastically embraced numerous planetary systems (Russell 1943). Definitive evidence, however, would be much more elusive, for it turned out that Russell's declaration was 50 years premature.

Even as the nebular hypothesis has been elaborated in ever more subtle form, attempts to pin down the abundance of planetary systems proved very difficult. Through the 1960s and 1970s the search was dominated by the astrometric method,

Table 7.1 Estimates of frequency of planetary systems, 1920–1961

Author	Argument	Number of planetary systems in galaxy	Number of habitable planets in galaxy
Jeans (Jeans 1919, 1923)	Tidal theory	Unique	1
Shapley (1923)	Tidal theory	"Unlikely"	"Uncommon"
Russell (1926)	Tidal theory	"Infrequent"	"Speculation"
Jeans (1941)	Number of stars	10^2	–
Jeans (1942a, b)	Improved tidal	One in six stars	Abundant
Russell (1943)	Companions	Very large	$>10^3$
Page (1948)	Weizsäcker	$>10^9$	$>10^6$
Hoyle (1950)	Supernovae	10^7	10^6
Kuiper (1951)	Binary star statistics	10^9	–
Hoyle (1955)	Stellar rotation	10^{11}	–
Shapley (1958)	Nebular hypothesis	10^6–10^9	–
Huang (1950)	Stellar rotation	10^9	10^9
Hoyle (1960)	Stellar rotation	10^{11}	10^9
Struve (1961)	Stellar rotation	$>10^9$	–

Adapted from Dick (1996, 199)

whereby the proper motions of stars are studied for the gravitational effects of planetary systems. In the 1960s Peter van de Kamp and others made claims for planetary systems around other stars (Van de Kamp 1963). In the 1980s another method for determining planetary effects on stars—this time utilizing their line-of-sight "radial velocities"—came into use. At the same time the Infrared Astronomical Satellite spacecraft discovered circumstellar disks, initially interpreted as proto-planetary disks (now believed to be debris disks left over after planet formation). But it was only in 1995 that the radial velocity method proved unambiguously successful, when the Swiss astronomers Michel Mayor and Didier Queloz discovered a planet around the star 51 Pegasi (Mayor and Queloz 1995). The American astronomers Geoff Marcy and Paul Butler confirmed the discovery almost immediately, and after that the floodgates were opened for more discoveries (Marcy and Butler 1998). They came not only from the radial velocity method, but also from the "photometric method," whereby milli-magnitude dips in stellar brightness were measured as a planet passed in front of its parent star. It was this method that the Kepler spacecraft used beginning in 2009, discovering more than 2,000 planetary candidates by 2012. Of these almost 900 are Earth- or Super-Earth-sized, 1,200 are Neptune sized, and about 250 are Jupiter sized or larger. 48 planet candidates were found in the habitable zones of their stars, and it is estimated that at least 5% of all Sun-like stars host Earth-sized planet candidates.

7.1.3 Origins of Life

Even as the idea of abundant planetary systems was being revived in the 1950s, work was also progressing on the biological question of the origins of life, a crucial factor in the question of extraterrestrial life (Fry 2000). In the 1920s the Russian biochemist Aleksandr Ivanovich Oparin (Oparin 1924, 1936). And the British biologist J. B. S Haldane had independently suggested that life originated on Earth by chemical evolution in a hot dilute soup under conditions of a primitive Earth atmosphere. The experiments of Harold Urey and Stanley Miller in 1953 showed how amino acids could be produced under just such conditions, believed at the time to be highly "reducing" atmosphere, rich in hydrogen compounds such as methane and ammonia (Miller and Urey 1953). Their success set off numerous experiments around the world in chemical evolution as related to the origins of life. The major thrust of NASA's exobiology program, begun in the early 1960s, was to undertake such experiments on the origin of life, as well as to research life detection methods for spacecraft headed to Mars (Dick and Strick 2004).

Since the original Miller-Urey experiments, a better appreciation of the difficulties of the many steps in the origin of life—as well as uncertainty about the nature of the primitive Earth atmosphere—has somewhat tempered optimism among biologists. Whereas astronomers focus on the enormous size of the universe and the likelihood of planets emerging from an abundance of stars, biologists point to the extremely complex steps in the origin and evolution of life. Thus a dichotomy of opinion has developed between astronomers and biologists, further widened by the biologists' recognition that the evolution of life beyond Earth might lead to forms of life and intelligence very different from the humanoid form and alien to the human concept of intelligence.

Over the past quarter century theories of the origin of life have proliferated, with various implications for exobiology. Furthermore, the discovery of life in extreme environments—around deep sea hydrothermal vents, in deep underground rock, and in conditions of great salinity and acidity—has fostered a new appreciation for the tenacity of life, and broadened our idea of the conditions under which life might originate on another planet, or on Earth. As the possibilities of panspermia have become more widely accepted, spurred on by the Mars rock controversy (discussed in the next section) and by the realization that material does transfer between planets, some researchers believe that so-called "exogenous delivery" of organic compounds may be the key to the origin of life on Earth.

The question of the origin of life on Earth and in space shared many philosophical issues. Old problems such as chance, necessity, and the nature of life— already recognized in the terrestrial realm—were magnified in the extraterrestrial realm. The crucial question for exobiology was whether life would arise wherever it could, or whether the Earth was a fluke. The contingency or necessity of life would be one of the greatest scientific and philosophical questions of the extraterrestrial life debate. The two points of view are classically represented by the French biologist and Nobelist Jacques Monod on the one hand, and the Belgian-American biochemist and Nobelist Christian de Duve on the other. In his classic work *Chance and Necessity* (Monod 1971, 144–146) argued "the universe was not pregnant with life, nor the biosphere with man. Our number came up in the Monte Carlo game." Nor was Monod the only one to favor chance; the astronomer Fred Hoyle agreed that the chance of a random shuffling of amino acids producing a workable set of enzymes was miniscule, and went one step further in asserting that life must have been assembled by a "cosmic intelligence," though not necessarily the supernatural intelligence of Christianity (Hoyle 1983). de Duve, on the other hand, argued just the opposite, declaring Monod wrong and viewing life as a "cosmic imperative," while evolutionary biologist Richard Dawkins argued that "climbing Mt. Improbable" was not impossible (de Duve 1995; Dawkins 1997).

7.1.4 Search for Extraterrestrial Intelligence

All these questions in the origin of life arena are multiplied when it comes to the nature of consciousness, mind and intelligence. In many ways defining "intelligence" remains more problematic than defining "life," with many different possible approaches undertaken in a very large literature (Sternberg 2000; Sternberg 2002). To frame it another way, there is no "general theory of intelligence" or even of human brain function, much less a general theory of intelligence in a cosmic context. Carl Sagan argued in his *Dragons of Eden* that "once life has started in a relatively benign environment and billions of years of evolutionary time are available, the expectation of many of us is that intelligent beings would develop. The evolutionary path would, of course, be different from that taken on Earth... But there should be many functionally equivalent pathways to a similar end result. The entire evolutionary record on our planet, particularly the record contained in fossil endocasts, illustrates a progressive tendency toward intelligence" (Sagan 1977, 230).

That conclusion embodies many assumptions that others have questioned. Evolutionists such as George Gaylord Simpson and Theodosius Dobzhansky, for example, had already argued just the opposite (Simpson 1964; Dobzhansky 1972), and Harvard evolutionist Ernst Mayr also differed strongly with Sagan, arguing that intelligence (by his definition) had emerged only once on Earth (Mayr 1985; Mayr 1988). Outspoken Harvard evolutionist Stephen Jay Gould (1989, 301) agreed with the non-prevalence of *humanoid* intelligence, arguing in an entire book on the Burgess Shale fossils of the Cambrian explosion that if we "Wind back the tape of life to the early days of the Burgess Shale; let it play again from an identical starting point, and the chance becomes vanishingly small that anything like human intelligence would grace the replay." By contrast evolutionary paleobiologist Simon Conway Morris (Conway Morris 1998, Conway Morris 2003) has argued from the same evidence, and others, that evolutionary convergence applies not only to morphology, but also to intelligence, if only the conditions are present. He is, however, skeptical that the proper conditions often obtain, summarizing his position in the subtitle of his 2003 book *Life's Solution: Inevitable Humans in a Lonely Universe*. In this he reached the same conclusion as had Peter Ward and Donald Brownlee (Ward and Brownlee 2000), who famously argued that complex life and thus intelligence in the universe will be rare, not from a lack of convergence but because so many factors must come together in order for it to exist.

These problems are leapfrogged to some extent by the radio search for extraterrestrial intelligence, or, to put it more accurately, the search for extraterrestrial technology. In 1959 the physicists Giuseppe Cocconi and Philip Morrison, both at Cornell, proposed a search in the radio region of the spectrum using the 21-cm hydrogen line (Cocconi and Morrison 1959). The radio astronomer Frank Drake independently undertook the first search of such signals at the National Radio Astronomy Observatory in 1960. It was in the context of a meeting in 1961 in the wake of this search that the so-called Drake equation was formulated. A general equation embodying the various factors of star and planet formation, the likelihood of the origin and evolution of life and intelligence, and the lifetimes of technical civilizations, it came to serve in the last third of the century as a paradigm for discussion of the issues (Dick 1996, 431–454). Although almost everyone acknowledges that the parameters of the equation are not well known, resulting in values ranging from one planet in our galaxy with intelligence (our own) to 100 million or more, this uncertainty has not prevented its use as a basis for discussion of the abundance of technological civilizations in the galaxy. Many radio searches have been undertaken worldwide since 1960, all unsuccessful.

7.1.5 Birth of a New Discipline

In the 1950s and 1960s these four scientific fields—planetary science, the search for planetary systems, origin of life studies, and SETI—converged to give birth to the field of exobiology (Dick 1996). At first quite separate in terms of researchers, techniques, and goals, these fields over four decades gradually became integrated, in large measure because of the scientific and public desire to search for life beyond Earth. NASA served as the most important patron for the new field. By 1963 NASA's life sciences expenditures (including exobiology) had reached $17 million. The $100 million spent on the Viking biology experiments was closely related to origin of life issues, since an informed search for life required a definition of life and a knowledge of its origins. Even though exobiology saw a slump in the 1980s in terms of space missions in the aftermath of the Viking results, NASA kept the program more than alive with a grant program of about $5–$10 million per year, funding research on such broad topics as deep ocean hydrothermal vents and their associated archaea, the primitive Earth atmosphere, the Gaia hypothesis, mass extinctions, exogenous delivery of organic compounds, and the RNA world (Dick and Strick 2004). At the same time NASA also operated the largest exobiology laboratory in the world at its Ames Research Center in California.

In 1995 a deep organizational restructuring at NASA precipitated a rebirth of the field under a new name, "astrobiology." NASA's strategic plan for 1996 used he term astrobiology for the first time anywhere in a NASA document (though it had been sporadically used elsewhere as much as 50 years earlier). Astrobiology under NASA would focus on three key questions. It was "the study of the living Universe" to be sure, but in particular it was seen as providing the scientific foundation for studying the origin and distribution of life in the universe, the role of gravity in living systems, and the study of the Earth's atmosphere and ecosystems. In 1998 an astrobiology 'roadmap' laid out three specific questions: How does life begin and evolve? Does life exist elsewhere in the universe? And what is life's future on Earth and beyond? Specific goals were set to answer these questions (Des Marais et al. 2008).

The contrast between the exobiology and astrobiology programs was quite striking. They both shared the core concerns of origin of life research and the search for life beyond Earth. But astrobiology placed life in the context of its planetary history, encompassing the search for planetary systems, the study of biosignatures, and the past, present and future of life. Astrobiology added new techniques and concepts to exobiology's repertoire, raised multidisciplinary work to a new level, and included the study of the history of Earth's life and present organisms. Today astrobiology is a robust field, a worldwide effort supported especially by NASA, but also by other international research-funding agencies.

All of this did not occur without skepticism, extending even to the period 50 years ago when exobiology was born. In 1964 George Gaylord Simpson, pointing to the long history of the debate, wrote that "There is even increasing recognition of a new science of extraterrestrial life, sometimes called exobiology—a curious development in view of the fact that this 'science' has yet to demonstrate that its subject matter exists!" Simpson noted that this supposed new science was very expensive, and called exobiology "a gamble at the most adverse odds in history," resembling "more a wild spree more than a sober scientific program" (Simpson 1964, 775). Simpson concluded with a plea "that we invest just a bit more of our money and manpower, say one-tenth of that now being gambled on the expanding space program," on studying the systematic and evolution of earthly organisms—that is to say, his own field! An interesting case of the rhetoric of science, clearly Simpson had an ulterior motive in declaring that exobiology was not a science. But with Isaac Asimov's article in the *New York Times Magazine* the following year entitled "A Science in Search of a Subject" (Asimov 1965), the phrase was too good to ignore as a kind of mindless meme deployed innumerable times in the course of the following decades, despite the article's positive assessment of exobiology (Strick 2004).

Even a minimal consideration of this idea suffices to show it is a misrepresentation of science, even if admittedly a catchy phrase. One could say the search for gravitational waves, or the Higgs boson, or planetary systems, are, or were, "sciences without a subject." But this hardly seems a productive way of approaching the problem. Every science is looking for a subject until it finds it (planetary systems), thinks it may have found it (the Higgs boson), or does not find it (gravitational waves, at least so far). From an epistemological point of view, the methods of astrobiology are as empirical as in any historical science such as astronomy or geology (Cleland 2001; Cleland 2002), though it is true that astrobiological observations and experiments are often especially difficult, and the inferences more tenuous. With the broad array of research now being undertaken in astrobiology, the "science without a subject" meme has outlived its usefulness.

Although Simpson criticized the pioneer in the field, Joshua Lederberg, by claiming that exobiology was not strictly biology because its techniques differed (Wolfe 2002), certainly astrobiologists today would be surprised to learn they are not doing science; from their point of view their endeavors constitute not only science, but cutting-edge science. While more than one practitioner early on heralded astrobiology or its equivalent as a new scientific discipline (Shklovskii 1965; Billingham 1981), these claims may have been premature (Dick 1996, 475–478). Moreover, being labeled a discipline may be good or bad in terms of "Balkanization" and isolation from broader parent fields, such as was contemplated, but did not happen, in the case of radio astronomy in relation to astronomy as a whole (Sullivan 2009, 435–438). An historical comparison of discipline formation in other fields such as biochemistry (Kohler 1982), molecular biology (Abir-Am 1992), and geophysics (Good 2000) would help illuminate the problem for astrobiology.

7.1.6 Cosmic Evolution as the Context for Astrobiology

The concerns of astrobiology—the origins and evolution of life, intelligence and culture—are embedded in the larger process of cosmic evolution, the 13.7 billion

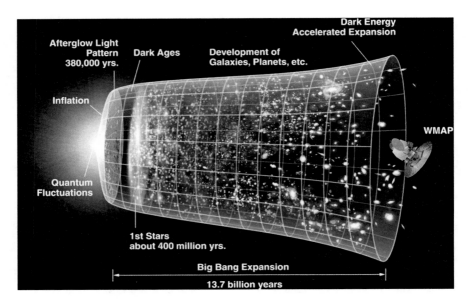

Fig. 7.1 The Master Narrative of the Universe, 13.7 billion years of cosmic evolution, as depicted by the Wilkinson microwave anisotropy probe (*WMAP*) program, which narrowed the estimated age of the universe to within 100 million years. The current model has the universe beginning with the Big Bang, stars forming within the first few hundred million years, followed by the development of galaxies, planets and life. The concerns of astrobiology must be seen within this framework, which encompasses physical, biological and cultural evolution. Courtesy NASA/WMAP Science team

year Master Narrative of the Universe (Fig. 7.1). The concept has its roots in the 18th and 19th centuries, but only became widely accepted and a major driver for research programs in the last half of the 20th century (Dick 2009; Zakariya 2010). I have argued elsewhere that the outcome of cosmic evolution may result in a physical, biological or postbiological universe, in other words, a physical universe composed of planets, stars and galaxies in which life is a fluke; a biological universe full of carbon-based life; or a postbiological universe in which cultural evolution has resulted in a universe full of artificial intelligence (Dick 2003). These outcomes determine the long-term destiny of humanity, and because the scope of astrobiology as set down in the Astrobiology Roadmap applies not only to the past and present, but also the future, the destiny of humanity falls within the purview of the philosophy of astrobiology

7.1.7 The Biological Universe as Worldview

The 20th century view of a universe full of life may perhaps best be seen as a cosmology in its own right, a "biophysical cosmology" that asserts the importance of both the physical and biological components of the universe. Like all cosmologies, it makes a claim about the large-scale nature of the universe, and its claim is that life is not only a possible implication, but also a basic property of the universe. Over the last four decades some scientists have come to question why the laws of nature and the physical constants appear to be "biofriendly," giving rise to what has been termed the "anthropic principle." The principle has many variants, all having to do with the apparent fine-tuning of the physical constants for life (Carter 1974; Barrow and Tipler 1986; Carr 2007). The phrase is a spectacular misnomer, and the term "biocentric principle" is much preferred, since in the context of astrobiology the universe appears to be friendly to life, and the very question to be answered is whether humans are the only intelligent life (Davies 2007).

The prospect of a fine-tuned universe has given rise to the idea of an ensemble of universes, termed a "multiverse," as an explanation for why we happen to be in a universe particularly suited for life (Carr 2007). Whether or not we invoke the multiverse, the physicist Freeman Dyson has suggested that the prospects are bright for a future-oriented science, joining together in a disciplined fashion the resources of biology and cosmology (Dyson 1988). In such a "cosmic ecology," life and intelligence would play a central role in the evolution of the universe, no less than its physical laws.

Like other cosmologies the biophysical cosmology redefines our place in the universe. And most importantly, like other cosmologies in the 20th and 21st centuries the biophysical cosmology has become increasingly testable; this is the role and the importance of modern astrobiology and SETI programs. Viewed in this light, the transition from the physical world to the biological universe is one of the great revolutions in Western thought, no less profound that the move from the closed world to the infinite universe described by the French historian of science Alexandre Koyré almost a half century ago (Koyré 1957). That transition has already occurred to some extent in the minds of most people. Whether the biological universe exists in reality, and what its effect will be on culture when and if it extraterrestrial life is actually discovered, remains to be seen. Its potential cultural impact is discussed in Section III of this volume.

7.2 Lessons Learned from the Twentieth Century Extraterrestrial Life Debate

Now that historians have completed surveys of the extraterrestrial life debate (Dick 1982; Crowe 1986; Guthke 1990; Dick 1996; Dick and Strick 2004), we can begin to study the possible lessons learned from that history. In this section we make that attempt in three overlapping areas: (1) the problematic nature of evidence and inference, and its relation to scientific preconceptions; (2) the role of theory in raising expectations, interpreting observations, and generating conclusions; and (3) an evaluation of the success or failure of some of the debate's most general arguments, including the Principles of Plenitude and Mediocrity and

"Goldilocks-type" arguments that life occurs under such tight constraints that it is rare in the universe. Another widespread general argument, the argument from analogy, we reserve for section III of this volume because of its overriding use and importance.

Whether or not there are lessons to be learned from history is a subject of some contention among historians. It is, of course, a dangerous game, with some (politicians in particular) reading into history whatever lessons they want to learn based on their own ideology. My attitude is more optimistic: lessons may be ambiguous, but they are there and can be debated and deployed. After all, not without reason does there exist a National Archives in the United States with the words "What is Past is Prologue" scrolled along the top of its impressive facade, a building whose function is duplicated in all civilized countries of the world. Not without reason did the Columbia Accident Investigation Board devote an entire chapter to history in its official report on the Space Shuttle's demise, and conclude that "history is not just a backdrop or a scene-setter, history is cause" (Columbia Accident Investigation Board 2003, 195). And not without reason does every high school, college and university teach history, ever hopeful that at the very least it will provide context, if not lessons, for students as they enter a complex world.

My optimism in this regard holds despite the fact that many thinkers—from Samuel Taylor Coleridge to Georg Wilhelm Friedrich Hegel, from Aldous Huxley to scholars today—have concluded that the main lesson of history is that the lessons of history are either misused or never learned. Thus Coleridge: "If men could learn from history what lessons it might teach us! But passion and party blind our eyes, and the light which experience gives is a lantern on the stern, which shines only on the waves behind us!" (Coleridge 1831). Hegel: "What experience and history teach is this—that people and governments never have learned anything from history, or acted on principles deduced from it" (Hegel 1832). Aldous Huxley: "That men do not learn very much from the lessons of history is the most important of all the lessons of history" (Huxley 1959, 222). Or, as a recent author put while contemplating Herodotus's ancient message about intercultural understanding: "it goes unheeded, as it always has and it always will, because history teaches us that we do not learn from history, that we fight the same wars against the same enemies for the same reasons in different eras, as though time really stood still and history itself as moving narrative was nothing but artful illusion" (Marozzi 2008, 95). With such cautions in mind, we nevertheless proceed to examine possible lessons to be learned from the history of the extraterrestrial life debate, in the (perhaps misguided) hope that scientists are more receptive to lessons learned than politicians.

7.2.1 Evidence, Inference and Preconceptions

Evidence, inference and interpretation are problems in all areas of science, not to mention broader areas such as law, where 5–4 votes are not uncommon on the

Supreme Court, based on interpretation of the best available evidence. Ask any three people to describe in detail *any* event they have just witnessed, much less something unexpected and emotional like a UFO event, and the likely outcome is three different answers. Scientists are trained in gathering evidence and making conclusions from that evidence, so one would hope their record would be better. Sometimes it is, but often not, especially when pushed to the limits of science, as is certainly the case in astrobiology. The episode surrounding the canals of Mars centered around the beginning of the 20th century is the most infamous example (Crowe 1986; Dick 1996). But we need not set our gaze back that far. Here we examine four episodes in the second half of the 20th century that we have already mentioned: William Sinton's claim of spectroscopic evidence for supposed Martian vegetation; Peter van de Kamp's claim of planets around Barnard's star; the ambiguities of the Viking experimental results; and the controversy surrounding the Mars rock ALH 84001. While many more exemplars could be used, these will suffice to illustrate the problems of evidence, inference and preconceptions.

7.2.1.1 Vegetation on Mars?

As the favorable close approaches of Mars neared in 1954 and 1956, interest in the red planet, driven by interest in the Martian vegetation hypothesis based on seasonal changes, was increasing. As the 1956 opposition approached, Harvard astronomer William Sinton planned a direct search for vegetation by spectroscopic methods. Keenly aware that previous tests for infrared reflectivity characteristic of plants had been negative, Sinton's own search had a new element: it depended on the fact that organic molecules have absorption bands at about 3.4 microns in the infrared part of the spectrum—beyond Gerard P. Kuiper's work that had been done in the 1–2.5 micron region leading to the discovery of carbon dioxide on Mars (Kuiper 1949). Sinton still used a lead sulfide photoconductive cell, as had Kuiper, but now cooled to 96 °K with liquid nitrogen to increase its sensitivity to 3.6 microns. The difficulties of the observations can be appreciated from the fact that the sensitive area of this cell was only 0.16 mm^2, and the diameter of Mars was less than 1 mm. Nevertheless, after four nights of observations Sinton believed he had enough evidence for his conclusion that the probability was "very high that an organic spectrum is required to account for the data" (Sinton 1957, 237).

Sinton was very much aware of previous visual evidence for vegetation in the form of seasonal changes in the size and shape of the Martian dark areas. In fact he saw the dip at 3.4 microns as "additional evidence for vegetation," and concluded that "this evidence, together with the strong evidence given by the seasonal changes, makes it seem extremely likely that plant life exists on Mars" (Sinton 1957, 237). Thus, his claim of infrared absorption did not constitute direct visual confirmation, but depended on the interpretation of spectrograms, an interpretation undoubtedly affected by preconceived ideas. The result caused considerable excitement, especially when Sinton confirmed it with equipment ten times more sensitive on the 200-inch Palomar telescope during the 1958 opposition. Although the image of Mars was only 2 mm, this time Sinton separated the dark areas from the bright areas on Mars and confirmed his previous conclusion of absorption bands near 3.5 microns (Fig. 7.2). Again he concluded that "the observed spectrum fits very closely ... that of organic compounds and particularly that of plants" (Sinton 1959, 1237). In addition a 3.67 micron absorption band was confirmed, which Sinton attributed to carbohydrate molecules in plants, analogous to tests on plants on Earth.

Although Sinton's results were widely hailed and cited in the literature, they were open to interpretation: not only were other biological interpretations possible, by 1963 researchers had done extensive work on infrared reflection spectra of terrestrial compounds and were critical of Sinton's interpretation (Rea, Belsky, and Calvin 1963). And by 1965 Sinton himself suggested that two of the Sinton bands were due to heavy "deuterated" water (HDO) in the Earth's atmosphere, with the remaining band still possibly organic (Rea, O'Leary, and Sinton 1965). In the end, Sinton's refined methods had been mitigated by refined problems. Whereas V. M. Slipher a half-century before believed he had found oxygen and water vapor on Mars only to find that his results were contaminated by the Earth's own

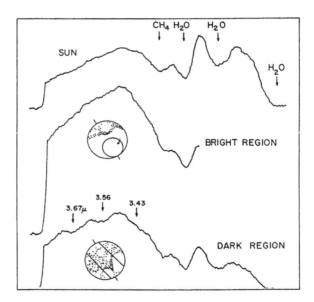

Fig. 7.2 Sinton's infrared spectroscopic evidence for vegetation on Mars, obtained on the Palomar 200-inch telescope. The top curve shows a solar spectrum, with superimposed absorptions by methane and water in the Earth's atmosphere. The middle curve shows a spectrum of a bright desert area of Mars, where no vegetation was expected. The bottom spectrum, obtained when the spectrograph slit was placed over one of the dark areas of Mars, shows three apparent absorption features (indicated by *arrows*) that were interpreted as due to vegetation. The evidence turned out to be spurious; the absorptions were actually due to deuterated water (*HDO*) in the Earth's atmosphere, as Sinton himself published six years later. With permission, from Sinton (1959), 1234. Copyright 1959 AAAS

atmosphere, Sinton's results too were contaminated, this time by heavy water, and despite his attempt to separate analysis of Martian dark areas from its bright areas.

As with the canals of Mars, the search for Martian vegetation demonstrates again differences in approach and world view among scientists, with one extreme much more likely to go out on a limb and to extrapolate than the other. Some astronomers probing the physical conditions on Mars presented their data and left it at that. Others used their data—indeed, were probably first inspired to gather their data—in the service of the question of extraterrestrial life. Still others rendered no opinion at all. In his book *Physics of the Planet Mars*, the astronomer Gerard de Vaucouleurs only rarely mentioned the problem of life because "It is our belief that such a problem is still, to a large extent, beyond the limits of our positive knowledge and can only be the subject—either way—of vague speculations in which general 'principles' of a metaphysical nature have always to be taken as a guide" (de Vaucouleurs 1954, 19).

To many, such a cautious attitude was not satisfying. They undoubtedly realized that the stakes in the debate extended far beyond Mars: as Kuiper wrote "If life truly exists on the only two planets of the solar system that are at all suitable to sustain it, it is tempting to conclude that, after enough time has elapsed, it will develop spontaneously wherever conditions permit. Since planetary systems are presumed to be very numerous, life would then be no exception in the universe" (Kuiper 1952, 404).

At the dawn of the Space Age, then, the canal controversy had receded, and much was known about the physical conditions of the planet Mars. Vegetation of some sort was still a very real possibility, dependent to some extent on what one saw as the limits to the adaptability of life. Vegetation did not have the popular appeal of intelligence, but to the scientist it was still a holy grail that held the promise of revealing the secrets of life. That promise was to play no small role in making Mars and important target for interplanetary probes of the space age.

7.2.1.2 Organics on Mars? The Viking Experiments

The culmination of the twentieth century search for life in the solar system was the landing of two Viking spacecraft on the surface of Mars in 1976, surely one of the great adventures in the history of science and technology (Ezell and Ezell 1984). The Viking project, initiated in 1968 after the demise of the Mars Voyager project and now managed by NASA's Langley Research Center, was an example of "big science" at its best in terms of budget, staff, goals and results. The cost of the Viking spacecraft, including the orbiters, landers and support (but not launch vehicles) was $930 million. Although the usual funding hurdles had to be overcome and many critics answered, in the end two Viking "orbiters" arrived at the planet on June 19 and August 7, 1976. After suitable reconnaissance, as the United States celebrated its Bicentennial back on planet Earth, two Viking landers set down on Mars in July and September. Under the guidance of project scientist Gerald A. Soffen, thirteen teams with a total of 78 scientists undertook thirteen separate investigations, including three mapping experiments from the orbiter, one atmospheric experiment, one radio and radar experiment, and eight surface experiments. The total costs for development and execution of these experiments was another $227 million. The results increased knowledge of Mars far beyond all previous investigations combined, finally providing definitive answers to age-old questions, including the issues of temperature, atmospheric composition and pressure so crucial to life.

From beginning to end, though the various science teams grappled with the myriad problems of meteorology, seismology, chemistry, imaging and physical properties of the planet Mars, the Viking biology experiments were the driving force behind the project, as evidenced by both budget and public, Congressional and even scientific interest. $59 million was spent on the Viking biology package and another $41 million on the molecular analysis experiment that was relevant to the question of life because of its ability to detect organic molecules. Harold P. Klein of NASA's Ames Research Center headed the Viking biology science team; Klaus Biemann of MIT headed the separate molecular analysis team. While the results of several of the teams were relevant to the question of Martian biology, these two out of the thirteen teams were most directly relevant.

The Viking biology package (Fig. 7.3) embodied in one piece of technology the most sophisticated thinking of the 20th century on the subject of extraterrestrial life in the solar system. The assumptions behind its experiments, the results obtained, and the ensuing controversies over the interpretation of these results are therefore of considerable importance. The diverse ideas about the nature of Martian life led to three different biology experiments aboard Viking, each representing a different approach to the problem of life. Indeed, biology team leader Klein later stated that had it not been for the constraints of 15 kg weight and about 1 cubic foot volume for the biology package, even more of the approaches conceived during the previous two decades would have been included on the spacecraft. The idea was that the three experiments, singled out and recommended by the Space Science Board of the National Academy of Sciences in 1968, would test for life using different philosophies, environmental conditions, and detectors.

One approach, which came to be known as the "labeled release" experiment, was developed by Gilbert Levin, who had spent much of the 1950s trying to improve methods for the detection of bacterial contaminants in city water supplies, and believed his method could be applied to the search for life on Mars. He was awarded a NASA contract for his "Gulliver" concept in 1961, and was reporting on his experimental apparatus already in the early 1960s. Levin's approach assumed that any Martian microorganisms, like those on Earth, would assimilate (eat) simple organic compounds, decompose them, and produce gases such as carbon dioxide, methane or hydrogen as end products. For this reason a dilute aqueous solution of seven such organic compounds, radioactively labeled for detection purposes, was added to the incubation chamber containing the Mars soil sample. The experiment tested for the expected "labeled release" of the gas produced as any organisms ate the organics and breathed out the decomposition products. The output was in the form of radioactive disintegrations, measured by a carbon-14 detector, in counts per minute.

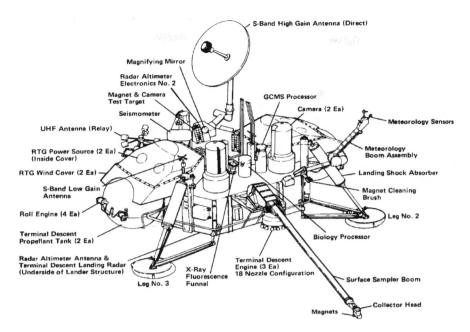

Fig. 7.3 The Viking lander, a complex machine incorporating eight experiments, landed on the surface of Mars in July, 1976, followed by another lander in September. The biology processor (labeled to the *lower right*) was within a small canister of volume 0.03 m^3. Nearby is the gas chromatograph mass spectrometer (*GCMS*), which detected no organic molecules at the two landing sites, down to parts per billion. Its conclusions have recently been called into question, and the biology experiment results are being reevaluated in the light of new evidence of the nature of the martian surface. (For scale: diameter of the lander body is ~3 m.) Courtesy NASA

The second biology test, the "gas exchange" experiment, was developed by Vance Oyama of NASA Ames Research Center, a veteran of life detection experiments on Apollo lunar samples. The gas exchange experiment tested for life under two different conditions. In the first mode, it was assumed that any organism in the dry Martian environment would be stimulated to metabolic activity by the addition of slight water moisture, and would give off a gas that could be detected by chromatography in the area immediately above the sample. In the second "wet nutrient" (or chicken soup) mode a rich nutrient of 19 organic compounds was added as an additional stimulus to metabolic activity, the products to be detected in the same manner. In both cases, the liquid added did not come into contact with the soil, but was added underneath the cell in which the soil "incubated." Water vapor gradually seeped up through the porous bottom of the incubation chamber, creating gradations of moisture through the soil. Experiments were also undertaken without the addition of any moisture.

The "pyrolytic release" experiment (also called the carbon assimilation experiment) was headed by Norman Horowitz of Caltech. Horowitz, a member of the WESTEX group in 1959, had cooperated with Levin's project in the early 1960s, but after Mariner IV showed that liquid water could not exist on the planet, he split with Levin and became convinced that it was best to test for Martian organisms under conditions known to exist on Mars when the experiment was designed. Thus to the small sample of Martian soil Horowitz proposed in his experiment to add only carbon dioxide and carbon monoxide, gases known to exist in the Martian atmosphere and now radioactively "tagged" for detection purposes. It was assumed that any organism on Mars would have developed the ability to assimilate these gases and convert them to organic matter. After 120 hours of incubation, the soil chamber was to be heated to 635 °C to pyrolyze the organic matter and release the volatile organic products, thus the name "pyrolytic release." A radiation counter yielded disintegrations per minute.

All three experiments sought to detect metabolic activities. Of the experiments Oyama's "wet nutrient" mode was the most Earth-like approach, in that it added rich terrestrial organics to stimulate any Martian organisms. Horowitz's was the most Mars-like, making few assumptions about Martian life except that it would be carbon based. Levin's, with his weak organic nutrient, fell in between. Levin and Oyama's experiments attempted to detect life by the decomposition of organics into gas during metabolism (a universal property of terrestrial organisms), while Horowitz sought to synthesize organic matter, which he would then pyrolyze in order to be able to detect. For detection purposes both Levin and Horowitz made use of standard techniques of radioactive carbon-14 as a "tracer," a method that did not change the chemistry, but provided a means of distinguishing atmospheric carbon from metabolized carbon. Oyama made use of the well-known method of gas chromatography for detection, as did Biemann (in conjunction with a mass spectrometer) for the organics experiment, which had nothing to do with metabolism. Ignorant of the nature of Martian life, the fondest hope of all the experimenters was that at least one of the experiments—hopefully their own—would turn up something.

Summer 1976 finally brought the day that Lowell, Kuiper and a host of scientific ghosts would have savored: the landing of two spacecraft on the surface of Mars to test for life in situ. They would not have been disappointed: Viking 1 landed successfully on the Chryse plain on July 20, and the first results of the biology experiments returned from Viking were exciting, to say the least. Although no visible life forms walked across the field of view of the camera, once the soil samples were collected on July 28, the biology experiments quickly began to return major surprises. Levin's experiment evolved gas into the chamber after the nutrient was added, then the reaction tapered off. Horowitz's pyrolytic release test was also positive, and Oyama's gas exchange experiment evolved not only CO_2 but also oxygen, the latter a reaction never before seen in tests on terrestrial or lunar soils. Because of the speed and course of the latter reaction, Oyama's experiment was not believed to be biological in nature. In short, two of the three biology experiments gave "presumptive positive results" for biology, and the third gave evidence of an oxidizing material in the surface at the Viking site. There was only one problem: in another unexpected finding, Biemann's organic analysis showed no organic molecules present to the level of a few parts per billion, a result Klein later called the most surprising single discovery of the mission. As Klein has subsequently recounted, these first results caused the carefully laid out experimental strategy to be abandoned, as the scientists attempted to discover whether chemical or biochemical reactions were taking place (Klein 1977; Dick and Strick 2004).

By eight and a half months after the first Viking had landed, 26 biological experiments had been carried out, and the first relatively complete results were reported, along with other Viking experiments, in *the Journal of Geophysical Research*. By then, shortly before the biological experiments were terminated in May 1977, Klein's considered judgment was that the positive result from Horowitz's pyrolytic release experiment was probably non-biological in origin, while Levin's labeled release experiment remained ambiguous (Klein 1977). Ironically, the gas exchange experiment of Oyama—the Viking scientist most optimistic about Martian life—showed no evidence at all for biological activity. Oyama and most of his colleagues concluded that the spontaneous evolution of oxygen was due to a chemical reaction involving "superoxides" such as hydrogen peroxide, perhaps by the effect of solar radiation on the small amount of water vapor in the upper atmosphere of Mars. "It's like the three bears," Klein later said. "Not too much water, not too little water, just the right amount of water in its atmosphere to produce something like this. This is one of the big mysteries, and any future missions to Mars have to find out what this stuff is" (Klein 1977, 4677–4680; Dick 1996, 155).

In the end, there was not complete consensus among the experimenters themselves. Writing for *Scientific American*, Horowitz concluded that although "it is not easy to point to a nonbiological explanation for the positive results" of his pyrolytic release experiment, "it appears that the findings of the pyrolytic-release experiment must also be interpreted nonbiologically," mainly because the reaction was less sensitive to heat that one expected from a biological process (Horowitz 1977, 61). Levin, however, did not agree; for decades he continued to argue forcefully that a biological interpretation of his data was still possible (Levin and Straat 1976; DiGregorio, Levin, and Straat 1997).

Clearly sensitive to their own assumptions, the Viking biologists continued to ponder the strategy of their experiments. What if their assumptions about Martian life, on which the biology experiments were based, were not correct? With this in mind Klein concluded his summary of Viking biology results with the astonishing remark that "we must not over look the fact, in assessing the probabilities of life on Mars, that all of our experiments were conducted under conditions that deviated to varying extents from ambient Martian conditions, and while we have accumulated data, these and their underlying mechanisms may all be coincidental and not directly relevant to the issue of life on that planet" (Klein 1977, 4679; Dick 1996, 157).

Ten years later, contemplating the experiments conducted for some ten months on the surface of Mars, Horowitz remained convinced that they not only proved the absence of life on Mars, but by extension "Since Mars offered by far the most promising habitat for extraterrestrial life in the solar system, it is now virtually certain that the earth is the only life-bearing planet in our region of the galaxy" (Horowitz 1986, 146). Although most scientists were not ready to make that quantum leap, it is also fair to say that they were much less optimistic about life on Mars in the aftermath of Viking. The Viking results were impressive enough that most scientists shifted the focus of their biological Martian interests to either past Martian history, or to different Martian environments such as rocks, polar caps, subsurface soil, volcanic regions, and the ancient river valleys.

That, however, is not quite the end of the Viking story. Though a consensus seemed to have been reached for several decades that life (indeed even organics) had not been found on the Martian surface at the two Viking landing sites, the issue was reopened especially after NASA's Phoenix lander discovered perchlorates on Mars in 2008. Some well-known and indisputably reputable scientists argued that such perchlorates would have destroyed any organics present in the Martian soil when it was heated during the Viking experiments (Navarro-Gonzalez et al. 2010). The issue remains open among prominent researchers today, with Levin more than ever convinced he discovered life on Mars (Levin 2011; Bianciardi et al. 2012). Surely, the extended discussion of the Viking results provides a cautionary note on the need for sensitivity to the preconceptions that enter into the design of experiments, and the difficulties of interpretation of the resulting evidence.

7.2.1.3 The Mars Rock

As the twentieth century approached its end, it appeared that the Viking landers had written the last chapter in the search for life on Mars. But almost exactly 20 years after the Viking landings, the world was startled with the announcement that organic molecules, possibly biogenic minerals, and even microfossils may have been found in a meteorite that originated on Mars. The result was controversial, though one might have thought that the inconclusive evidence would be balanced to some extent by the fact that the Martian meteorite could now be examined, not with the limited resources of a spacecraft on the surface of Mars, but with the full power of analytical techniques in many laboratories on Earth. A new era in Martian life studies had begun.

Meteorites had long been associated with the question of extraterrestrial life, but those meteorites were a special variety known as carbonaceous chondrites, and their parent body had not been identified. Only in the post-Viking era was a new category of extremely rare meteorites identified, and a case slowly built that they had originated on Mars. Known as SNC meteorites after the locations of their three types (shergottites, nakhlites and chassignites), they were also stony meteorites, but "achondrites," because they exhibited none of the millimeter-size embedded mineral spheres characteristic of chondrites. They were known to have come from Mars not only because of their chemical composition, but also because the gases trapped in them were precisely the same composition and proportions as those of the Martian atmosphere, as determined by the Viking landers. Thus, although Viking did not unambiguously find life itself, ironically it enabled the identification the SNC meteorites as Martian in origin.

The surprising announcement in the summer of 1996 centered on the Martian meteorite known as Allan Hills 84001, believed to have fallen on the ice fields of the Antarctic 13,000 years ago. The first meteorite found in the Antarctic during an NSF-sponsored search season in 1984 (thus the name ALH 84001), it was not identified as Martian in origin until 1994. One of only 12 such meteorites identified at the time, the 4.5 pound (1.9 kg) softball-sized rock was by far the oldest of the 12, estimated to have formed about 4.5 billion years ago, from a period when Mars was warmer and had water and an atmosphere. It was hypothesized that a meteorite impact on Mars fractured the rock about 3.6 billion years ago, and that another impact about 16 million years ago launched the rock into space, where it eventually intercepted the Earth.

The evidence, announced by a NASA team led by David McKay of NASA Johnson Space Center in Houston, consisted of four parts (McKay et al. 1996). None of these parts, the participants pointed out, were conclusive in themselves, but taken together they could be interpreted as biogenic. First, the multidisciplinary science team reported, the fractured surfaces of the rock contained large complex organic compounds in the form of polycyclic aromatic hydrocarbons (PAHs). This was already a step beyond what the Viking landers had found, but even though the NASA team undertook analysis that showed to their satisfaction that the PAHs were not contamination from Earth, this was not proof of life, since organic molecules could have originated by non-biogenic processes on Mars. But then the plot thickened: in the fractures the team also discovered carbonates and magnetite, minerals that are produced (among other ways) by certain "magnetotactic" bacteria on Earth. Finally, using a high resolution scanning electron microscope the team suggested the existence of microfossils in the carbonates and other mineral grains (Fig. 7.4); at only 20–100 nm they were 100 times smaller than the smallest known bacteria on Earth.

Less than two months later, a British team of scientists led by Colin Pillinger of the Open University announced independent evidence of possible traces of life, both in ALH 84001 and in a much younger Martian meteorite known as Elephant Moraine 79001 (EETA 79001, again named after the location of its discovery in the Antarctic). The latter meteorite was only 175 million years old, and was blasted from Mars only 600,000 years ago. This was so recent, geologically speaking, that it held open the possibility that life might still exist on Mars.

As in past controversies over Martian life, the stakes were high and the skeptics numerous. One of the chief objections came from Ralph Harvey of Case Western and Harry Y. McSween of the University of Tennessee, who had reported in *Nature* shortly before the NASA announcement that their analysis of the same meteorite showed that the carbonates formed not as a result of microbial life, but during the asteroid impact when carbon dioxide combined with the rock at temperatures of 1,200 °F. Such temperatures are inimical to life; if this method of carbonate formation was confirmed it would cast severe doubt on the claims of past Martian life. Others, however, argued for low-temperature formation of the carbonates, one that did not rule out life.

More general questions of inference from evidence were also asked. For example, do four independent but (critics said) weak arguments—from the morphology of the nanostructures, carbonate globules, the presence of magnetite and polycyclic aromatic hydrocarbons—add up to a strong argument for biogenesis? The

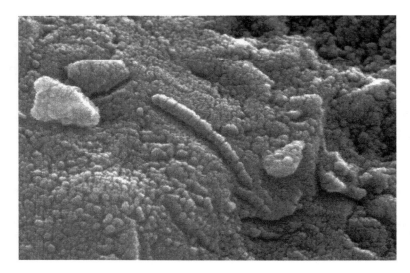

Fig. 7.4 High-resolution scanning electron microscopic image showing an unusual tube-like structure less than 1/100th the width of a human hair, found in Martian meteorite ALH 84001 and interpreted by some to be evidence of fossil life on Mars. Such morphological evidence was challenged by paleobiologist J. William Schopf, among other critics. Courtesy NASA

authors of the discovery paper thought so, ending their paper with the argument "Although there are alternative explanations for each of these phenomena taken individually, when they are considered collectively, particularly in view of their spatial association, we conclude that they are evidence for primitive life on early Mars" (McKay et al. 1996, 930). Critics, including paleobiologist J. William Schopf, thought not, arguing that "spatial association" held no persuasive value at all, and citing Carl Sagan's dictum "extraordinary claims require extraordinary evidence" (Dick and Strick 2004, 191).

Definitive proof of past life on Mars would come only by sectioning thin sections of the microfossils to search for cell walls, DNA or other structures unambiguously linked to life. In the years following the announcement many teams did precisely that, but with still ambiguous results. Though there is now consensus that Martian nanofossils have likely not been found, some (including most scientists who made the original claims) have not given up. Thus, the pattern is similar to that of the claims of Gilbert Levin about extant life on Mars based on the Viking experiments.

7.2.1.4 Planetary Systems?

If the evidence from the relatively nearby solar system proved problematic, the evidence for other much more distant solar systems would be even more difficult, if of an entirely different type. It is some measure of the difficulty of the search for planetary systems that the Space Age did not bring immediate advances in the problem. Unlike solar system studies, where planetary spacecraft brought immediate and revolutionary progress in our knowledge of the planets, no such prospect was in store for planetary systems. It is true that increased knowledge of our own planetary system provided voluminous data for the refinement of theories of the origin of the solar system, which by the usual gross analogies could be applied to other solar systems. But, although substantial, these refinements changed little the fortunes of planetary systems. Perhaps the largest impact of the Space Age on planetary systems science was the infusion of funds from space agencies such as NASA, which displayed an interest in both observational and theoretical aspects of the subject almost from the beginning, but with delayed results.

We should therefore not be surprised that, while most astronomers in the second half of the 20th century were optimistic about other planetary systems, observational proof of their existence through the 1970s remained entirely dependent on the old astrometric technique. That technique, the results of which remained elusive in many cases, created a public and scientific sensation with the announcement in the 1960s of the detection of several planetary systems. The promise and limitations of this technique, and the difficulties of tackling a problem at the limits of science, may best be seen in the famous case of Barnard's star. The central figure in the case is astronomer Peter van de Kamp, who had begun his search for low mass companions at the Sproul Observatory of Swarthmore College in Pennsylvania in 1937. Such is the long-term nature of the problem of determining perturbations in stellar motions that only 25 years later was Van de Kamp beginning to announce results with planetary companions.

Barnard's star was a star of 9.5 magnitude, so-called after Barnard's discovery in 1916 of its enormous proper motion of about 10.3 arcseconds per year. This meant that it was a close star (the closest known after the Alpha Centauri system), and it was immediately placed on observational programs, including the parallax program at Sproul in 1916–1919. In 1938 van de Kamp had placed it back on the Sproul parallax program with his arrival as Director in 1937, and by 1944 he announced a low-mass companion stellar in nature. Over the next 20 years, as Kuiper and others were predicting an abundance of planetary systems based on their own work, and as theory once again made plausible abundant planetary systems, van de Kamp patiently collected data on Barnard's star and other nearby stars.

There is no doubt that van de Kamp was sensitive to the question of whether low-mass companions were "stars or planets," at least since his 1944 article on the subject, an article undoubtedly stimulated by the observational claims of 1943. In a progress report on "Planetary Companions of Stars" in 1956, van de Kamp pointed out that while numerous unseen objects had been detected over the last two decades with masses 0.05 of the Sun's or greater, it was "extremely likely" that these objects were stars. "There are tentative indications of unseen companion objects with about 0.01 solar masses or more, and these may be planetary companions. However, definitive interpretation can hardly be reached at present, partly due to limitations of accuracy," he wrote in that year (Van de Kamp 1956, 1040). In particular, van de Kamp pointed out that the 1943 claims for a planetary companion of 70 Ophiuchi had not been confirmed as of 1952; and he held out hope only for the claim in 1943 for a companion of 61 Cygni of 0.016 solar masses, well into the planetary range, as confirmed by two other observers. As for his own program, now almost two decades old, van de Kamp claimed only that the companion to Lalande 21185 was probably a star of low luminosity, and that no satisfactory explanation existed for some perturbations seen in the motion of Barnard's star. Almost two decades after the start of his observational program at Sproul, no one could accuse van de Kamp of rushing to judgment on planetary companions.

All of this was to change in the 1960s. First, in 1960 Sarah Lippincott, van de Kamp's colleague at Sproul, announced that the companion of Lalande 21185 had a mass of only 0.01 that of the Sun. Although this was at within planetary range (recall that the 1943 claims for planetary companions to 61 Cygni and 70 Ophiuchi gave them about this same mass), Lippincott's technical article in the *Astronomical Journal* made no mention of the word "planet," and perhaps for this reason her announcement did not raise much of a stir. But van de Kamp's 1963 article with the mundane title "Astrometric Study of Barnard's Star from Plates Taken with the 24-inch Sproul Refractor" created a sensation (Van de Kamp 1963). In it he announced the discovery of a companion to Barnard's star with a mass of only 0.0015 the mass of the Sun, only 1.6 times the size of Jupiter, which he specifically characterized as a planet. He further found the distance of the planet from Barnard's star to be similar to that of Jupiter from the Sun, and its surface temperature about 60 K compared to 120 K for Jupiter.

Van de Kamp's claim was based on 25 years of photographic observations, using three types of photographic emulsions, and 50 observers, yielding 2413 plates. To the extent that the public was aware of such details, they might have been persuaded by this alone that such an intensive scientific effort must have yielded a definitive result. But they would have not been aware of the subtleties of the technique, which included taking into account a variety of insidious errors that could affect the results. Having taken into account all known sources of error as best as he could, van de Kamp found a perturbation in the motion of Barnard's star with a period of about 24 years (Fig. 7.5). In order to come up with an actual mass for the companion, he further had to carry out a "dynamical interpretation" calculation, using the mass of Barnard's star. By Kepler's law, once this mass was known, and the period of the orbiting body, the mass of the latter could be calculated. It was here that van de Kamp finally came to the figure of 0.0015 times the mass of the sun for his new planet: "The orbital analysis leads, therefore, to a perturbing mass of only 1.6 time the mass of Jupiter. We shall interpret this result as a companion of Barnard's star, which therefore appears to be a planet, i.e., an object of such a low mass that it would not create energy by the conventional nuclear conversion of hydrogen into helium" (Van de Kamp 1963, 521).

Like the announcements 20 years before, the reaction to van de Kamp's result was swift. From *Time* magazine to popular science magazines and more sober scientific journals, countless reports of van de Kamp's results hailed the discovery of another planetary system. Independent verification of the result, on the other hand, was more difficult, since the observations were very specialized and required

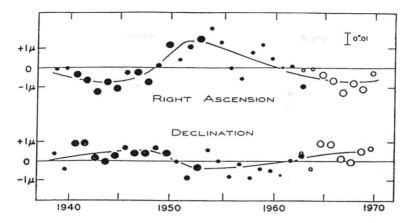

Fig. 7.5 Peter van de Kamp's evidence for a planet around Barnard's star made use of the classical astrometric method for planet detection. Van de Kamp reported that the star underwent minute, periodic gravitational perturbations of a few hundredths of an arcsecond over three decades. The two plots show the star's measured east–west and north–south relative positions (in microns) on photographic plates taken with a 24-inch refractor. The data was proven spurious, but only after several decades. From van de Kamp (1963), reproduced by permission of the American Astronomical Society

decades to reach a result—van de Kamp had been at it for a quarter century. It is not surprising, therefore, that van de Kamp himself was the first to reinforce his own result. In 1969, with five more years of photographic measures of Barnard's star, he reiterated his claim that a planetary companion existed around that star, with a slightly revised mass of 1.7 times that of Jupiter. In the same year, Van de Kamp proposed an alternate analysis of his data that held out the possibility that two planets orbited Barnard's star, with periods of 26 and 12 years, and masses of 1.1 and 0.8 times Jupiter.

But trouble was around the corner, and the 1970s saw serious questions raised about van de Kamp's momentous result. In 1973, John Hershey, one of van de Kamp's own students, found that changes made to the Sproul telescope, in particular a change of lens cell in 1949, caused jumps in the data at that point and may have affected the results for Barnard's star. In the same year George Gatewood of the Allegheny Observatory in Pittsburgh and Heinrich Eichhorn of the University of South Florida concluded "with disappointment," based on an independent analysis of 241 photographic plates, that no perturbations existed in the motion of Barnard's star. Attempting to explain their result, they pointed to the disadvantages of van de Kamp's analysis technique, to changes in the optical system of van de Kamp's telescope over the extended period of time of his study; and to the fact that his claimed perturbation was just "on the verge of significance," a status similar to claimed measurements of parallax before Bessel (Gatewood and Eichhorn 1973, 776). A similar analysis by Gatewood published the following year gave the same null result for Lippincott's 1960 claim of a planetary companion around

Lalande 21185. Two other studies of van de Kamp's data in 1973 were more favorable to his claim of one or more planetary companions, but a decade after the first announcement by the van de Kamp group, planetary systems were once again in trouble.

Van de Kamp understandably did not take lightly this negation of the main result of his work of 25 years. In order to take the objections into account, especially the finding that changes in the instrument might have affected positional measurements, van de Kamp re-measured his plates on a new machine, and included only material from 1950 onwards. He confirmed the existence of the shorter period planet, with a mass now 0.4 Jupiter, while the second planet was "less well determined." In 1977, 60 years after the discovery of Barnard's star, van de Kamp took the occasion to reassert his belief in the reality of its planetary companions. In addition to the now familiar scientific defense, the article concluded with a Rembrandt etching on the appearance of Christ to Thomas, with a caption "blessed are they that have not seen, and yet have believed," suggesting a religious invocation of faith undoubtedly seen by some as not readily transferable to the scientific realm (van de Kamp 1977, 521). Van de Kamp's last paper based on new data, published in 1982, again supported the conclusion of two planets around Barnard's star, a conclusion he never relinquished.

The Barnard's star episode was only the most notorious of several claims made for planetary companions by the mid-1970s, all subject to the same limitations of technique and inference. Although the assault on Barnard's star continued to receive the greatest attention, in the field of astrometric perturbations it was not unique, and thus could not be written off as a fluke. In an extensive 1975 review of the subject of unseen astrometric companions van de Kamp could list 17 "well-established perturbations" of stars by unseen companions, including Barnard's star and three others with possible planetary companions. Another 14 stars, including the famous 61 Cygni, were listed with "perturbations of provisional, suspected, or uncertain nature" (van de Kamp 1975, 312–313).

Not everyone was convinced that even those stars showing well-established perturbations necessarily harbored planets, for this depended on theoretical ideas about the cutoff point for stable hydrogen burning in stars. While CalTech geochemist Harrison Brown supported the idea of numerous planets by an extension of the "luminosity function" (the distribution of the stars with their visual magnitudes) to low masses, S. S. Kumar, for example, argued that all of the objects claimed as planets were probably very low mass "degenerate" objects that he termed "black dwarfs" (Kumar 1967). The dividing line between stars and planets remained the subject of vigorous discussion, and the accompanying search for what came to be known as "brown dwarfs" (objects not massive enough to sustain nuclear fusion) remained almost as elusive as the search for planets themselves.

Numerous other instances could be examined, including the controversy surrounding the claims of extrasolar planets in the 1980s and 1990s (subsequently demonstrated to be true), the claims of protoplanetary systems found by the Infrared Astronomical Satellite (IRAS) and the Hubble Space Telescope, and the recent reaction to the claims of arsenic life in the Halomonadaceae bacterium. In fact, it is safe to say, controversies about the interpretation of evidence are the central core of science, the rule rather than the exception.

All these examples also bear on the role of preconceptions, manifesting themselves especially in the form of prior assumptions feeding into the design of experiments, the interpretation of observations and the definitions of life and intelligence. For example, the Viking experiments were designed with metabolic conceptions of life, as opposed to other possibilities such as Darwinian processes, metabolism, energy and thermodynamics, complexity theory, cybernetics, or some new insight. This was in part due to the constraints of spacecraft investigation; nevertheless with the perspective of four decades we can look back on these experiments and ask how they might have been done differently. Certainly had scientists known about perchlorates on the surface of Mars, the experiments would have had a different design.

But what are the lessons learned from these representative case studies? Scientists need hardly be reminded to be careful with the interpretation of data. But the general public can never be reminded too often that this is the nature of science. The lesson is not that science should be mistrusted or abandoned because of its imperfections, but that despite the difficulties history shows that something approaching "the truth" eventually emerges as Nature is continuously interrogated at increasingly more subtle levels, even if the final outcome may take decades. The lesson is both pedestrian and profound: with all its personal and cultural biases, science is the best way we have of interrogating nature, and even in extremely difficult areas such as astrobiology progress can be made, even if we have to resort to analogies, which much of microbial astrobiology does. As scientists know—but as much of the public doesn't seem to understand—science, including astrobiology, is a series of trial and error observations, subject to constant interpretation and re-interpretation, and therefore to constantly changing ideas, asymptotically approaching the truth—until the truth may suddenly change with a new way of looking at things. In this it in no way resembles religion, as is sometimes charged. The difficult and expensive nature of science, especially space exploration, makes experimental and observational iterations very extended. But in the end, once again, most scientists would say the objective truth is out there (though this is a deep philosophical problem), and remarkably the human mind can eventually ferret it out.

7.2.2 The Role of Theory

The role of theory in science is a huge and complex subject. We confine ourselves here to one example that demonstrates that theory can serve both as catalyst and as hindrance: the case of the origin of solar systems. As we have seen, during the first half of the 20th century, the primary and most widely accepted theory of the origin of solar systems was the Chamberlin-Moulton and the Jeans-Jeffreys tidal theory, whereby solar systems arise by material tidally pulled out during the close encounters of stars. This theory lowered expectations for observing other solar systems, since encounters between stars were believed to be extremely rare. Only in the 1940s, when the tidal theory was replaced by a revived nebular hypothesis, which postulated solar systems as a common byproduct of stellar evolution, did the expectations radically change, providing the backdrop first to Van de Kamp's work, and to the real discoveries of exoplanets 30 years later. So theory can affect both the undertaking and the interpretation of observations (Dick 1996).

In the absence of decisive observational evidence for planetary systems, one might expect that theories of solar system formation would play an especially important role, at least in determining the plausibility of such systems. This had indeed been the case for the nebular hypothesis, which favored abundant planetary systems because the formation of planets from a rotating gaseous disk was assumed to be a universal process. But as we have briefly seen in the first part of this chapter, that hypothesis was under serious attack in 1900, and for the first two decades of the twentieth century the new theory, to the limited extent that it addressed the issue at all, gave conflicting indications about the possibility of other planetary systems. Developed by T. C. Chamberlin, chairman of the Geology Department at the University of Chicago, and F. R. Moulton, an astronomy graduate student at the same university, the Chamberlin-Moulton hypothesis sought to surmount the technical weaknesses of the nebular hypothesis by proposing instead that solar systems were formed by the close encounter or actual collision of stars in space. According to this hypothesis, the close encounter caused material to be ejected from the Sun. The passing intruder then caused the ejected material to form spiral arms. These arms contained knots of denser material that condensed into nuclei, which in turn grew into planets and satellites by the capture of planetesimals, cold particles in the nebula. The spiral nebulae recently observed in the heavens, they believed, might be evidence of such collisions and of solar systems in formation. Curiously, rarity or abundance of planetary systems does not seem to have been an issue for Chamberlin or Moulton. To the extent that their rarity or abundance was an issue at all, it oscillated between the twin pillars of the "planetesimal hypothesis:" the spiral nebulae, which implied abundance, and stellar encounters, which implied rarity. With the gradual realization that spiral nebulae were too large to represent planetary systems in formation, the stellar encounter aspect of the theory was free to gain the upper hand—and with it the implication of the rarity of planetary systems.

This, in fact, is precisely what occurred, not in America but in Britain, where in the tradition of William Whewell and A. R. Wallace, the scientific community seemed more skeptically inclined toward planets and life. It was at the hand of the British mathematical physicist and astronomer James Jeans that the question of other solar systems would become closely linked with the rarity of planets and life in the universe. Jeans, a 1903 graduate of Trinity College, Cambridge, had done important work on atomic theory and statistical mechanics prior to 1914. After 1914 he turned from the microscopic to the macroscopic, from atoms to astronomy, and specifically to cosmogony. Jeans' attention was at first devoted to the stability of rotating bodies, on which subject he published two lengthy papers in 1915 and 1916. This work he applied to cosmogony in 1916 with reference to tidally distorted masses, in other words, determining how a rotating astronomical body would be affected by tidal forces raised by another passing astronomical object, as would happen in the case of a close stellar encounter. In a paper read before the Royal Astronomical Society in 1916 and published in the Society's Memoirs in the following year, Jeans dealt not only with the origin of solar systems, but also with binary star formation and spiral nebulae. In contrast to the binaries and spirals, Jeans concluded that the solar system might well have been formed from a tidally distorted mass, in particular by another star approaching our sun. Unlike the Chamberlin-Moulton hypothesis, however, Jeans' analysis showed that neither spiral nebulae nor planetesimals played a role in planet formation, and he thus emphasized that for solar systems "the origin which seems most probable is not that of the planetesimal hypothesis" (Jeans 1917, 48). Instead, his analysis showed that rather than the streams of gas torn from the Sun condensing into numerous small cold planetesimals that in turn accreted to form the planets, a single cigar-shaped filament of hot gas would be ejected and condense directly into the planets. As the theory was later elaborated, he pointed out that the largest planets would form near the center where the filament was thickest, and the smaller ones at each end, giving the distribution of planets observed in our solar system.

The central question in determining whether this mathematical conclusion could really occur in nature was the frequency of close stellar encounters. It is clear at the outset of the paper that Jeans was already thinking in these more general terms, not only with regard to the origin of our solar system, but in connection with the frequency of planetary systems. In his earliest statement on what would become a lifelong contentious issue, he wrote: "We have absolutely no knowledge as to whether systems similar to our solar system are common in space or not. It is quite possible, for aught we know to the contrary, that our system may have been produced by events of such an exceptional nature that there are only a very few systems similar to ours in existence. It may even be that our system is something quite unique in the whole of space" (Jeans 1917, 46).

Jeans' analysis showed that the issue of abundance was very sensitive to the assumptions made about a variety of parameters, including the density of stars in the universe, the velocity of the stars in space, the age of the stars and of the universe, and the size and mass of the stars at time of encounter. All of these parameters were subject to change in the discussion that ensued over the next three decades. For now, using the best estimates known in 1916 and assuming stellar masses and velocities similar to the Sun, Jeans found that at most 1 star in 4,000 might have experienced a "non-transitory" encounter at the distance of Jupiter in a lifetime of 10 billion years, the upper limit that he placed on the age of the universe. If the encounter distance were a hundred times greater and the other parameters adjusted accordingly, one star in three might have experienced such an encounter, and "we may, without postulating anything very improbable, suppose our system to have experienced an encounter as close as this…" (Jeans 1917, 46–47). However, Jeans clearly did not think all these conditions would ensue at one time, and in the end he labeled these occurrences as "somewhat improbable" and systems similar to our own "somewhat rare," but in general the entire process not "impossible or very improbable." Given the number of parameters and their uncertainty, Jeans' waffling is not surprising. But he emphasized that no reasonable choice of parameters was likely to alter the result that only very few stars can have experienced non-transitory encounters. And most importantly, Jeans stressed, the theory violated no quantitative criterion.

In his classic work *Problems of Cosmogony and Stellar Dynamics* (1919), Jeans discussed the problem in more detail, and ended with results even more pessimistic: only one encounter in 30 billion years, a situation so improbable in the present universe as to cast doubt on the validity of the close encounter hypothesis. Pointing out that the parameters were not well known, Jeans concluded that while tidal breakup by a passing star was hardly a likely event, its improbability was not grounds for rejecting the tidal theory. In whatever case one adopted, the solar system seemed to be very exceptional, "and for aught we know may be unique" (Jeans 1919, 290). In his 1923 lecture "The Nebular Hypothesis and Modern Cosmogony," Jeans carried his train of thought one step further, arguing that it was just possible, though not probable, that only the earth could support life in the universe. "Astronomy does not know whether or not life is important in the scheme of nature, but she begins to whisper that life must necessarily be somewhat rare" (Jeans 1923, 30).

In the hands of Jeans, this whisper soon grew to a crescendo. In both his technical and popular publications by the late 1920s, Jeans spread his view far and wide. The numbers varied somewhat, but always present was the basic scenario that the stars are sparsely scattered in space, close encounters exceedingly rare, and the conditions for life very exacting. "All this suggests," Jeans inevitably concluded, "that only an infinitesimally small corner of the universe can be in the least suited to form an abode of life" (Jeans 1930, 335). In his popular works this view of the disruptive approach of stars was vividly drawn, and the rarity of such approaches and their ensuing solar systems was an integral part of this picture—clear even to the public.

For two decades the Jeans tidal theory—with contributions by Sir Harold Jeffreys was widely accepted, and when the beginning of the end came in 1935 it was once again because of problems with physical principles. This time it was the Americans' turn again, in the form of Henry Norris Russell, who criticized the tidal hypothesis because it could not account for the present orbits of the planets. Russell could not see how a close stellar encounter would remove the planets so far from the Sun and give them most of the angular momentum of the system rather than the Sun, which was a thousand times more massive (Russell 1935). He also could not see how the planets could condense out of the high-temperature matter ejected from the Sun, an objection given definitive form by Russell's student Lyman Spitzer four years later. In their discussion the possibility of other planetary systems played no role, but their fatal objections left science without a workable theory of the origin of the solar system, and by association placed in limbo the idea that such systems were rare.

The 19th century view of abundant planetary systems was thwarted for decades by the tidal theory of Jeans and Jeffreys. Far from the teleological view of R. A.

Proctor or J. E. Gore, Jeans' colleague Sir Arthur Eddington asked "How many acorns are scattered for one that grows to an oak? And need she be more careful of her stars than of her acorns? If indeed she has no grander aim than to provide a home for her greatest experiment, Man, it would be just like her to scatter a million stars whereof one might haply [sic] achieve her purpose" (Eddington 1929, 179). To have provided a theoretical underpinning for this startlingly different worldview was no small part of the legacy of James Jeans.

But alas, this world view had no more claim to objective truth than the 19th century belief in abundant planetary life, for if the early observational claims for planetary systems at the turn of the century had yielded no definitive result, by 1940 neither had theory solved the problem—nor could it—especially with the departure of spiral nebulae as confirming evidence. The discredited nebular hypothesis had been superseded by the planetesimal hypothesis of Chamberlin and Moulton and then the tidal theory of Jeans and Jeffreys, only to have Russell and Spitzer overturn the latter, leaving only the void. Reviewing the collisional and nebular hypotheses in 1938, Lick Observatory Director Emeritus Robert G. Aitken still saw the development of planetary systems as an "exceptional event." "Exceptional" did not mean unique to Aitken, who pointed out that even if only one star in a million had planets, there would still be 30,000 solar systems in the Milky Way Galaxy—and two million galaxies were within the range of current telescopes (Aitken 1938).

As we saw at the beginning of this chapter, the 15 years between 1943 and 1958 saw a remarkable turning point in the fortunes of planetary systems. It had begun with Russell's criticism of the Jeans-Jeffreys tidal theory, but it was fueled by the revival of a modified nebular hypothesis, developments in fields as diverse as double star astronomy and the measurement of stellar rotation periods, and—most surprising of all—by insistent claims that planetary systems, or their effects, had been actually observed (Table 7.1). Moreover, broader events in the field of cosmology conspired toward change also, events that Jeans himself could not ignore.

The implications of the revolution in cosmology of the 1920s and 1930s—a greatly enlarged Galaxy, the existence of innumerable "island universes" full of stars, a universe expanding in space and expanded in time—are evident in Jeans' review of the subject of life on other worlds published in 1942. Having given a dim view of the chances of life on Mars and Venus, Jeans turned to the realm of the stars, and the origin of planetary systems. He pointed out that under present conditions in the universe the frequency of stellar encounters would be only 1 in 10^{18} years, so that for stars two billion years old, one star in 500 million might have planets. So far this was his old argument. But it was a sign of the times that he went on to say that though this seemed like a small fraction, in a universe with 10 billion galaxies, each with 100 billion stars, this minute fraction still represented 2 million stars that might have planetary systems! Statistics—and the new cosmology—had caught up with Jeans, even if only 2,000 of these systems might be located in our own galaxy. Straining the definition of "rare," Jeans was forced to conclude that "although planetary systems may be rare in space, their total number is far from insignificant" (Jeans 1942a, 83).

Later that year, however, Jeans' view had undergone a much more radical change. In a letter to *Nature* of June 20, 1942, reacting to recent claims of serious dynamical problems arising for the tidal theory assuming the Sun was about its present size at time of encounter, Jeans asserted that the Sun was most likely comparable in size to the present orbit of Uranus or Neptune when an encounter took place. In a last-ditch effort to save the tidal theory from dynamical objections, Jeans was forced to increase greatly the size of the Sun at the time of supposed planetary formation, a concession that greatly increased its cross-section and by analogy the cross section of other suns. Not only did this address the dynamical objections in Jeans' opinion, it also led to another conclusion: that the total chance of planet formation was now 1 in 6 with such a size for the Sun. Thus, "there is no longer any need to strain the probabilities to account for the existence of the planets" (Jeans 1942b, 695). And the final conclusion is one hardly expected from Jeans: "A far larger proportion of the stars than we have hitherto imagined must be accompanied by planets; life may be incomparably more abundant in the universe than we have thought" (Jeans 1942b, 695) The whole exercise demonstrated the fragility of the argument, and the dangers of using equations whose parameters were not well-determined. For 25 years Jeans had epitomized the concept of the rarity of life in the universe. Now in the last years of his life he recanted, and his death in 1946 left no substantial heirs to his theory.

Jeans' turnabout was just the beginning, and the cracks opening in the tidal theory in 1941–1942 were to become a breach through which the floodwaters of change would rush in the following year, when strong and independent observational claims were made for the existence of two planetary systems around nearby stars. Many astronomers were quick to draw general conclusions, especially in light of the new observations; as Henry Norris Russell wrote in 1943, "On the basis of this new [observational] evidence, it therefore appears probable that among the stars at large there may be a very large number which are attended by bodies as small as the planets of our own system. This is a radical change—indeed practically a reversal—of the view which was generally held a decade or two ago" (Russell 1943, 19). Table 7.1 indicates how completely the change was, due in no small part to a change in theory.

Examples of the role of theory in the extraterrestrial life debate could be multiplied, but the lesson in this case seems to be that theories can play both positive and negative roles in getting at the truth, especially when empirical evidence is lacking. One thinks of Harold Urey's theory of a reducing atmosphere on the primitive Earth, based on his speculation that the solar nebula was largely hydrogen. This led to Stanley Miller's famous experiment in which amino acids were produced under such a simulated primitive Earth atmosphere, a result that in turn gave much hope to those who believed extraterrestrial life might be common. The nature of the primitive Earth atmosphere has since been called into question, leaving scientists oscillating between optimism and pessimism—one might almost say between hope and despair—undoubtedly also influenced by predispositions to one side or the other.

7.2.3 Testing History: Plenitude, Mediocrity, Anthropocentrism and Rare Earth

History demonstrates a mixed record of success for general arguments in the extraterrestrial life debate, including the uniformity of nature, the Principle of Plenitude, the Principle of Mediocrity, the Goldilocks argument for rare Earth, large number arguments, and the Fermi paradox. In closing we focus here briefly on three related arguments: plenitude, rare Earth, and the Principle of Mediocrity, with the idea that our current understanding of the number of exoplanets, the conditions for the origins of life, the conditions for life on planetary and satellite surfaces, and "The Great Silence" can be used to evaluate the validity of arguments in astrobiology over the last 50 years and more. The arguments of Ward and Brownlee (2000) in their book *Rare Earth*, makes such an evaluation in the light of history an important endeavor.

That general arguments can sometimes be illuminating is illustrated in the case of the principle of plenitude, which posits the fecundity of God or Nature and states that whatever God or Nature can do, they will do. In his classic volume *The Great Chain of Being* the historian Arthur O. Lovejoy put it this way: that "no genuine potentiality of being can remain unfulfilled, that the extent and abundance of the creation must be as great as the possibility of existence and commensurate with the productive capacity of a 'perfect' and inexhaustible 'Source,' and that the world is better, the more things it contains" (Lovejoy 1936 1960, 52). He posited this argument as the chief argument of the entire plurality of worlds debate through the 19th century, a claim now seen to be too simplistic.

That does not mean, however, that ideas of plenitude have not played a significant role throughout history, as Chap. 1 of this volume demonstrates (Crowe and Dowd 2013). Nor does it mean that the same principle is not invoked even today, consciously or unconsciously, implicitly or explicitly. Despite skepticism prior to 1995, the last two decades have shown that a Principle of Plenitude does indeed apply in the case of exoplanets, which we now know exist in abundance around normal Sun-like stars. That the idea of plenitude has its limits is indicated by the fact that, although some planets are also found around exotic stars such as pulsars and in binary systems, they do not exist around all classes of stars. General metaphysical ideas such as plenitude must be mediated by sober physical reality and empirical findings. Whether or not this kind of mediated general argument carries over to life and intelligence remains to be seen—indeed that is the very question to be answered by more scientific means. But the Principle of Plenitude remains a kind of guiding argument for the optimists, and so far a useful one.

Other arguments, such as the "rare Earth" and related "Goldilocks" genre, do not fare so well when evaluated in terms of history. For example the natural philosopher William Whewell, who coined the term "scientist" in the mid-19th century, used numerous Goldilocks arguments to "prove" that other worlds could not exist—that the Earth was indeed rare. His treatise *Of a Plurality of Worlds: an Essay*, which appeared anonymously in 1853, was the most learned, radical, and influential anti-pluralist treatise of the century (Crowe 1986). As with all participants in the debate, Whewell had his predispositions in the matter of extraterrestrial life. When it came to the compatibility of other worlds with Christianity, unlike Thomas Paine and others Whewell argued that it was other worlds, not Christianity, that should be rejected. To the argument that all the vast space must have some purpose, he countered that geology reveals human existence on Earth to be but a short "atom of time" compared to the age of the Earth; therefore why could not intelligence be confined to the "atom of space" that was the Earth?

Moreover, although the universe was indeed vast (about 3,000 light years by his estimate), Whewell argued that the possible locales for inhabitants had been vastly overrated. With arguments that were plausible at the time he held that (1) all nebulae are gaseous rather than stellar systems, meaning that not nearly as many stars existed as some believed; (2) the analogy between the Sun and the stars may be less strong than many thought—all stars may not be similar to our Sun; (3) many stars are double stars, unsuitable for planets; and (4) there is no evidence of planets around other stars. In short, Whewell saw the analogies as greatly exaggerated in the case of other worlds. No longer was the Copernican implication that the planets were Earths a sufficiently precise argument; greater attention had to be given to the details of their physical conditions. But he was wrong that all nebulae are gaseous, and while there was no evidence at the time of other planets, today they are known to exist in abundance, even around double stars. Whewell was indeed correct that not all stars are similar to the Sun, but this argument was obliterated by the fact that we now know we live in a vastly larger universe that he conceived, harboring billions of stars even in a single galaxy.

All of Whewell's conclusions were based on uncertain evidence, but his arguments narrowing the number of habitats and claiming that everything had to be just right for life on those habitats were misguided. Today, with our knowledge of the extremities of life in deep sea hydrothermal vents, deep underground environments, and in conditions of extreme salinity, acidity, pressure, radiation, and almost every variable one can imagine, Goldilocks arguments seem even less relevant as a guiding principle. Such "rare Earth" type of arguments are now known to be fallacious in light of new knowledge, and it seems likely that the modern rare Earth hypothesis (Ward and Brownlee 2000) will suffer a similar fate, as the Kepler spacecraft is already beginning to indicate. Of course, the rare Earth declaration depends on the definition of "rare," and on the definition of "Earths," especially with the discovery of the category of planets called "Super Earths." But with cases in our solar system such as Europa, Callisto, and Ganymede, the very concept of a "habitable zone" has been revolutionized.

Related to the rare Earth argument is the insidious role of anthropocentrism, deployed consciously or unconsciously. Fifty years after Whewell's treatise, at the beginning of the 20th century, A. R. Wallace—the co-founder with Darwin of the theory of natural selection—used arguments similar to Whewell's to demonstrate the Earth had a favored place in the universe, that the sole purpose of the universe was to produce humans, and that humans were the only life in the universe. His influential work *Man's Place in the Universe: A Study of the Results of Scientific Research in Relation to the Unity or Plurality of Worlds* (Wallace 1903a) incorporated many of the biological problems that would be elaborated in ever more subtle form throughout the century. Although Wallace's book in some ways marks a signal advance in the debate about other worlds, its failure is marked by the dominance of the anthropocentric worldview over all other arguments. Convinced of the nearly central position of the Sun in the universe, Wallace first sought—and found—the significance of this fact in the uniqueness of life, and then adduced arguments in favor of the view that life was found beyond the Earth neither in our solar system nor in others. Fifty years after Whewell's treatise, Wallace confidently concluded that "Our position in the material universe is special and probably unique, and ... it is such as to lend support to the view, held by many great thinkers and writers today, that the supreme end and purpose of this vast universe was the production and development of the living soul in the perishable body of man" (Wallace 1903b, 474). Although professing a scientific approach, Wallace's book serves as a lesson on the limits of science when worldviews dominate empirical evidence (Dick 1996). It is a lesson the twenty-first century should take to heart.

The same lesson, however, needs to be applied to the other side of the argument. During the 20th century the tug-of-war between anthropocentrism and other worlds was profoundly affected by radical changes in astronomical worldview. While it was still possible as the century began for scientists to argue for an anthropocentric universe based on the Earth's privileged physical position in the cosmos, by 1930 advances in astronomy had destroyed this argument. The resultant world view—an expanding universe of enormous dimensions in which the solar system was at the periphery of one galaxy among millions—tipped the scales strongly toward the presumption of other worlds for the rest of the century. "The assumption of mediocrity" became an underlying current of thought favoring other inhabited worlds, superseding the assumption of uniqueness that had opposed it. The hopes for anthropocentrism at the beginning of the century, and its rapid demise thereafter, constitute one of the profound shifts in twentieth century thought. In this sense the proponents of extraterrestrial life therefore champion not only a scientific theory, but an entire philosophy. But a Principle of Mediocrity cannot be taken for granted any more than can an anthropocentric worldview. Only observation will provide the answer.

Whatever the scientific merits of the extraterrestrial life debate, the emotional issue of human status is inextricably linked to all discussions of inhabited worlds. Pluralism and anthropocentrism have long been locked in a deadly battle that has not been completely decided by the dawn of the twenty-first century. Committed anthropocentrists, whether for religious or other reasons, are likely to be the staunchest foes of pluralism, no matter what the evidence, and pluralists—whether they liked it or not—contributed significantly to the demise of anthropocentrism. For all the appeal to scientific argument, the continuing battle between anthropocentrism and other worlds pervades modern discussions of extraterrestrial life, and carries the Darwinian debate on the status of humanity into the universe at large.

7.3 Overview of Part II

The chapters that follow in this section provide only the briefest glimpse at the richness of the history of the 20th century extraterrestrial life debate and the many approaches to it. Danielle Briot (2013) focuses on a single individual, Gavriil Adrianovich Tikhov, a Soviet astronomer whose work is not well known, but that is of pioneering interest for current work in the field. As we mentioned above, by mid-century Tikhov claimed to have created the science of "astrobotany" by his comparison of the reflection spectra of terrestrial plants with observations of Martian surface spectral features. Tikhov was correct in his negative detection of chlorophyll characteristic of plants, foreshadowing the work of Gerard Kuiper and William Sinton (if not the latter's conclusion), but, like them, he was mistaken in his belief that Martian vegetation existed at all. While this undermined the claims of a science of astrobotany for Mars, Briot points out that Tikhov's method was valid in two ways: today observations of Earthshine (which Tikhov also pioneered) have provided the biosignature of Earth's atmosphere, and our own planet's biosignature in turn provides the basis for eventually determining the biosignatures of the many planets now being discovered beyond our solar system.

Astronomer Chris Impey (2013) details the development of the field of extra-solar planet detection, including biosignatures. His chapter is of interest not only for the modern history of the field over the last two decades, but also because of his wider philosophical claims, namely, that the Copernican Principle, also known as the Principle of Mediocrity has been robust enough to lead us in a direction that is now confirmed empirically. "Our situation on a rocky planet that orbits a middle-weight star on the outskirts of an unexceptional spiral galaxy appears not be unusual or unique," he writes. He concludes that a billion habitable locales in our Milky Way galaxy is a conservative estimate, and this must be multiplied by a hundred billion to arrive at the number of potentially habitable locales in the observable universe. Twenty years ago we knew of no other planets outside our solar system; today we know of thousands, with the number increasing exponentially. Only the future will tell whether these locales are actually inhabited.

Douglas Vakoch (2013) represents another approach to the history of the modern debate touched on in this chapter: an investigation of how individual conceptual frameworks affect views on extraterrestrial life. In particular he shows how the modern evolutionary synthesis affected the opinions of four leading evolutionary biologists: Theodosius Dobzhansky, George Gaylord Simpson, and less directly, H. J. Muller and Ernst Mayr. In contrast to astronomers, he shows how increasing acceptance of the evolutionary synthesis led to a consensus by 1980 among biologists, anthropologists and paleontologists (at least the relatively few who addressed the subject) that complex life was rare in the universe. Such a conclusion, though based on a relatively small sample and still open to discussion and interpretation, points the way toward a research program in which personal, cultural, and conceptual factors may be examined with an eye toward their role in belief in extraterrestrial life.

Finally, Aaron Gronstal's (2013) chapter provides a transition to a subject that will be explored in more detail in Section III of this volume: the societal impact of the discovery of extraterrestrial life. His approach is unusual in focusing on the discovery of microbial rather than intelligent life. He makes the crucial point that the discovery of microbial life could have more immediate impact on Earth than the discovery of extraterrestrial intelligence, which is likely to be extremely distant and limited in its ability to communicate with us. The impact of microbial life, he argues, would likely not affect our theological and philosophical world-views. Rather, it could have an immediate economic and technological impact in the areas of biotechnology and medicine. Drawing an analogy with the economic benefits of extremophile research over the last few decades, Gronstal concludes that microbial ecosystems on other worlds could provide similar economic value. On the other side of the coin, the traditional concerns about back contamination of Earth by extraterrestrial microbes must be heavily weighed against any economic benefits. Wiping out Earth life in the process of trying to create new medicines would not be an optimal outcome. The cost-benefit debates over the discovery and uses of microbial life beyond the Earth are thus likely to be heated and protracted.

References

Abir-Am, Pnina G. 1992. "The Politics of Macromolecules: Molecular Biologists, Biochemists and Rhetoric." *Osiris* 7: 164–191.

Aitken, Robert G. 1938. "Is the Solar System Unique?" *Astronomical Society of the Pacific Leaflet* 112: 98–106.

Asimov, Isaac. 1965. "A Science in Search of a Subject." *New York Times Magazine*, 23 May, 52–58.

Barabashev, Nicholas P. 1952. *A Study of the Physical Conditions of the Moon and Planets*. Kharkov, Russia: Kharkov Univ. Press.

Barrow, John D., and Frank J. Tipler. 1986. *The Anthropic Cosmological Principle*. Oxford: Oxford Univ. Press.

Bianciardi, Giorgio, Joseph D. Miller, Patricia A. Straat, and Gilbert V. Levin. 2012. "Complexity Analysis of the Viking Labeled Release Experiments." *International Journal of Aeronautical and Space Sciences* 13: 14–26. Accessed July 30, 2012. http://ijass.org/On_line/admin/files/2)(014-026)11-030.pdf.

Billingham, John. 1981. *Life in the Universe*. Cambridge, MA: MIT Press.

Briot, Danielle. 2013. "The Creator of Astrobotany, Gavriil Adrianovich Tikhov. In *Astrobiology, History, and Society: Life Beyond Earth and the Impact of Discovery*, ed. Douglas A. Vakoch. Heidelberg: Springer.

Brush, Stephen. 1996. *Nebulous Earth: The Origin of the Solar System and the Core of the Earth from Laplace to Jeffreys*. Cambridge: Cambridge Univ. Press.

Carr, Bernard, ed. 2007. *Universe or Multiverse?* Cambridge: Cambridge Univ. Press.

Carter, Brandon. 1974. "Large Number Coincidences and the Anthropic Principle in Cosmology." In *Confrontation of Cosmological Theories with Observational Data*, ed. Malcolm S. Longair, 291–298. Dordrecht: Reidel.

Chamberlin, Thomas C., and Forest R. Moulton. 1900. "Certain Recent Attempts to Test the Nebular Hypothesis." *Science* 12: 201–208.

Cleland, Carol E. 2001. "Historical Science, Experimental Science, and the Scientific Method." *Geology* 29: 978–990.

Cleland, Carol E. 2002. "Methodological and Epistemic Differences between Historical Science and Experimental Science." *Philosophy of Science* 69: 474–496.

Coblentz, William W., and C.O. Lampland. 1924. "New Measurements of Planetary Radiation." *Science* 60: 295.

Cocconi, Giuseppi, and Philip Morrison. 1959. "Searching for Interstellar Communications." *Nature* 184: 844–846.

Coleridge, Samuel Taylor. 1831. In *Specimens of the Table Talk of Samuel Taylor Coleridge,* ed. Henry N. Coleridge. Remark made 18 December 1831. Accessed July 30, 2012. http://www.gutenberg.org/cache/epub/8489/pg8489.html.

Columbia Accident Investigation Board. 2003. *Report,* vol. 1.

Conway Morris, Simon. 1998. *The Crucible of Creation.* Oxford: Oxford Univ. Press.

Conway Morris, Simon. 2003. *Life's Solution: Inevitable Humans in a Lonely Universe.* Cambridge: Cambridge Univ. Press.

Crowe, Michael J. 1986. *The Extraterrestrial Life Debate, 1750–1900: The Idea of a Plurality of Worlds from Kant to Lowell.* Cambridge: Cambridge Univ. Press.

Crowe, Michael J., and Matthew F. Dowd. 2013. "The Extraterrestrial Life Debate from Antiquity to 1900." In *Astrobiology, History, and Society: Life Beyond Earth and the Impact of Discovery,* ed. Douglas A. Vakoch. Heidelberg: Springer.

Davies, Paul C.W. 2007. *Cosmic Jackpot: Why Our Universe is Just Right for Life.* Boston: Houghton Mifflin.

Dawkins, Richard. 1997. *Climbing Mount Improbable.* New York: W. W. Norton.

Des Marais, David J., Joseph A. Nuth III, Louis J. Allamandola, et al. 2008. "The NASA Astrobiology Roadmap." *Astrobiology* 8: 715–730.

DeVaucouleurs, Gerard. 1954. *Physics of the Planet Mars: An Introduction to Areophysics.* London: Faber and Faber.

Dick, Steven J. 1982. *Plurality of Worlds: The Origins of the Extraterrestrial Life Debate from Democritus to Kant.* Cambridge: Cambridge Univ. Press.

Dick, Steven J. 1996. *The Biological Universe: The Twentieth Century Extraterrestrial Life Debate and the Limits of Science.* Cambridge: Cambridge Univ. Press.

Dick, Steven J. 1998. *Life on Other Worlds.* Cambridge: Cambridge Univ. Press.

Dick, Steven J. 2003. "Cultural Evolution, the Postbiological Universe and SETI." *International Journal of Astrobiology* 2: 65–74. Reprinted as Steven J. Dick. 2009. "Bringing Culture to Cosmos: The Postbiological Universe." In *Cosmos & Culture: Cultural Evolution in a Cosmic Context,* ed. Steven J. Dick and Mark Lupisella, 463–488. Washington, DC: NASA.

Dick, Steven J. 2009. "Cosmic Evolution: History, Culture and Human Destiny." In *Cosmos & Culture: Cultural Evolution in a Cosmic Context,* ed. Steven J. Dick, and Mark Lupisella, 25–59. Washington, DC: NASA.

Dick, Steven J., and Mark Lupisella, eds. 2009. *Cosmos & Culture: Cultural Evolution in a Cosmic Context.* Washington, DC: NASA. Also available online. Accessed July 29, 2012. http://history.nasa.gov/SP-4802.pdf.

Dick, Steven J., and James E. Strick. 2004. *The Living Universe: NASA and the Development of Astrobiology.* New Brunswick, NJ: Rutgers Univ. Press.

DiGregorio, Barry E., Gilbert Levin, and Patricia A. Stratt. 1997. *Mars: The Living Planet.* Berkeley, CA: Frog Books.

Dobzhansky, Theodosius. 1972. "Darwinian Evolution and the Problem of Extraterrestrial Life." *Perspectives in Biology and Medicine* 15: 157–175.

Duve, Christian de. 1995. *Vital Dust: Life as a Cosmic Imperative.* New York: Basic Books.

Dyson, Freeman. 1988. *Infinite in All Directions.* New York: Harper and Row.

Eddington, Arthur S. 1929. *The Nature of the Physical World.* New York: Macmillan.

Ezell, Edward C., and Lin Ezell. 1984. *On Mars: Exploration of the Red Planet, 1958–1978.* Washington, DC: NASA.

Fry, Iris. 2000. *The Emergence of Life on Earth: A Historical and Scientific Overview.* New Brunswick, NJ: Rutgers Univ. Press.

Gatewood, George, and Heinrich Eichhorn. 1973. "An Unsuccessful Search for a Planetary Companion of Barnard's Star (BD + 43561)." *Astronomical Journal* 78: 769–776.

Good, Gregory. 2000. "The Assembly of Geophysics: Scientific Disciplines as Frameworks of Consensus." *Studies in the History and Philosophy of Modern Physics* 31: 259–292.

Gould, Stephen J. 1989. *Wonderful Life: The Burgess Shale and the Nature of History.* New York: W. W. Norton.

Gronstal, Aaron L. 2013. "Extraterrestrial Life in the Microbial Age." In *Astrobiology, History, and Society: Life Beyond Earth and the Impact of Discovery*, ed. Douglas A. Vakoch. Heidelberg: Springer.

Guthke, Karl S. 1990. *The Last Frontier: Imagining Other Worlds from the Copernican Revolution to Modern Science Fiction.* Ithaca, NY: Cornell Univ. Press.

Hegel, Georg W. F. 1832. *Lectures on the Philosophy of History.* Introduction, section 8. Accessed July 30 , 2012. http://www.marxists.org/reference/archive/hegel/works/hi/history2.htm#II.

Horowitz, Norman H. 1977. "The Search for Life on Mars." *Scientific American* 237: 52–61.

Horowitz, Norman H. 1986. *To Utopia and Back: The Search for Life in the Solar System.* New York: W. H. Freeman.

Hoyle, Fred. 1983. *The Intelligent Universe: A New View of Creation and Evolution.* New York: Holt, Rinehart and Winston.

Huxley, Aldous. 1959. "Case of Voluntary Ignorance." In *Collected Essays.* New York: Harper.

Impey, Chris. 2013. "The First Thousand Exoplanets: Two Decades of Excitement and Discovery." In *Astrobiology, History, and Society: Life Beyond Earth and the Impact of Discovery*, ed. Douglas A. Vakoch. Heidelberg: Springer.

Jeans, James. 1917. *The Motion of Tidally Distorted Masses.* Memoirs of the Royal Astronomical Society. London: Royal Astronomical Society.

Jeans, James. 1919. *Problems of Cosmogony and Stellar Dynamics.* Cambridge: Cambridge Univ. Press.

Jeans, James. 1923. *The Nebular Hypothesis and Modern Cosmogony.* Oxford: Oxford Univ. Press.

Jeans, James. 1930. "Life and the Universe." In *The Universe Around Us.* Cambridge: Cambridge Univ. Press.

Jeans, James. 1942a. "Is There Life on Other Worlds?" *Science* 95: 589.

Jeans, James. 1942b. "Origin of the Solar System." *Nature* 149: 695.

Klein, Harold P. 1977. "The Viking Biological Investigation: General Aspects." *Journal of Geophysical Research* 82: 4677–4680.

Kohler, Robert E. 1982. *From Medical Chemistry to Biochemistry.* Cambridge: Cambridge Univ. Press.

Koyré, Alexandre. 1957. *From the Closed World to the Infinite Universe.* Baltimore: Johns Hopkins Univ. Press.

Kuiper, Gerard P., ed. 1949. *The Atmospheres of the Earth and Planets: Papers Presented at the Fiftieth Anniversary Symposium of the Yerkes Observatory, September, 1947.* Chicago: University of Chicago Press.

Kuiper, Gerard P., ed. 1952. *The Atmospheres of the Earth and Planets*, (2nd edition). Chicago: University of Chicago Press.

Kuiper, Gerard P. 1955. "On the Martian Surface Features." *Publications of the Astronomical Society of the Pacific* 67: 271–282.

Kumar, Shiv S. 1967. "On Planets and Black Dwarfs." *Icarus* 6: 136–137.

Levin, Gilbert V., and Patricia A. Straat. 1976. "Viking Labeled Release Biology Experiment: Interim Results." *Science* 194: 1322–1329.

Levin, Gilbert V. 2011. Private communication.

Lovejoy, Arthur O. 1936. *The Great Chain of Being.* Cambridge, MA: Harvard Univ. Press.

Marcy, Geoff, and Paul Butler. 1998. "Detection of Extrasolar Giant Planets." *Annual Review of Astronomy and Astrophysics* 36: 57–97.

Marozzi, Justin. 2008. *The Way of Herodotus: Travels with the Man Who Invented History.* Cambridge, MA: Da Capo Press.

McKay, David, Everett K. Gibson, Jr., et al. 1996. "Search for Past Life on Mars: Possible Relic Biogenic Activity in Martian Meteorite ALH84001." *Science* 273: 924–930.

Mayor, Michel, and Didier Queloz. 1995. "A Jupiter-Mass Companion to a Solar-Type Star." *Nature* 378: 355–359.

Mayr, Ernst. 1985. "The Probability of Extraterrestrial Intelligent Life." In Ernst Mayr. 1988. *Toward a New Philosophy of Biology: Observations of an Evolutionist*, 67–74. Cambridge, MA: Harvard Univ. Press.

Mayr, Ernst. 1988. *Toward a New Philosophy of Biology: Observations of an Evolutionist.* Cambridge, MA: Harvard Univ. Press.

Miller, Stanley, and Harold C. Urey. 1953. "Organic Compound Synthesis on the Primitive Earth." *Science* 130: 245–251.

Monod, Jacques. 1971. *Chance and Necessity*. New York: Vintage Books.

Navarro-González, Rafael, Edgar Vargas, Jose de la Rosa, Alejandro C. Raga, and Chris P. McKay. 2010. "Reanalysis of the Viking Results Suggests Perchlorate and Organics at Mid-latitudes on Mars." *Journal of Geophysical Research* 115: E12010.

Oparin, Alexander I. 1924. *Proiskhozhdenie zhinzy.* Moscow. trans. by Ann Synge as "The Origin of Life." In John D. Bernal. 1967. *The Origin of Life*, Appendix I, 199–234. Cleveland and New York: World Publishing Company. Reprinted in David W. Deamer and Gail R. Fleischaker. 1994. *Origins of Life: The Central Concepts*, 31–71. Boston: Jones and Bartlett Publishers.

Oparin, Alexander I. 1936. *Vozhiknovenie zhizny na aemle.* trans. as 1938. *The Origin of Life.* London. Republished 1952. New York: Dover.

Petit, Edward, and Seth B. Nicholson. 1924. "Radiation Measures on the Planet Mars." *Publications of the Astronomical Society of the Pacific* 36: 269–272.

Rea, D. G., T. Belsky, and Melvin Calvin. 1963. "Interpretation of the 3-to 4- Micron Infrared Spectrum of Mars." *Science* 141: 923–937.

Rea, D. G., Brian O'Leary, and William M. Sinton. 1965. "The Origin of the 3.58 and 3.69-Micron Minima in the Infrared Spectrum." *Science* 147: 1286–1288.

Reuyl, Dirk, and Erik Holmberg. 1943. "On the Existence of a Third Component in the System 70 Ophiuchi." *Astrophysical Journal* 97: 41–45.

Russell, Henry Norris. 1935. *The Solar System and Its Origin.* New York: Macmillan.

Russell, Henry Norris. 1943. "Anthropocentrism's Demise." *Scientific American*, July, 18–19.

Russell, Henry Norris, Raymond Smith Dugan, and John Quincy Stewart. 1945. *Astronomy: A Revision of Young's Manual of Astronomy.* Boston: Ginn and Company.

Sagan, Carl. 1977. *The Dragons of Eden: Speculations on the Evolution of Human Intelligence.* New York: Random House.

Shklovskii, Ioseph S. 1965. "Multiplicity of Inhabited Worlds and the Problem of Interstellar Communications." In *Extraterrestrial Civilizations*, ed. G. V. Tovmasyan. Erevan: Akademii Nauk Armyanskoi SSR. English trans., 1965. Jerusalem: Israel Program for Scientific Translations, 5–13.

Simpson, George G. 1964. "The Non-Prevalence of Humanoids." *Science* 143: 769–775. Reprinted in *This View of Life: The World of an Evolutionist*, 253–271. New York: Harcourt, Brace & World.

Sinton, William M. 1957. "Spectroscopic Evidence for Vegetation on Mars." *Astrophysical Journal* 126: 231–239.

Sinton, William M. 1959. "Further Evidence of Vegetation on Mars." *Science* 130: 1234–1237.

Slipher, Edward C. 1927. "Atmospheric and Surface Phenomena on Mars." *Publications of the Astronomical Society of the Pacific* 39: 209–216.

Spinrad, Hyron, Guido Münch, and L.D. Kaplan. 1963. "The Detection of Water Vapor on Mars." *Astrophysical Journal* 137: 1319–1321.

Sternberg, Robert J., ed. 2000. *Handbook of Intelligence.* Cambridge: Cambridge Univ. Press.

Sternberg, Robert J. 2002. "The Search for Criteria: Why Study the Evolution of Intelligence." In *The Evolution of Intelligence*, ed. R. J. Sternberg, and J. C. Kaufman, 1–8. Mahwah, NJ: Lawrence Erlbaum Associates.

Strick, James E. 2004. "Creating a Cosmic Discipline: The Crystallization and Consolidation of Exobiology, 1957–1973." *Journal of the History of Biology* 37: 131–180.

Sullivan, Woodruff T., III. 2009. "Cosmic Noise: A History of Early Radio Astronomy." Cambridge: Cambridge Univ. Press.

Tikhov, Gavriil A. 1955. "Is Life Possible on Other Planets?" *Journal of the British Astronomical Association* 65: 193–204.

Tikhov, Gavriil A. 1960. *Principal Works: Astrobotany and Astrophysics, 1912–1957*. Washington, DC: U.S. Joint Publications Research Service.

Trumpler, Robert J. 1927. "Observations of Mars at the Opposition of 1924." *Lick Observatory Bulletin* 387: 19–45.

Vakoch, Douglas A. 2013. "Life Beyond Earth and the Evolutionary Synthesis." In *Astrobiology, History, and Society: Life Beyond Earth and the Impact of Discovery*, ed. Douglas A. Vakoch. Heidelberg: Springer.

van de Kamp, Peter. 1956. "Planetary Companions of Stars." *Vistas in Astronomy* 2: 1040–1048.

van de Kamp, Peter. 1963. "Astrometric Study of Barnard's Star from Plates Taken with the 24-Inch Sproul Refractor." *Astronomical Journal* 68: 515–521.

van de Kamp, Peter. 1975. "Unseen Astrometric Companions of Stars." *Annual Reviews of Astronomy and Astrophysics* 13: 295–333.

van de Kamp, Peter. 1977. "Barnard's Star 1916–1976: A Sexagesimal Report." *Vistas in Astronomy* 20: 501–521.

Wallace, Alfred R. 1903a. *Man's Place in the Universe*. New York: McClure, Phillips and Co.

Wallace, Alfred R. 1903b. "Man's Place in the Universe." *The Independent* 55: 473–483.

Ward, Peter, and Donald Brownlee. 2000. *Rare Earth: Why Complex Life is Uncommon in the Universe*. New York: Springer.

Wolfe, Audra. 2002. "Germs in Space: Joshua Lederberg, Exobiology, and the Public Imagination, 1958–1964." *Isis* 93: 183–205.

Zakariya, Nasser. 2010. *"Towards a Final Story: Time, Myth and the Origins of the Universe."* PhD dissertation, Harvard University.

Chapter 8
The Creator of Astrobotany, Gavriil Adrianovich Tikhov

Danielle Briot

Abstract The Russian astronomer, Gavriil Adrianovich Tikhov (1875–1960), was one of the main pioneers of astrobiology and was the creator of astrobotany. From 1906 to 1941, he began his career as an astronomer in the Pulkovo Observatory, near Saint Petersburg, and then moved to Alma-Ata (Kazakhstan) until the end of his life. He specialized in many different fields of astronomy: besides astrobotany and astrobiology, he studied the Sun, the planets, the comets, the blue color and polarization of the sky, the Earthshine, variable stars, as well as interstellar absorption. He designed new instruments and wrote more than 230 scientific papers. As early as 1914, on the basis of observations of Earthshine, he concluded that Earth seen from space has to be seen with a pale blue color. Tikhov's main research was focused on the search for extraterrestrial life, particularly the presence of vegetation on Mars. At this time, many astronomers believed in the existence of canals on the Martian surface. Seasonal variations of color on the surface of Mars were often interpreted as changes of vegetation, as on Earth. From 1909, he observed Mars during favorable configurations, that is to say Mars oppositions, first using filters and then with a spectrograph. However, he failed to detect chlorophyll in Mars' spectra. So he decided to go looking for plants with no chlorophyll, especially plants growing in extreme environments like on Mars and to study and measure their reflectance spectrum. Tikhov, who was one of the first scientists to use the word "astrobiology," coined the word "astrobotany" in 1945. In 1947 he founded a Department of Astrobotany at Alma-Ata Observatory, where students, biologists, botanists, and physicists joined to study the reflectance spectra of plants growing in conditions similar to those found on Mars. Expeditions were organized in very cold or very dry places with this aim in view. After he died, the Department of Astrobotany was dismantled. After nearly half a century of lack of interest, the work of G.A. Tikhov appears really modern. Nowadays, in order to prepare for the detection of life in remote extrasolar planets, astronomers observe

D. Briot (✉)
Observatoire de Paris (GEPI), 61 avenue de l'Observatoire, Paris, France
e-mail: Danielle.Briot@obspm.fr

D. A. Vakoch (ed.), *Astrobiology, History, and Society*, Advances in Astrobiology and Biogeophysics, DOI: 10.1007/978-3-642-35983-5_8,
© Springer-Verlag Berlin Heidelberg 2013

Earthshine to detect the spectrum of terrestrial chlorophyll, and specifically the Vegetation Red Edge (VRE) in the near infrared. Although the VRE is only a few percent, it is higher when continents covered by vegetation are facing the Moon, and lower in the case of oceans. In order to study the daily variation of chlorophyll spectra in the Earthshine according the part of Earth facing the Moon, we developed an observational program from the scientific station Concordia in Antarctica.

8.1 Introduction

The history of astrobiology is fascinating, particularly because some ideas developed by its original pioneers were forgotten for a long period. With the discovery of extrasolar planets and studies of possible conditions of life on these planets, we are rediscovering and applying many of these original ideas. The work of Gavriil Adrianovich Tikhov is a beautiful example of the reuse of "old" original ideas.

8.2 Biography and Early Works

Gavriil Adrianovich Tikhov (1875–1960), whose name is sometimes written in the Latin alphabet as Tichow or Tikhoff, was born in a small village near Minsk, in Belarus. After he completed his higher studies at the University of Moscow in 1897, he traveled to France to take some courses at the Sorbonne University. There his professors were Emile Picard for mathematics, Paul Appell for analytical mechanics, Henri Poincaré for theoretical physics, Charles Wolf for astronomy, and Gaston Bonnier for botany, all being very famous scientists. He met astronomers at the Paris Observatory, and he collaborated with Jules Janssen, then director of the Meudon Observatory, which consequently played a crucial part in his career. He went up in a balloon to observe the Leonid meteor shower, and went up twice to the observatory established by Janssen on top of the Mont Blanc, the highest mountain in Western Europe, with an altitude of 4,810 m at its highest point, to observe the Sun.

Some time after he returned to Russia, he became an astronomer at the Pulkovo Observatory, near Saint Petersburg, in 1906. During the First World War, in 1917 he was mobilized in the army and became a pilot watcher. Then he came back to the Pulkovo Observatory.

During the Russian revolution in 1917 and the civil war, the conditions of life became very difficult. Battles took place on the field of the observatory, the first one in 1917 between Cossacks and Communists, and the second one in 1919 between Communists and White Russians, continuously for one week. Food, clothes, shoes, and wood for heating were lacking. A letter from an anonymous Russian astronomer from Pulkovo published in *Popular Astronomy* in October 1921 (Anonymous 1921) describes the tragic way of life during this time. In 1941, just before the siege of Leningrad (Saint Petersburg) by German troops, which took nearly 900 days and completely destroyed the Pulkovo Observatory, he went away to participate in an expedition to Alma-Ata Observatory (Kazakhstan) to observe a total solar eclipse. He used for that a special four-camera astrograph, and the observations were very successful. After the war and until his death in 1960, Tikhov worked in Alma-Ata, where an Institute of Astronomy and Physics was created.

His range of research was very broad. He studied the Sun, planets, comets, the blueness of the sky and its polarization, radial velocities of stars, Earthshine, variable stars, interstellar absorption, and, of course, astrobotany and astrobiology. He also developed new astronomical instruments and was especially interested in photographic photometry. In total, he wrote more than 230 publications. In this chapter we will mainly report on Tikhov's work on Earthshine, astrobotany, and astrobiology.

8.3 Study of Earthshine

Tikhov studied Earthshine in 1914. Earthshine, also called Ashen Light, is the faint reddish light that can be seen at night within the thin crescent Moon, just at the beginning or at the end of the Moon's cycle. It corresponds to the Earth's light backscattered by the non-Sunlit Moon. It should be pointed out that the "phases of the Earth" as seen from the Moon are the reverse of the phases of the Moon as seen from the Earth. So a lighted Earth faces the Moon when the Moon phase is near the New Moon. During the periods close to the New Moon, the Sun light arriving on the Earth is reflected in the direction of the Moon, where it is reflected again, and finally comes back to the Earth, where it can be observed from the non-lighted part (night). During this path, light crosses the Earth's atmosphere three times. Because of the larger size of Earth and the larger albedo of Earth (~0.37) compared to that of the Moon (~0.12), the Earthshine on the Moon is approximately 40 times more intense than moonlight on Earth. Since Greek philosophers, several interpretations have been suggested about the origin of the faint light seen in the dark part of the Moon, most notably either a translucent or luminous Moon, or a Moon lighted by stars or by Venus. The first correct explanation of Ashen Light or "secondary" light is most often attributed to Galileo in the *Sidereus Nuncius* (*The Sidereal Messenger*) (Galilei 1610). However, the same correct explanation can be seen in some manuscripts of Leonardo da Vinci, the *Codex Leicester* written between (Vinci 1506–1509), that is to say one century before, but not yet published at that time.

As early as 1914, long before any space mission, Tikhov concluded from his study of Earthshine that if we were able to see the Earth space, it would have a pale blue color because of Rayleigh scattering (Tikhoff 1914). It is remarkable to note that a "pale blue dot" is exactly the term used to describe a photograph of the planet Earth taken in 1990 by the Voyager 1 spacecraft from a distance of about 6 billion km, at the request of Carl Sagan. This is now a very famous expression that has become the symbol of the image of a terrestrial exoplanet that we expect to see in a near future. More a decade later than Tikhov's observations, Earthshine has been regularly observed since 1926 by the French astronomer André Danjon and collaborators (Danjon 1928, 1954; Dubois 1947).

8.4 Birth of Astrobotany

The most interesting part of Tikhov's research was undoubtedly his investigations on Mars and the search for life. In the early twentieth century, many astronomers, but not all, believed that there were canals on Mars. The word "canal" was first used by Angelo Secchi (1858, p. 73), who referred to dark elongated structures on the surface of Mars. It is difficult to know if by "canals" Secchi meant natural or artificial structures. Since 1877, nearly two decades later, the same word was used extensively by Schiaparelli. However, let us note that Schiaparelli mentioned, at least in some papers, that the he did not know what precisely defined objects correspond to the so-called "canals" (Schiaparelli 1882a, b). The dark color of some parts of the planet was first explained by the presence of "seas," but when some "canals" were seen across the "seas," the dark color of "seas" was attributed to some vegetation. Moreover the changes of Mars' colors according to various seasons of Mars suggest a vegetation similar to Earth's, where vegetation colors change with the seasons.

At these times, before the era of space instruments and very large telescopes, more information could be obtained when Mars was as close as possible to the Earth. Considering that the Earth and Mars are rotating around the Sun, the distance between Earth and Mars is the smallest when the Sun, the Earth and Mars are collinear. Such a configuration is named a Mars opposition, and happens approximately every 2 years and 49 days. However, because planetary trajectories are not exactly circular, but elliptical—particularly Mars' trajectory—some oppositions are more favorable than the others, when the Earth-Mars distance is shorter. These are the perihelic oppositions, that is to say when the Earth and Mars are at their closest distances to the Sun. Perihelic oppositions occur every 15 or 17 years. The most favorable oppositions occur every 15, 32, 47, 79, and 205 years. So it was very important that astronomers attempting to solve the mysteries of Mars should carefully prepare and carry out observations at these moments. In 1909, the distance between Mars and Earth was a little more than 58 million km. During the opposition of 1909, Tikhov made some observations of Mars using colored filters that he made himself. During the next oppositions in 1918 and 1920, he used a spectrograph with the aim of detecting the chlorophyll spectrum, but without obtaining the expected result. The reflectance spectrum of vegetation presents a small "bump" in green wavelengths, making grass appear green to our eyes, and yielding a particularly sharp edge in the near infrared at about 725 nm, the so-called Vegetation Red Edge (VRE). This feature is due to photosynthetic pigments. Tikhov could not to identify chlorophyll in his Martian spectra. Considering the very hard physical conditions on Mars, he thought that some plants could grow on Mars without chlorophyll. He described the probable landscape of Mars as a low blue vegetation growing in a red soil, with the plants being some kind of mosses or lichens. His aim was to search for terrestrial plants growing in Mars-like living conditions, and to study their reflectance spectrum, to determine when chlorophyllian pigments can be unusual or even missing. He began to compare visible and near infrared spectra of deciduous species and those of evergreen species, and he studied various conifers from different countries.

Tikhov coined the word "astrobotany" for the first time in 1945 during a communication about his observations of Mars. He was also among the first ones to use the word "astrobiology." The study of reflectance spectra of various plants became so important that in 1947, he founded in Alma-Ata a Department of Astrobotany with young students, biologists, botanists, and physicists. The purpose was to study the reflectance of plants growing in specific places, such as high mountains or polar regions, where the environmental conditions are extreme and could resemble those on Mars. Some expeditions were sent to the Pamir Mountains, where the mean temperature is −1 °C, and where the variations of ground temperature can reach 60 °C during a day, and 102° in a year. The relative humidity in this region hardly reaches 9–15 %. Moreover there are some geysers where the water temperature equals 71 °C. Other expeditions were organized in the Zailiskiy Alatau Mountains, near Alma-Ata, in the cold desert of Central Tien-Shan, to the Ob mouth, in Yakutia, and up to the shores of Arctic Polar Sea.

Very numerous plants were studied by spectroscopy, during various seasons and under various physical conditions, always to demonstrate that vegetation was possible on Mars. Tikhov counted more than 200 species of plants growing in Siberia in a climate as harsh as that of some parts of Mars. We briefly recall some of his studies and results:

- He showed that, under very low temperatures, the absorption band of chlorophyll can decrease and even completely disappear, sometimes for the same plant.
- He studied the color of plants and demonstrated that plants growing at very low temperatures can have other colors than green.
- He demonstrated and studied the fluorescence of plants in the infrared and studied the heat that plants can produce by this process, and how this can help the plants be more adaptable.
- He observed that plants can adapt to very hard climatic conditions, low temperatures, and lack of oxygen, changing their optical properties and increasing or decreasing their solar radiation absorption.
- He studied how to resolve the problem of the lack of an ozone layer on Mars.
- He studied the conditions of the primitive Earth and the evolution of plants (paleobotany), and he made hypotheses about the possible evolution of the climate of Mars.

These results culminated in many interesting publications (e.g., Tikhov 1947, 1955, 1960). Additional information can also be found in Omarov and Tashenov (2005).

8.5 From Astrobotany to Astrobiology and Cosmobiology

Assuming that primitive microorganisms have in common remarkable properties of adaptation when put in different environmental conditions, astrobotanists started to establish a wide range of observations and research fields. They studied the extreme conditions required for life on Earth, drawing from the studies of microbiologists and investigating whether the physical conditions on various planets would be opportune for the emergence of primitive life. The field of astrobotany thus broadened to form astrobiology, then cosmobiology. Tikhov used the word cosmobiology as a generalization for research about life on other planets, and particularly planets orbiting around other stars than our Sun.

We have to note that Tikhov did not claim that life exists on Mars, but he thought that life develops inevitably under favorable conditions. He noted that if conditions on Mars were the same as on Earth, it would be much easier to demonstrate the presence of life on Mars, but in this case many observations about optical properties of plants under very difficult climatic conditions would never be realized. In (Tikhov 1949), he published a book entitled *Astrobotany* and in (Tikhov 1953) another named *Astrobiology*. Unfortunately, following the death of Tikhov in 1960, the Department of Astrobotany of Alma Alta was dissolved. At the beginning, attempts were made to maintain Tikhov's house, the observatory, and the so-called "astrobotany garden," but a few years later they were all very sadly destroyed, despite numerous protests (see Tejfel 2010). The name of Tikhov has been given to a lunar crater, to a Martian crater, and to an asteroid (2251).

8.6 Contemporary Studies

At the same time, other astronomers also investigated the hypothesis of vegetation on Mars, studying chlorophyll spectra, such as Slipher (1924), Millman (1939) and Kuiper (1949, p. 339). Sinton (1957a, b, 1958) studied and compared reflectance spectra of several plants and concluded there is evidence of vegetation on Mars. However, his studies concerned the near infrared spectrum where vegetation reflectance curves could be confused with those of soils, and unlike Tikhov, he investigated only a few plants. It should be noted that there were other astronomers who never believed in the reality of Martian "canals" and in the presence of vegetation on this planet.

To estimate the importance of Tikhov in his own time, we reviewed the astronomical bibliography that existed then. From 1899 to 1968 the Astronomischer Rechen-Institut in Heidelberg, Germany published every year a bibliographic book *Astronomischer Jahresbericht*, containing as exhaustive as possible a thematic recording of astronomical literature throughout the world during that year. The language used is nearly always German. These books were replaced by *Astronomy and Astrophysics Abstracts* from 1969 to 2000. Authors and keywords are listed in each volume. The study of various keywords is a very good index to determine the areas of astronomy that are or are not considered as important during that year. The word "astrobotany" appears as a keyword in 1949, for Tikhov's book so entitled, and again from 1951 to 1954, whereas the keyword "astrobiology" appears only once in 1953. So the word "astrobotany" appears as a keyword before the word "astrobiology." It is important to note that these words are only keywords and not headings.

8.7 Present Studies

The work of Tikhov is particularly interesting for astrobiologists. At the present time, astronomers—without or before knowing Tikhov's research—independently use again his methods in astrobotany as well as in observations of Earthshine. Of course we know now that there is no vegetation on Mars, but we can reasonably imagine that vegetation is present on other celestial objects.

In 1995, the discovery of the first extrasolar planet 51 Pegasi b by Michel Mayor and Didier Queloz was tremendous news for the world of astronomers, as well as for people in general. This discovery provided the answer to a philosophical question that humankind has wondered for more than 2,000 years. This first planet discovered was very rapidly followed by many others. A new very promising domain of astronomy is being developed very quickly, both in theoretical studies and in instrumentation. Many ground and space observational programs are planned and carried out.

In the near future, we can expect to see the image of a terrestrial planet located in the habitable zone. What kind of life could be detected on such a planet? How is it possible to detect life at the distance of an extrasolar planet that appears as merely a dot in the sky? So far, we don't know anything about extraterrestrial life and the form it may take. The only possibility is to study and prepare for the detection of life similar to the only life we know, that is terrestrial life. So we search for evidence of life on the only planet known to harbor life, that is our Earth. To observe Earth under the same conditions as an exoplanet, we use Earthshine, where each point on the Moon reflects the light of the Earth facing the Moon. This idea was suggested by Jean Schneider, from Paris-Meudon Observatory, in 1998, without knowing any previous similar idea. Actually, as early as 1912, Arcichovsky suggested to look for chlorophyll absorption features in the Earthshine spectrum to calibrate this pigment in the spectrum of other planets, but the spectral resolution of spectrometers at that time was not sufficient for that purpose. This approach was completely forgotten until 1998.

No known animal life could be detected at the distance of an extrasolar planet, in contrast to vegetation that can cover vast areas. Very probably, the vegetation on extrasolar planets would be very different from that on Earth. For example, Kiang

(2008) studies the colors and characteristics that plants would have on extrasolar planets, particularly according to the temperature of the central star of the planetary system. However, today Earth's vegetation is the only target we can study and observe. A very distinctive signature of chlorophyll is the Vegetation Red Edge (VRE), around 0.7 μm, in the green vegetation reflectance spectra. A small bump is also present at 0.5 μm, which explains the green color of plants seen by our eyes. However, the VRE in the near infrared is much larger. If our eyes could see in the near infrared, plants would be seen as "infrared," something like red and bright. Near infrared photos of landscapes show grass and trees very bright, looking as if they are snow covered. The VRE is due to the darkness in the red part of the visible spectrum because of absorption by chlorophyll, and the high reflectance in the near infrared. More information about the reflectance of plants can be obtained, for example, in Ustin et al. (2009).

So Earthshine has been observed since 1999 to detect atmospheric biomarkers and especially the Vegetation Red Edge (VRE) in the near infrared spectrum of chlorophyll. The first results were obtained simultaneously by Arnold et al. (2002) and Woolf et al. (2002). A review of studies and results obtained from 2001 to 2006 is presented by Arnold (2008). Observations showed that the vegetation signature is detectable in an integrated Earth spectrum, however, this signature is weak, only a few percent (0–5 % range). It depends on many factors including the ratio between ocean and continents in view from the Moon, or the cloud cover above the vegetation area during observation. By using observations at the New Technology Telescope (NTT) at the European Southern Observatory, Hamdani et al. (2006) showed that the chlorophyll Vegetation Red Edge is larger when continents are facing the Moon, instead of the Pacific Ocean. These observations detect also in the red side of the Earth reflectance spectrum, the presence of O_2 and H_2O absorption bands, and in the blue side the Huggins and Chappuis ozone (O_3) absorption bands. The higher reflectance observed in the blue is due to Rayleigh scattering, which explains the blue color of our planet, as discovered nearly a century ago by Tikhov (1914).

At mean or low latitudes, it is well known that Earthshine observations are possible during twilight, i.e., just after sunset or just before sunrise. So observations last a short time, and roughly speaking, for one telescope, only two enlightened parts of the Earth face the Moon: either the part located at the West of the observing telescope for evening observations (beginning of the lunar cycle), or the part of the Earth located at the East of the observing telescope for morning observations (last days of the lunar cycle). However, there are other possibilities. If observations are made from a site located at high latitudes, conditions of Earthshine observations are different. Six to eight times a year, around equinoxes, Earthshine can be observed for several hours, and even—at very high latitudes—during a full Earth rotation (total diurnal cycle). Observations around the equinox in March correspond to the last days of the lunar cycle, and observations around the equinox in September correspond to the first days of the lunar cycle. Note that Antarctica being located in the southern hemisphere, the equinox of March corresponds to the autumnal equinox and the equinox of September corresponds to the spring equinox. During these long observing windows, different landscapes successively face the Moon, while the Earth rotates. Consequently, the Earthshine corresponding to various parts of our Earth can be studied: continents with vegetation or oceans. In order to obtain more detailed results, we performed observations of Earthshine from the French-Italian scientific Concordia station, established in Dome C, in Antarctica, where the polar night last 3 months. The geographical coordinates are 75° South latitude, 123° East longitude, with an altitude of 3,220 m. The mean air temperature is −50.8 °C, and the lowest is −84.4 °C. A dedicated instrumentation called LUCAS (*LUmière Cendrée en Antarctique par Spectroscopie*) was designed and built. After a period of testing, this instrument was adapted to any special needs generated by the extreme climatary conditions that prevail at Dome C. The feedback we got from the first time we observed was very important for detecting, analyzing, and correcting instrumental problems due to extreme temperature and physical conditions. Successful Earthshine observations were carried out, with runs lasting up to 8 h. As in Tikhov's time, long distance missions again provide a way to detect vegetation on other planets.

8.8 Conclusion

Tikhov's heritage is considerable. Unfortunately, it is largely misunderstood and quite unknown for several reasons. The first one is that during Tikhov's life, there was very little communication between countries located on opposite sides of the Iron Curtain. The papers of Tikhov and his collaborators were published in Russian, and in scientific journals from Alma-Ata or Moscow. Another reason concerns the subject of his studies. When it was realized that canals of Mars did not exist, almost all research about plant life on Mars were forgotten. Moreover, the stopping of activity of the Department of Astrobotany after Tikhov's death led to his research sinking into total oblivion.

Studies of pigment absorption features in order to investigate the presence of vegetation on other planets, as well as observations of Earthshine, represent good examples of scientific topics that were first carefully investigated, then abandoned, and finally came back to the forefront again. Concerning observations of Earthshine, some geophysical applications also exist, according to the recommendations made by the NASA Navigator Program: "Continued observations of Earthshine are needed to discern diurnal, seasonal, and interannual variations."

As astrobiology has expanded and as the search for extrasolar life has become a hot topic, it can be seen how modern the studies of Gavriil Adrianovich Tikhov were, even though they were conducted more than a half century ago. Of course, techniques have progressed considerably in the meantime. However, many fundamental ideas about astrobiology were developed in the first part of the twentieth century. Tikhov understood very early that astrobiology was an interdiciplinary science and so founded a team composed of scientists specialized in such different fields as astronomy and botany. He was really a remarkable pioneer.

Tikhov used to recall an ancient maxim, which was appreciated in Pulkovo Observatory: "If the results of a research correspond to what it is expected, it is a pleasure, if not, it gets interesting" (Tikhov 1960, p. 156).

Acknowledgments It is a pleasure to thank my brother Alain Briot for having introduced me to Tikhov's life and research. I am very much indebted to Jean Schneider and Florence Raulin for helpful discussions and their precious encouragement. Thanks are also due to Patrick Rocher for calculations of Mars oppositions and to Stéphane Jacquemoud and Patrick François for their careful reading of the manuscript.

References

Anonymous. 1921. "An Astronomer's Life in Russia." *Popular Astronomy* 29: 524–525.
Arcichovsky, Vladimir M. 1912. "Auf der Suche nach Chlorophyll auf den Planeten." *Annales de l'Institut Polytechnique Don Cesarevitch Alexis a Novotcherkassk* 1(17): 195–216.
Arnold, Luc, Sophie Gillet, Olivier Lardière, Pierre Riaud, and Jean Schneider. 2002. "A Test for the Search for Life on Extrasolar Planets: Looking for the Terrestrial Vegetation Signature on the Earthshine Spectrum." *Astronomy and Astrophysics* 392: 231–237.
Arnold, Luc. 2008. "Earthshine Observation of Vegetation and Implication for Life Detection on Other Planets: A Review of 2001–2006 Works." *Space Science Review* 135: 323–333.
Danjon, André. 1928. "Recherches de photométrie La lumière cendrée et l'albedo de la Terre," chapter 14. In vol. 2, 165–180. Annales de l'Observatoire de Strasbourg.
Danjon, André. 1954. "Albedo, Color and Polarization of the Earth." In *The Earth as a Planet: The Solar System*, vol. 2, ed. Gerard P. Kuiper, 726–738. Chicago: University of Chicago Press.
Dubois, J. 1947. "Sur l'albedo de la Terre." *Bulletin Astronomique* 13: 193–196.
Galilei, Galileo. 1610. *Sidereus Nuncius*. Venice: Thomas Baglioni.
Hamdani, Slim, Luc Arnold, Cédric Foellmi, Jérôme Berthier, Malvina Billeres, Danielle Briot, Patrick François, Pierre Riaud, and Jean Schneider. 2006. "Biomarkers in Disk-averaged Near-UV to Near-IR Earth Spectra Using Earthshine Observations." *Astronomy and Astrophysics* 460: 617–624.
Kiang, Nancy Y. 2008. "The Color of Plants on Other Worlds." *Scientific American* 298(4): 48–55.
Kuiper, Gerard P. 1949. "Survey of Planetary Atmosphere." In *The Atmospheres of the Earth and planets*, ed. Gerard P. Kuiper, 305–345. Chicago: University of Chicago Press.
Mayor, Michel, and Didier Queloz. 1995. "A Jupiter-mass Companion to a Solar-type Star." *Nature* 378: 355–359.
Millman, Peter M. 1939. "Is There Vegetation on Mars?" *The Sky* 3(10): 10–11.
Omarov, Tuken B., and Bulat T. Tashenov. 2005. "Tikhov's Astrobotany as a Prelude to Modern Astrobiology." In *Perspectives in Astrobiology*, vol. 366, ed. Richard B. Hoover, Alexei Yu. Rozanov, and Roland Paepe, 86–87. NATO science series, series 1: Life and Behavioural Sciences.
Schiaparelli, Giovanni V. 1882a. "On Some Observations of Saturn and Mars." *The Observatory* 5: 221–224.
Schiaparelli, Giovanni V. 1882b. "Découvertes nouvelles sur la planète Mars." *L'Astronomie* 1: 216–221.
Secchi, Angelo. 1858. "Schreiben des Herrn Professors Secchi, Directors der Sternwarte des Collegio Romano, an den Herausgeber." *Astronomische Nachrichten* 49: 73–74.
Sinton, William M. 1957a. "Spectroscopic Evidence for Vegetation on Mars." *Astrophysical Journal* 126: 231–239.
Sinton, William M. 1957b. "Vegetation on Mars?" *Sky and Telescope* 16: 275.

Sinton, William M. 1958. "Spectroscopic Evidence of Vegetation on Mars." *Publications of the Astronomical Society of the Pacific* 70: 50–56.

Slipher, Vesto Melvin. 1924. "Observations of Mars in 1924 Made at the Lowell Observatory: II. Spectrum Observations of Mars." *Publications of the Astronomical Society of the Pacific* 36: 261–262.

Tejfel, Victor. 2010. "Gavriil Adrianovich Tikhov (1875–1960): A Pioneer in Astrobiology." *Highlights in Astronomy* 15: 720–721.

Tikhoff, Gavriil A. 1914. "Etude de la lumière cendrée de la Lune au moyen des filtres sélecteurs." *Mitteilungen der Nikolai-Hauptsternwarte zu Pulkovo* 62(6): 15–25.

Tikhov, Gavriil A. 1947. *Bulletin Astronomy and Geodesy Society URSS* 1(8): S3–S13 (in Russian).

Tikhov, Gavriil A. 1949. *Astrobotany*. Alma-Ata: Academy of Sciences Kazakh.

Tikhov, Gavriil A. 1953. *Astrobiology*. Moscow: Molodaya Gvardiya (Young Guard).

Tikhov, Gavriil A. 1955. "Is Life Possible on Other Planets?" *Journal of the British Astronomical Association* 65: 193–204.

Tikhov, Gavriil A. 1960. *L'énigme des planets*. Moscow: Editions en langues étrangères.

Ustin, Susan L., Anatoly A. Gitelson, Stéphane Jacquemoud, Michael E. Schaepman, Gregory P. Asner, John A. Gamon, and Pablo J. Zarco-Tejada. 2009. "Retrieval of Qualitative and Quantitative Information About Plant Pigment Systems from High Resolution Spectroscopy." *Remote Sensing of Environment* 113(S1): S67–S77.

Vinci, Leonardo da. 1506–1509. *Codex Leicester*, sheet 2A, folio 2r.

Woolf, Neville J., Paul S. Smith, Wesley A. Traub, and Kenneth W. Jucks. 2002. "The Spectrum of Earthshine: A Pale Blue Dot Observed from the Ground." *Astrophysical Journal* 574: 430–433.

Chapter 9
Life Beyond Earth and the Evolutionary Synthesis

Douglas A. Vakoch

Abstract For many astronomers, the progressive development of life has been seen as a natural occurrence given proper environmental conditions on a planet: even though such beings would not be identical to humans, there would be significant parallels. A striking contrast is seen in writings of nonphysical scientists, who have held more widely differing views. But within this diversity, reasons for differences become more apparent when we see how views about extraterrestrials can be related to the differential emphasis placed on modern evolutionary theory by scientists of various disciplines. One clue to understanding the differences between the biologists, paleontologists, and anthropologists who speculated on extraterrestrials is suggested by noting who wrote on the subject. Given the relatively small number of commentators on the topic, it seems more than coincidental that four of the major contributors to the evolutionary synthesis in the 1930s and 1940s are among them. Upon closer examination it is evident that the exobiological arguments of Theodosius Dobzhansky and George Gaylord Simpson and, less directly, of H. J. Muller and Ernst Mayr are all related to their earlier work in formulating synthetic evolution. By examining the variety of views held by nonphysical scientists, we can see that there were significant disagreements between them about evolution into the 1960s. By the mid-1980s, many believed that "higher" life, particularly intelligent life, probably occurs quite infrequently in the universe; nevertheless, some held out the possibility that convergence of intelligence could occur across worlds. Regardless of the final conclusions these scientists reached about the likely prevalence of extraterrestrial intelligence, the use of evolutionary arguments to support their positions became increasingly common.

D. A. Vakoch (✉)
SETI Institute, Mountain View, CA, USA
e-mail: vakoch@seti.org

D. A. Vakoch (ed.), *Astrobiology, History, and Society*, Advances in Astrobiology and Biogeophysics, DOI: 10.1007/978-3-642-35983-5_9,
© Springer-Verlag Berlin Heidelberg 2013

9.1 Introduction

The notion of extraterrestrial beings of bizarre yet somewhat humanoid forms existed well before science fiction movies became popular. In Christiaan Huygens's *The Celestial Worlds Discover'd*, we can see two poles of thought about life beyond Earth that are reflected in more recent works. That monograph, published posthumously in 1698, depicts possible denizens of other planets as in some ways very similar and also potentially markedly different from humankind.[1] After explaining why "Planetarians" would be upright beings with hands, feet, and eyes, he claimed that their form could still be quite alien:

> Nor does it follow from hence that they must be of the same shape with us. For there is such an infinite possible variety of Figures to be imagined, that both the Oeconomy of the whole Bodies, and every part of them, may be quite distinct and different from ours (Huygens 1968, 74).

Huygens was neither the first nor the last astronomer to speculate on extraterrestrial morphology.[2] But his position *is* representative of his profession. For many astronomers, the progressive development of life has been seen as a natural occurrence given proper environmental conditions on a planet. And even though such beings would not be identical to humans, they have argued, there would be significant parallels. A striking contrast is seen in writings of nonphysical scientists. Members of this latter group hold more widely differing views. But within this diversity, reasons for differences become more apparent when we see how views about extraterrestrials can be related to the differential emphasis placed on modern evolutionary theory by various scientists.

One clue to understanding the differences between the biologists, paleontologists, and anthropologists who speculated on extraterrestrials is suggested by noting who wrote on the subject. Given the relatively small number of commentators on the topic, it seems more than coincidental that four of the major contributors to the evolutionary synthesis in the 1930s and 1940s are among them. Upon closer examination it is evident that the exobiological arguments of Theodosius Dobzhansky and George Gaylord Simpson and, less directly, of H. J. Muller and Ernst Mayr are all related to their earlier work in formulating synthetic evolution. By examining the variety of views held by nonphysical scientists, we can see that there were significant disagreements between them about evolution into the 1960s. Within the next two decades, many but by no means all believed that "higher" life,

---

[1] One early reviewer of *The Celestial Worlds Discover'd* argued on the basis of analogy that stars are circled by inhabited worlds: "yet from the Analogy that is between the Sun and Stars, we may judge of the planetary Systems about them, and of the Planets themselves too, which probably are like the planetary Bodies about the Sun, (that is) that they have Plants and Animals, nay, and Rational ones too, as great admirers and Observers of the Heavens as any on Earth" (Anonymous 1699, 337).

[2] For more in-depth analysis of Christiaan Huygens's views of extraterrestrial life, see the first chapter of this volume by Crowe and Dowd (2013).

particularly intelligent life, probably occurs quite infrequently in the universe. Those arguing that extraterrestrial intelligence could plausibly exist were increasingly likely to make their case based on convergent evolution. While different scientists came to divergent conclusions about the likelihood of intelligence beyond Earth, the use of evolutionary arguments became increasingly common.

9.2 Early Critiques of Darwin's Theory of Evolution

To understand the 20th-century synthesis of evolution, it is useful to recall the main features of Charles Darwin's theory as seen in the first edition of *The Origin of Species*. His basic position can be summarized in two concepts: variation and natural selection. Darwin limited himself to minute differences between organisms that could be passed on to subsequent generations. Because each organism would be distinctly equipped for the "struggle for existence," those best suited to their environments would have the greatest chance of surviving to reproduce offspring that share some of their characteristics. Darwin (1968, 131) succinctly stated the relationship between this process of natural selection and variation: "This preservation of favourable variations and the rejection of injurious variations, I call Natural Selection."

In subsequent years, the efficacy of natural selection was questioned and rejected by many. Fleeming Jenkin (1867), for example, contended that any small beneficial variations would be diluted quickly in a population including many other organisms not similarly adapted. In later editions of *The Origin*, Darwin relied more heavily on "sports," individuals varying markedly from their forebears. This caused some critics to charge that Darwin had shifted to a position very similar to an older view that periodically new species abruptly appear.

Ironically, the mathematical analysis of heredity that was to play an important role in formulating the evolutionary synthesis began as an argument against the transmission of small variations from one generation to the next. When Francis Galton examined the "swamping effect" that Fleeming Jenkin described, he concluded that any variations from the mean type of a species would be lost in following generations. Thus, in the long run organisms would tend to have common characteristics. Deviations from the norm were, by Galton's analysis, transient. His protégé, Karl Pearson, came to the opposite conclusion. Pearson argued against the assumption that the fate of variations should be measured against a fixed ancestral type. Rather, he said that variations from an organism's ancestors could cause lasting changes in future generations.

In contrast to Pearson, others argued that evolution could only be accounted for through large-scale mutations. Supporting their views with Gregor Mendel's newly discovered paper, William Bateson, Hugo de Vries, and Wilhelm Johannsen proposed saltatory accounts of evolution. Mendel's early work focused on the inheritance of discontinuous characteristics. For example, for some of his experiments he used pea plants that had either pure yellow or pure green peas. When these were crossed, he did not obtain peas of an intermediate hue, but only of the same pure yellow of one of the parents. This emphasis on inheritance of discrete characteristics supported the views of those who explained evolution in terms of gross mutations. Moreover, many were skeptical of the existence of natural selection. For example, as late as 1915 Johannsen saw no reason to assume natural selection played a role: "Selection of differing individuals creates nothing new; a shift of the 'biological type' in the direction of selection has never been substantiated" (Johannsen 1915, 609).

9.3 The Evolutionary Synthesis

In the second and third decades of the twentieth century, there was a return to gradualistic evolution. The inadequacies of Darwin's original formulation were overcome by reconceptualizing variation and natural selection. From the combination of experimental and theoretical approaches to understanding these processes, the evolutionary synthesis was born.

A major emphasis of the evolutionary synthesis was to explain natural selection in mathematical terms. Especially through the work of R. A. Fisher, J. B. S. Haldane, and Sewall Wright, inheritance at the level of populations was explained through statistical models. Despite the highly theoretical nature of their contributions, their work was not divorced from experimentation. Fisher's work in quantifying variation and natural selection typified this synthesis of mathematics and empirical research. Using Muller's experiments, he showed how variation by micromutation could be estimated. The result was an indication of the rate at which variations entered populations. Next, he was able to specify the degree of selection by environmental factors. Either by comparing the differential rate of increase of two or more populations or by measuring changes of gene frequency within single populations, he was able to propose a statistical model of natural selection.

For all of Fisher's interest in natural populations, he was still a mathematician with little training in biology. At the other end of the mathematical/experimental continuum was H. J. Muller. By exposing genes to mutation-inducing X-rays, Muller was able to show the influence of environment on variation. But before the various stands of the evolutionary synthesis could be braided together, populations had to be understood both statistically and as they occur in nature. Theodosius Dobzhansky, George Gaylord Simpson, and Ernst Mayr were particularly adept at this.

When we consider Theodosius Dobzhansky's background, it is easy to understand why he made such an important contribution to the evolutionary synthesis. His early training with Sergei Chetverikov emphasized population genetics. In 1927 he went to the United States to work with Muller's mentor, T. H. Morgan. By combining Morgan's stress on experimentation with the Russian statistical approach, Dobzhansky did pioneering work in the genetics of free-living populations. This is evident even in his early work on variations of *Drosophila* in isolated mountain ranges (Lewontin et al. 1981). More influential, however, was his *Genetics and the Origin of Species*, first published in 1937 (Dobzhansky 1951).

Among those stimulated by this book was George Gaylord Simpson. As a paleontologist, his contacts with colleagues within his profession contributed little to his training in evolutionary theory. Paleontologists in the 1930s were more concerned with descriptive systematics than with the foundations of evolution. Consequently, Simpson (1978, 114–115) relied on the writings of people outside his discipline, including Fisher, Haldane, Wright, and Dobzhansky. After the 1930s, he also had personal contacts with Mayr and Dobzhansky (Mayr 1980a, 455). The high degree to which he assimilated populational approaches is evident in his 1944 *Tempo and Mode in Evolution*. His conclusions were in marked contrast to the Mendelians whose position was dominant a few years earlier. He acknowledged the importance of variation, but rejected macromutations:

> Single mutations with large, fully discrete, localized phenotypic effects are most easily studied; but paleontological and other evidence suggests that these are relatively unimportant at any level of evolution (Simpson 1944, 94).

His view of natural selection was diametrically opposed to that of Johannsen. According to Simpson (1944, 96), "Selection is a truly creative force and not solely negative in action. It is one of the crucial determinants of evolution."

A third major figure in the history of the evolutionary synthesis began by studying neither bones nor fruit flies, but rather birds. Unlike most other ornithologists of his day, however, Ernst Mayr worked in population genetics. Though Fisher, Haldane, and Wright had little influence on his early work, he was quickly attracted to the Russian school because of its emphasis on naturally occurring populations and taxonomy (Mayr 1980b, 421–422). Mayr's (1942, 67) central concern was speciation, which he thought could be discussed without recourse to large-scale mutations:

> Speciation is explained by the geneticist on the assumption that through the gradual accumulation of mutational steps a threshold is finally crossed which signifies the evolution of a new species.

Similarly, natural selection played a key role for Mayr (1942, 293): "Even genes with a small selective advantage will eventually spread over entire populations."

9.4 The Evolutionary Synthesis and Extraterrestrial Life

9.4.1 Simpson on the Nonprevalence of Humonoids

Now that we have seen how Darwin's notions of variation and selection were reformulated in the 1930s and 1940s by synthetic evolutionists, we are prepared to see the extent to which these ideas influenced those who speculated on the possibility of extraterrestrial life. An appropriate starting point is Simpson's article from 1964, "The Nonprevalence of Humanoids."[3] In addition to drawing on evolutionary factors

---

[3] For a related article see Simpson (1962). See Dick (2013) in this volume on Simpson's skepticism about exobiology being a science.

we have already seen, Simpson discussed other considerations affecting the probability of life beyond Earth. Simpson agreed with others who held that it is likely that rudimentary macromolecules will form from chemical processes, which should occur throughout the universe. But, Simpson said, this did not commit him to the conclusion that many others, particularly physical scientists, had reached: that therefore more complex forms of life will also evolve.

To go beyond chemical to biological activity, Simpson (1964, 772) said three processes were required: "mutation, recombination, and selection." (While two of these three are familiar from earlier discussions, recombination did not play as significant a role in the evolutionary synthesis.) The critical question for Simpson was whether or not these three factors interact in such a way as to make advanced forms of life a likely outcome of the origin of pre-biotic molecules. He argued that there are two ways to approach this issue: through the actual history of life on Earth and from theoretical considerations. On both counts Simpson was not optimistic that the development of extraterrestrial life would be a common occurrence.

According to Simpson (1964, 773), paleontological evidence gave no indication for the inevitability of higher forms of life: "The fossil record shows very clearly that there is no central line leading steadily, in a goal-directed way, from a protozoan to man." The reason for this can be understood by considering the mechanisms by which life arose. Variations are introduced through mutation, and individual differences are increased even more through recombination. Through interactions between the organisms and their environments, however, only a fraction of these variations will become established in the population. Given the combination of the numerous factors responsible for the evolution of any given species, Simpson (1964, 773) argued that terrestrial life is very likely to be unique:

> The existing species would surely have been different if the start had been different and if any stage of the histories of organisms and their environments had been different.... Man cannot be an exception to this rule. If the causal chain had been different, *Homo sapiens* would not exist.

9.4.2 Dobzhansky Against the Convergent Evolution of Extraterrestrial Life

Though the thrust and conclusion of Dobzhansky's argument was similar to Simpson's line of reasoning, Dobzhansky discussed explicitly two issues that Simpson dealt with only in passing: chance and convergence in evolution. Dobzhansky isolated the same three factors of mutation, sexual recombination, and natural selection as central to evolution. But only the first two, he said, operate randomly; selection works against chance. While acknowledging that selection is probabilistic, he maintained that because it relates the individual and its environment through a feedback mechanism, it is an antichance process.

Dobzhansky's speculations about extraterrestrial life were consistent with the emphasis on mutation and selection in the early days of the evolutionary synthesis.

In spite of mentioning recombination as a factor in terrestrial evolution, when he committed himself to determining the characteristics that all life should possess, he mentioned only selection and mutation:

> Despite all the uncertainties inevitable in dealing with a topic so speculative as extraterrestrial life, two inferences can be made. First, the genetic materials will be subject to mutation. Accurate self-copying is the prime function of any genetic materials, but it is hardly conceivable that no copy erors [sic] will ever be made. If such errors do occur, the second inference can be drawn: the variants that arise will set the stage for natural selection. This much must be a common denominator of terrestrial and extraterrestrial life (Dobzhansky 1972, 170).

A second issue Dobzhansky addressed was convergent vs. divergent evolution. He pointed out that in many instances on Earth, organisms of disparate ancestries can have similar characteristics. As an example he noted that fish and whales have similar forms because they both adapted to an aqueous environment. Some have held that because this sort of convergent evolution is so common on Earth, the process may be universal. Therefore, the argument goes, extraterrestrials may well resemble life on Earth. Dobzhansky argues against this belief on the grounds that in many cases similar environments have resulted not in convergent, but in divergent evolution (Dobzhansky 1972, 168–169).

Dobzhansky concluded that, given the number of discrete interactions between organism and environment in the evolutionary history of the human species, the probability of humans evolving on another Earth-like planet is virtually zero. Even assuming another planet equipped with all life forms that existed in the Eocene period, the re-evolution of humankind would involve the same mutations and the same selection on the roughly 50,000 genes that would have changed in *Homo sapiens* since then (Dobzhansky 1972, 173).

9.4.3 Muller, Mutation, and Intelligence

When H. J. Muller addressed the question of life beyond Earth, it is not surprising that he emphasized the role of mutation. What may seem more remarkable is that someone who played such an important role in the evolutionary synthesis still kept room for interplanetary convergence of intelligence. He agreed with Simpson and Dobzhansky about the importance of chance:

> Just what steps will be taken at a particular point is sometimes a matter of accident: of what mutation manages to take hold, and then what combination of mutations, until some novel structure of [sic] manner of functioning is thereby brought into being that acts as a key to open up an important new way of living (Muller 1963, 80).

Though Muller believed a wide range of morphologies was possible, he thought intelligence was the natural product of evolution (Muller 1963, 83). One possible explanation for this view of limited directedness may be the influence of one of his students, Carl Sagan (Carlson 1981, 389). Though Carl Sagan worked with him only one summer, Carl Sagan said he "always kept in touch with him" (Cooper

1980, 42–43). By the time Muller wrote the above article, the young Carl Sagan had also published about life beyond Earth.

9.4.4 Mayr and the Importance of Chance

Though Mayr claimed his analysis was very similar to Simpson's reasoning, there were significant differences. Most obvious is Mayr's lesser emphasis on mechanisms of evolution. Instead, he provided an extended summary of the history of the human species. This may simply be a reflection of the time Mayr was writing. Dobzhansky, Simpson, and Muller all wrote first about extraterrestrials in the early 1960s. Mayr's article was written two decades later. The evolutionary synthesis may have been so well accepted by then that a detailed justification of its basic tenets would have seemed superfluous. Nevertheless, throughout the piece his discussion emphasized the importance of chance. Though his primary concern was to discuss the likelihood of extraterrestrial intelligence, not merely multicelluar life, he reached the same conclusions as Simpson.

Mayr amplified Dobzhansky's argument against the convergent evolution of intelligence by addressing the multiple emergence of vision on Earth. A common argument has been that evidence for the widespread occurrences of convergent evolution can be seen in the independent evolution of eyes numerous times. Mayr said that his own studies had drawn him to conclude that eyes have developed at least 40 different times in unrelated lineages. In contrast, intelligence has evolved only once on Earth (Mayr 1985, 28).[4]

9.4.5 Divergent Views of Extraterrestrial Life: Outside and Within the Evolutionary Synthesis

Speculations in the 1950s and 1960s by those not intimately involved with the evolutionary synthesis were not as similar to one another as the views we have seen thus far. For example, in 1953 the anthropologist Loren Eiseley focused on the uniqueness of humankind. After examining mimicry among terrestrial organisms, he concluded that this could not be used to argue for extraterrestrials resembling life on Earth: "No animal is likely to be forced by the process of evolution to imitate, even superficially, a creature upon which it has never set eyes and with which it is in no form of competition" (Eiseley 1953, 84).

Even more fascinating is Eiseley's description of the opinion of cytologist Cyril D. Darlington. In Eiseley's (1953, 81) words, Darlington "dwells enthusiastically

---

[4] For a summary of Mayr's debate with Carl Sagan about the likelihood of extraterrestrial intelligence, see Garber (2013).

on the advantages of two legs, a brain in one's head and the position of surveying the world from the splendid height of six feet." Eiseley failed to mention where Darlington stated this, and I was not able to find any relevant passages. I was able to find a potential partial explanation for why a contributor to the evolutionary synthesis would hold a view so different from those of the other four key figures we have seen. First, note that Darlington was writing several years before the others, and thus the evolutionary synthesis may not have solidified. Second, he favorably noted Henry Fairfield Osborn's orthogenesis and Bernhard Rensch's directed evolution, which held that evolution is teleological (Darlington 1969, 22).

Another anthropologist, William Howells, concluded in 1961 that extraterrestrial intelligence probably exists. He repeatedly made comments contrary to the mainstream views of the evolutionary synthesis. Several times he suggested that evolution is a volitional process. For example, Howells (1961, 239) said "Intelligent creatures will have made a choice, early in evolution, of a nervous system which is more open to fresh impressions: a brain which can learn." He thought such "choices" would likely lead to intelligence very human in appearance.

Oceanographer and ecologist Robert Bieri's conclusions were similar to those of Howells, but the basis for his belief was more explicit. Bieri opened his article with a quote from geneticist G. W. Beadle (1959), against which he argued. In opposition to Beadle's assertion that there are an extraordinary number of evolutionary pathways open to life, Bieri (1964, 452, 457) stressed the limitations imposed by the properties of chemical elements and by the "forms of energy" available. Such constraints, Bieri wrote, are evident in the finite range of variability of terrestrial organisms. Because of these restrictions, organisms beyond Earth will conform to the same patterns imposed on life as we know it. After considering a number of characteristics that he thought would be universal, he concluded with his prediction of the form of extraterrestrial intelligence: "If we ever succeed in communicating with conceptualizing beings in outer space, they won't be spheres, pyramids, cubes, or pancakes. In all probability they will look an awful lot like us" (Bieri 1964, 457).

Bacteriologist Francis Jackson and co-author astronomer Patrick Moore seemed less decided. At one point in their 1962 book they said it would be absurd to imagine that humans are constructed on an ideal model that would be followed on other planets (Jackson and Moore 1962, 115). Yet a few pages later they included a sentence that gives the opposite sense: "It is by no means impossible that, on planets closely similar to the Earth, chemical and biological evolution might have followed a strikingly similar course, even occasionally to the production of men" (Jackson and Moore 1962, 124). There is no absolute contradiction in holding both of these views. However, it is noteworthy that Jackson and Moore were comfortable with either possibility.

As we examine works through the mid-1980s, we continue to see a variety of perspectives. Dale Russell, a paleontologist, was reluctant to generalize from evolution on Earth to extraterrestrial conditions. In only one sentence did he suggest that the existence of extraterrestrial life is by no means a foregone conclusion. Within the context of astrophysical considerations, he concluded, "It would seem that the origin of life is intrinsically a much more probable event than the origin of higher intelligence" (Russell 1981, 270).[5] Another paleontologist, C. Owen Lovejoy, was more definitive than Russell. Lovejoy thought intelligence beyond Earth could be quite common, but he distinguished this from the much rarer occurrence of cognition. He said that because cognition as exemplified in humans is the result of our specific evolutionary path, the combination of events making cognition possible is highly unlikely to occur on most planets where intelligent life is present (Lovejoy 1981, 327).

In spite of the increasing trend to view the possibility of extraterrestrials in light of synthetic evolutionary theory, there remained concerns about some of the principles of its founding fathers. Gerald Feinberg and Robert Shapiro, a physicist and a biochemist, rejected the conclusion of space scientists Roger MacGowan and Frederick Ordway "that the majority of intelligent extrasolar land animals will be of the two legged and two armed variety" (MacGowan and Ordway 1966, 240). Instead they pointed out, citing Simpson, that great divergences from terrestrial forms are possible through the joint action of mutation and natural selection. Yet they also maintained that "we will undoubtedly encounter [convergent evolution] on other worlds" (Feinberg and Shapiro 1980, 411). Paleontologist David Raup certainly understood the force of arguments against convergence toward humanoid forms elsewhere, but he countered that too little is known about the process of convergence to make any definitive claims. The evolution of other humanoids may be highly improbable, he wrote, but not necessarily impossible (Raup 1985, 36).[6]

Two other tendencies were also present among nonphysical scientists: hard-headed theorizing and more free-form speculation. In a manner somewhat reminiscent of the earlier evolutionary systematists, James Valentine approached the question by distinguishing between microevolution, involving selection within a

---

[5] Paleontologist Peter Ward and astronomer Donald Brownlee came to a similar conclusion in their more recent book *Rare Earth* (Ward and Brownlee 2000).

[6] More recently, while evolutionary paleobiologist Simon Conway Morris was certainly conversant with the evolutionary synthesis, he emphasized the ubiquity of convergence, contesting the view that historical contingencies make it impossible to predict the likely forms of life on other worlds: "Rerun the tape of the history of life, as S. J. Gould would have us believe, and the end result will be an utterly different biosphere. Most notably there will be nothing remotely like a human, so reinforcing the notion that any other biosphere, across the galaxy and beyond, must be as different as any other: perhaps things slithering across crepuscular mudflats, but certainly never the prospect of music, no sounds of laughter. Yet, what we know of evolution suggests the exact reverse: convergence is ubiquitous and the constraints of life make the emergence of the various biological properties very probable, if not inevitable. Arguments that the equivalent of *Homo sapiens* cannot appear on some distant planet miss the point: what is at issue is not the precise pathway by which we evolved, but the various and successive likelihoods of the evolutionary steps that culminated in our humanness" (Conway Morris 2003, 283–284). Recent supporters of Conway Morris's emphasis on convergence include anthropologists Kathryn Coe, Craig T. Palmer, and Christina Pomianek, who noted, "It is now time to take the implications of evolutionary theory a little more seriously, and convergence is the norm" (Coe, Palmer, and Pomianek 2011, 209). They also maintained that "evolutionary theory, theoretically, should apply anywhere to anything that is living" (Coe, Palmer, and Pomianek 2011, 215), in a line of reasoning similar to biologist Richard Dawkins's argument for "Universal Darwinism" (Dawkins 1983).

population, and macroevolution, dealing with evolution above the species level. He concluded that the microevolutionary details of life on another planet, e.g., their genetic materials, would probably be very different from their terrestrial counterparts. But macroevolution, he thought, should yield extraterrestrial patterns of "multicellular diversification" similar to those seen on Earth (Valentine 1981, 253).

Imagination reigned in Bonnie Dalzell's exhibit of possible alien creatures for the Smithsonian. By hypothesizing planets that vary from Earth in gravity and temperature, she created environments that would foster a wide variety of land-bound, aquatic, and aerial life (Dalzell 1974). The combination of her artistic talent and her background in paleontology seemed more heavily weighted toward the former. Anthropologist Doris Jonas and psychiatrist David Jonas, by contrast, considered not only the morphology but also the possible perceptual worlds of extraterrestrials. Though their work was not as informed by theory as that of some of the contributors to the evolutionary synthesis, their basic tenet was the same:

> One thing is for certain: we have no reason to assume that evolutionary forces on other planets will produce forms or intelligences that are the same as ours even though the basic raw materials must be similar. Whatever chance factors combine to produce any form of life, infinitely more must combine to produce an advanced form (Jonas and Jonas 1976, 9).

9.5 Conclusion

Some of the most incisive arguments for and against the possibility of extraterrestrial life have come from scientists who have only a passing interest in the question. Their views typically were more influenced by their professional work in their own disciplines than by more extended contacts with others interested in life beyond Earth. Thus, when trying to evaluate their positions, it is vital to understand the conceptual frameworks of the disciplines from which these speculations arose. One such framework that played a major role in the 20th and 21st centuries is modern evolutionary theory. By examining the extent to which this paradigm has made an impact in various fields over the past few decades, we can better understand the diversity of views about extraterrestrial life held by scientists from a variety of disciplines.

Acknowledgments   This chapter is an adaptation of Vakoch, Douglas A. 2013. "The Evolution of Extraterrestrials: The Evolutionary Synthesis and Estimates of the Prevalence of Intelligence Beyond Earth." In *Archaeology, Anthropology, and Interstellar Communication*, ed. Douglas A. Vakoch. Washington, DC: NASA.

References

Anonymous. 1699. "Review: An Account of Books." *Philosophical Transactions* 21: 335–342. Quoted in Gingras, Yves, and Alexandre Guay, "The Uses of Analogies in Seventeenth and Eighteenth Century Science." *Perspectives on Science* 19 (2): 154–191.

Beadle, G. W. 1959. "The Place of Genetics in Modern Biology." Eleventh Annual A. D. Little Memorial Lecture, Massachusetts Institute of Technology.

Bieri, Robert. 1964. "Humanoids on Other Planets?" *American Scientist* 52: 452–458.

Carlson, Elof Axel. 1981. *Genes, Radiation, and Society: The Life and Work of H. J. Muller.* Ithaca, NY: Cornell Univ. Press.

Coe, Kathryn, Craig T. Palmer, and Christina Pomianek. 2011. "ET Phone Darwin: What Can an Evolutionary Understanding of Animal Communication and Art Contribute to Our Understanding of Methods for Interstellar Communication?" In *Civilizations Beyond Earth: Extraterrestrial Life and Society,* ed. Douglas A. Vakoch and Albert A. Harrison, 214–225. New York: Berghahn Books.

Conway Morris, Simon. 2003. *Life's Solution: Inevitable Humans in a Lonely Universe.* Cambridge: Cambridge Univ. Press.

Cooper, Henry S. F., Jr. 1980. *The Search for Life on Mars: Evolution of an Idea.* New York: Holt, Rinehart and Winston.

Crowe, Michael J., and Matthew F. Dowd. 2013. "The Extraterrestrial Life Debate from Antiquity to 1900." In *Astrobiology, History, and Society: Life Beyond Earth and the Impact of Discovery,* ed. Douglas A. Vakoch. Heidelberg: Springer.

Dalzell, Bonnie. 1974. "Exotic Bestiary for Vicarious Space Voyagers." *Smithsonian Magazine* 5 (October): 84–91.

Darlington, C. D. 1969. *The Evolution of Man and Society.* New York: Simon and Schuster.

Darwin, Charles. 1968. *The Origin of Species by Means of Natural Selection: Or the Preservation of Favoured Races in the Struggle for Life.* New York: Penguin Books. First published in 1859.

Dawkins, Richard. 1983. "Universal Darwinism." In *Evolution from Microbes to Men,* ed. D. S. Bendall, 403–425. Cambridge: Cambridge Univ. Press.

Dick, Steven J. 2013. "The Twentieth Century History of the Extraterrestrial Life Debate: Major Themes and Lessons Learned." In *Astrobiology, History, and Society: Life Beyond Earth and the Impact of Discovery,* ed. Douglas A. Vakoch. Heidelberg: Springer.

Dobzhansky, Theodosius. 1951. *Genetics and the Origin of Species,* 3rd ed. New York: Columbia Univ. Press.

Dobzhansky, Theodosius. 1972. "Darwinian Evolution and the Problem of Extraterrestrial Life." *Perspectives in Biology and Medicine* 15 (2): 157–175.

Eiseley, Loren C. 1953. "Is Man Alone in Space?" *Scientific American* 189 (July): 80–86.

Feinberg, Gerald, and Robert Shapiro. 1980. *Life Beyond Earth: The Intelligent Earthling's Guide to Life in the Universe..* New York: William Morrow and Company.

Garber, Stephen J. 2013. "A Political History of NASA's SETI Program." In *Archaeology, Anthropology, and Interstellar Communication,* ed. Douglas A. Vakoch. Washington, DC: NASA.

Howells, William. 1961. "The Evolution of 'Humans' on Other Planets." *Discovery* 22 (June): 237–241.

Huygens, Christiaan. 1968. *The Celestial Worlds Discover'd.* London: Frank Cass and Co. Ltd. Facsimile reproduction of 1698 ed.

Jackson, Francis, and Patrick Moore. 1962. *Life in the Universe.* London: Routledge & Kegan Paul.

Jenkin, Fleeming. 1867. "The Origin of Species." *The North British Review* 46 (June): 277–318. Reprinted in David, L. Hull. 1973. *Darwin and His Critics: The Reception of Darwin's Theory of Evolution by the Scientific Community,* 303–344. Cambridge, MA: Harvard Univ. Press.

Johannsen, Wilhelm. 1915. "Experimentelle Grundlagen der Deszendenslehre. Variabilität, Vererbung, Kreuzung, Mutation," *Kultur Der Gegenwort III* 4. Quoted in Mayr, Ernst. 1980. "Prologue: Some Thoughts on the History of the Evolutionary Synthesis." In *The Evolutionary Synthesis: Perspectives on the Unification of Biology,* ed. Ernst Mayr and William B. Provine, 1–8. Cambridge, MA: Harvard Univ. Press, 7.

Jonas, Doris, and David Jonas. 1976. *Other Senses, Other Worlds*. New York: Stein and Day.

Lewontin, R.C., John A. Moore, William B. Provine, and Bruce Wallace, eds. 1981. *Dobzhansky's Genetics of Natural Populations I–XLIII*. New York: Columbia Univ. Press.

Lovejoy, C. Owen. 1981. "Evolution of Man and Its Implications for General Principles of the Evolution of Intelligent Life." In *Life in the Universe*, ed. John Billingham, 317–329. Cambridge, MA: The MIT Press.

MacGowan, Roger A., and Frederick I. Ordway, III. 1966. *Intelligence in the Universe*. Englewood Cliffs, NJ: Prentice Hall.

Mayr, Ernst. 1942. *Systematics and the Origin of Species: From the Viewpoint of a Zoologist.*. New York: Columbia Univ. Press.

Mayr, Ernst. 1980a. "G. G. Simpson." In *The Evolutionary Synthesis: Perspectives on the Unification of Biology*, ed. Ernst Mayr and William B. Provine, 452–63. Cambridge, MA: Harvard Univ. Press.

Mayr, Ernst. 1980b. "How I Became a Darwinian." In *The Evolutionary Synthesis: Perspectives on the Unification of Biology*, ed. Ernst Mayr and William B. Provine, 413–423. Cambridge, MA: Harvard Univ. Press.

Mayr, Ernst. 1985. "The Probability of Extraterrestrial Intelligent Life." In *Extraterrestrials: Science and Alien Intelligence*, ed. Edward Regis, Jr., 23–30. Cambridge: Cambridge Univ. Press.

Muller, H. J. 1963. "Life Forms To Be Expected Elsewhere Than on Earth." *Spaceflight* 5 (3): 74–85. Reprinted from H. J. Muller. 1961. "Life Forms to be Expected Elsewhere Than on Earth." *The American Biology Teacher* 23 (6): 331–346.

Raup, David M. 1985. "ETI Without Intelligence." In *Extraterrestrials: Science and Alien Intelligence*, ed. Edward Regis, Jr., 31–42. Cambridge: Cambridge Univ. Press.

Russell, Dale A. 1981. "Speculations on the Evolution of Intelligence in Multicellular Organisms." In *Life in the Universe*, ed. John Billingham, 259–275. Cambridge, MA: The MIT Press.

Simpson, George Gaylord. 1944. *Tempo and Mode in Evolution*. New York: Cambridge Univ. Press.

Simpson, George Gaylord. 1962. "Some Cosmic Aspects of Organic Evolution." In *Evolution und Hominisation*, ed. Gottfried Kurth, 6–20. Stuttgart: Gustav Fischer Verlag. Also reprinted in George Gaylord Simpson. 1964. "Some Cosmic Aspects of Organic Evolution." In *This View of Life*, 237–252. New York: Harcourt, Brace, and World, Inc.

Simpson, George Gaylord. 1964. "The Nonprevalence of Humanoids." *Science* 143: 769–775. Also reprinted in George Gaylord Simpson. 1964. "The Nonprevalence of Humanoids." In *This View of Life: The World of an Evolutionist*, 253–271. New York: Harcourt, Brace, and World, Inc.

Simpson, George Gaylord. 1978. *Concession to the Improbable: An Unconventional Autobiography*. New Haven, CT: Yale Univ. Press.

Valentine, James W. 1981. "Emergence and Radiation of Multicellular Organisms." In *Life in the Universe*, ed. John Billingham, 229–257. Cambridge, MA: The MIT Press.

Ward, Peter, and Donald Brownlee. 2000. *Rare Earth: Why Complex Life Is Uncommon in the Universe*. New York: Springer.

Chapter 10
The First Thousand Exoplanets: Twenty Years of Excitement and Discovery

Chris Impey

Abstract The recent "explosion" in the number of extrasolar planets, or exoplanets, is perhaps the most exciting phenomenon in all of science. Two decades ago, no planets were known beyond the Solar System, and now there are more than 770 confirmed exoplanets and several thousand more candidates, while the mass detection limit has marched steadily downwards from Jupiter mass in 1995 to Neptune mass in the early 2000s to Earth mass now. The vast majority of these exoplanets are detected indirectly, by their gravitational influence on the parent star or the partial eclipse they cause when they periodically pass in front of it. Doppler detection of the planet's reflex motion yields a period and an estimate of the mass, while transits or eclipses yield the size. Exoplanet detection taxes the best observatories in space, yet useful contributions can be made by amateur astronomers armed with 6-inch telescopes. The early discoveries were surprising; no one predicted "hot Jupiters" or the wild diversity of exoplanet properties that has been seen. It is still unclear if the Solar System is "typical" or not, but at current detection limits at least 10 % of Sun-like stars harbor planets and architectures similar to the Solar System are now being found. Over a hundred multiple planet systems are known and the data are consistent with every star in the Milky Way having at least one planet, with an implication of millions of habitable, Earth-like planets, and of which could harbor life. Doppler and transit data can be combined to give average density, and additional methods are beginning to give diagnostics of atmospheric composition. When this work can be extended to rocky and low mass exoplanets, and the imprint of biology on a global atmosphere can be measured, this might be the way that life beyond Earth is finally detected for the first time.

C. Impey (✉)
Department of Astronomy, University of Arizona, Tucson, AZ, USA
e-mail: cimpey@as.arizona.edu

D. A. Vakoch (ed.), *Astrobiology, History, and Society*, Advances in Astrobiology and Biogeophysics, DOI: 10.1007/978-3-642-35983-5_10,
© Springer-Verlag Berlin Heidelberg 2013

10.1 Introduction

The recent "explosion" in the number of extrasolar planets, or exoplanets, is perhaps the most exciting phenomenon in all of science. Two decades ago, no planets were known beyond the Solar System, and more than a few researchers had been burned by claims of detections that did not hold up, while many others had given up on the chase. When a planet with half the mass of Jupiter was found whipping around the star 51 Peg every 4 days, it was a stunning surprise (Mayor and Queloz 1995). We should, however, spare some surprise for the earlier discovery of planets around a pulsar, demonstrating that expectations are meant to be defied in astrobiology (Wolszczan and Frail 1992). Since 1995, the number of confirmed exoplanets has had a doubling time of 30 months. When the burgeoning number of candidates from NASA's Kepler satellite is included, the number of exoplanets soared through a thousand early in 2012. Alongside these growing numbers is the steady downward March of the detection limit from Jupiter mass in 1995 to Neptune mass in the early 2000s to Earth mass now. History has not prepared us for what we have learned about distant planets (Raulin Cerceau 2013). The pace of progress and discovery has been dizzying even for experts in the field.

10.2 The Detection Problem

For centuries, scientists and philosophers speculated about the existence of planets around other stars (see other contributions in this volume). Once the Copernican revolution displaced the Earth from the center of the universe and it became one rocky body orbiting a normal star, the Principle of Mediocrity suggested that other Solar Systems should exist. By extension, this heuristic suggests the existence of planets similar to ours, and fuels expectations of life on beyond Earth and hence the whole subject of astrobiology.

To understand the challenge of exoplanet detection, consider a scale model. If the Sun is a glowing ball of plasma ten feet across, the Earth is a large blue-white marble 400 yards away and Jupiter is a pale yellow sphere the size of a beach ball just over a mile away. On this scale, the Solar System is 20 miles across, while the nearest Sun-like star would be another ten-foot glowing plasma ball 50–100,000 miles away. Looking towards that nearest Sun-like star, a giant planet like Jupiter would reflect a billionth of the star's light and an Earth five times less, and both would be buried in the glare of the star, since their angular separation is less than the blurring of the star image seen through a telescope. Planets can also be detected by the reflex motion they induce on the star they orbit. In our Solar System, Jupiter causes the Sun to pirouette around its edge, a ten-foot wobble that would be imperceptible from thousands of miles away. The periodic Doppler motion induced on the star is also subtle, 11 m/s for Jupiter and 10 cm/s for the Earth, equivalent to a very slow walking speed. As fractions of the speed of light, these are four parts in a billion for Jupiter and a 100 times less for the Earth.

10.3 Failure and Frustration

Success in the search for exoplanets did not come easily or quickly. In the nineteenth century, unexplained motions of the binary star 70 Ophiuchus were attributed to a planet, and in the mid-twentieth century, Peter van de Kamp claimed to have detected a wobble of the nearby red dwarf Barnard's Star caused by a Jupiter-mass planet (van de Kamp 1969). That claim was discredited, although ironically, the Kepler team reported in 2012 the smallest exoplanets yet detected orbiting a red dwarf very similar to Barnard's Star. In 1988, Bruce Campbell and his collaborators published radial velocity evidence of a planetary companion to Gamma Cephei, though they used cautious language in their paper (Campbell et al. 1988). The interpretation of the evidence was called into question and it was 15 years before an exoplanet was confirmed in this system (Hatzes et al. 2003). An object times times Jupiter's mass was discovered in 1989 (Latham 1989), but uncertainty in its inferred properties mean that it might be a brown dwarf rather than a planet. In the second half of the 20th century, the stuttering progress in the search for exoplanets mirrored the development of astrobiology as a field (Dick 2013).

Pulsar timing provides an unusual window onto exoplanets. Pulsars are the collapsed, rapidly-spinning remnants of massive stars, and their rotation is so irregular that anomalies can be measured to a precision of one part in a trillion, allowing orbiting planets as slight as a tenth of the Earth's mass to be detected. A pulsar planet announced in 1991 received much publicity, but the claim was subsequently retracted. Yet the following year, Wolszczan and Frail (1992) found two Earth-mass planets around the millisecond pulsar PSR B1257+12, and that claim has stood the test of time, including the subsequent discovery of a third Moon-mass body. These radio astronomershad succeeded in find the first planet-mass objects beyond the Solar System, yet they experienced a strange kind of failure when the rest of the community seemed to relegate pulsar planets to the status of an exotic anomaly. Alan Boss (2009), Mike Perryman (2011) and Ray Jayawardhana (2011) have detailed at length the winding road that led to the first bona fide detection of a planet orbiting a main sequence star like the Sun. For interviews with many of the leading players, see Impey (2010).

10.4 The First Discoveries

The age of exoplanet discovery was formally ushered in on October 6, 1995, when Michel Mayor and Didier Queloz of the Geneva Observatory announced the discovery of an exoplanet half Jupiter's mass orbiting the G star 51 Pegasi (Mayor and Queloz 1995). While the exoplanet was unseen and only detected by the Doppler method, the 50 light years distant star was bright enough to be visible to the naked eye. News of the discovery must have been rather bittersweet for Geoff Marcy and Paul Butler, 5,000 miles away in California. Marcy had been running an experiment for 8 years and had removed 51 Pegasi from his sample due to an error in the star catalog. Within a week of the discovery, he and Butler had confirmed the Swiss discovery. They mined their data for similar objects and by the end of 1995 had found three more exoplanets (see Marcy and Butler 1998 for a snapshot of this rapidly emerging field of research). Mayor and Queloz had come to planet hunting from the perspective of binary star researchers, accustomed to orbital periods of hours or days, so they observed their candidates frequently. Marcy and Butler were taking data more sparsely and thought they had plenty of time to dig out planet signals, since Jupiter takes 12 years to orbit the Sun. Friendly competition drove them to push the limits of their spectrographs. These two groups have led the field in terms of Doppler detection ever since, with hundreds of new worlds to their names.

10.5 Surprise and Confusion

The detection of exoplanets had been anticipated for a long time, but the excitement of discovery was tinged almost immediately with confusion. 51 Peg b was a Jupiter-mass planet much closer its star than Mercury is to the Sun, whipping around a complete orbit in just over 4 days, at a scorching temperature of 1,000 °C (1,800 °F)! Discoveries were announced at a rate of about one a month for the first few years, accelerating to one a week in the early 2000s, and a current average of a new exoplanet daily. Properties of exoplanets are governed by an obvious observational truism: you can only detect planets that your technique allows you to detect. The statistical properties of the Doppler method sample have always been skewed in favor of high mass and short period, since those objects require less data of lower quality to be detected. But the first few dozen exoplanets were very surprising because they were so massive and so close to their parent stars, and most of them had orbital eccentricities larger than any of the planets in the Solar System. These "hot Jupiters" were unusual and completely unexpected.

In the absence of any other examples to test the paradigm with, planetary scientists had drawn as many inferences as possible from our Solar System. Locally, we see planets on nearly circular orbits within a few degrees of a single plane, with small rocky planets close to the Sun and rocky planets that have accreted large hydrogen and helium mantles far from the Sun. The underlying theory is based on the nebular hypothesis, which was proposed in 1734 by Emanuel Swedenborg and refined later in the 18th century by Immanuel Kant and Pierre-Simon Laplace. Problems with the nebular hypothesis were addressed by Victor Safronov in the 1970s and his work became the basis for the modern theory of planet formation. However, there had always been concerns that the theory might be overly tailored to the specific circumstances and history of one planetary system, making it a kind of "Just So" story. Planet formation is in some senses historical science, since its complexity cannot be captured either by a computer or by theory and evidence of the initial conditions might be unobtainable.

It was immediately clear that giant planets could not form so close to their stars; there simply isn't enough material at those distances in the proto-planetary disk and the temperature is too high (Lin et al. 1996). Rather, they migrate inwards due to interactions with each other and with material in the disk. This has to happen quickly since it only takes about a million years to grow by accretion from dust bunnies into planet embryos or planetesimals. These embryos are Moon- or Mars-sized in the inner regions and several times Earth's size beyond the snow line. One type of migration involves subtle resonance interactions between the embryo and gas in the disk and another happens after an embryo has grown to near Jupiter mass and it clears a gap in the disk, after which both the planet and the gap migrate to smaller distances. But that's not the whole story, as recent observations show that many hot Jupiters have highly included orbits and some even go around their stars in the opposite direction to the star's rotation! The details are complex, and planets interact violently with each other and can migrate in or out depending on the circumstances. Theory isn't yet mature enough to predict exoplanet properties.

10.6 Methods of Detection

For a decade after the discovery of 51 Peg b, the principle method for finding exoplanets was the Doppler method. It still yields the most confirmed exoplanets, but it has been "eclipsed" by Kepler in the number of exoplanet candidates. In a planetary system, planets and stars orbit a common center of gravity that is close to or inside the star but not as its center. The small reflex motion induced on the star by a massive planet is observable as a sinusoidal variation in velocity and that modest variation is detected with a series of high resolution spectroscopic observations. The radial velocity variation is inversely proportional to the square root of the orbital distance and proportional to the planet mass times the sin of the inclination angle of the orbit. Because of the uncertainty in inclination, a minimum mass is measured and for any sample of planet systems at random orientations the masses will on average be underestimated by a factor of two. Multiple planets can be detected with the same set of spectra; the most massive exoplanet is searched for first, then the best sinusoidal fit to the data is subtracted off, then a smaller signal is search for in the residuals. Each exoplanet contributes to the data as a harmonic of a particular strength and frequency, reminiscent of Kepler's harmony of the spheres.

The pioneering groups succeeded through exquisite experimental technique. Detecting a Jupiter mass planet involves measuring a long-term wavelength shift of a stellar absorption line by 0.1 % of its width. This requires a high dispersion spectrograph, high signal to noise spectra, and extremely accurate wavelength calibration. The second requirement is not too difficult to meet for the kind of bright stars targeted in the first radial velocity survey, many of which were visible to the naked eye like 51 Peg. The last requirement led to the innovation of passing light gathered by the telescope through an iodine cell, which imprinted a reference gird of thousands of narrow absorption lines on the spectrum. The first discoveries were made with a precision of 10 m/s and current accuracy is 1 m/s or slightly better. To see what a difference a decade can make, compare the *Annual Reviews of Astronomy and Astrophysics* summaries of Marcy and Bulter (1998) and Udry and Santos (2007).

When a planet passes in front of a star it dims it slightly and temporarily. For our Solar System seen edge-on from afar, Jupiter would dim the Sun by 1 % for 5 h every 12 years. The depth of the partial eclipse is just the ratio of the cross-sectional area of the planet to the cross-sectional area of the star. Observing exoplanet transits would seem like searching for needles in a haystack, but the prevalence of hot Jupiters improves the odds. For normal Jupiter-Sun systems with random orientations the odds of a transit alignment are one in a thousand but this rises to one in ten for systems with hot Jupiters. In 1999, the first transit was detected, of HD 209458b (Charbonneau et al. 2000). Since then, the number of exoplanets with transit detections has grown steadily to become about a third of the total sample. The combination of size from a transit and mass from radial velocity variation gives mean density, crucial extra information for characterizing an exoplanet. If the mean density is less than water, it is good evidence that the eclipsing object is a rocky, terrestrial planet.

The most compelling evidence of an exoplanet is an image showing separated from its star, with its orbit traced by multiple observations. This was very difficult to obtain because the reflected light from a giant planet is swamped by hundreds of millions of times brighter starlight. As with the radial velocity method, technical innovation opened the door for progress. Adaptive optics systems on large telescopes started to be able to correct for the distorting effects of the atmosphere on the incoming light wave front from a star. This allows a telescope to approach its diffraction limit, which is a linear function of diameter, and resolve or separate the dim light of the exoplanet from much brighter star. Imaging is most sensitive to large separations like 10 or 100 AU so is complementary to selection by the Doppler effect or by transits. It's also best done in the infrared when the contrast between the exoplanet and its star is time times better than at optical wavelengths. Exoplanets were first imaged a decade after they were first discovered (Chauvin et al. 2004), and the number successfully imaged is still only a few dozen. Rapid advances in achieving better contrast through adaptive optics led to the first image of multiple planets just a few years later (Marois et al. 2008).

The last and perhaps cleverest method for detecting exoplanets employs microlensing. When a star passes directly in front of another star, general relativity predicts a brightening of the background star by about 30 % as its light is magnified by the intervening star. No image splitting is seen because the gravity deflection angle is very small. If the foreground star has an orbiting exoplanet, it can cause a secondary brightening. Microlensing succeeded around the same time as imaging (Bond et al. 2004) and it has the potential to detect Earth-like planets (Gaudi et al. 2008). Unfortunately, the incidence rate of microlensing events is only one in a million and the events are not repeatable, limiting the amount that can be learned about these systems.

10.7 The Exoplanet Zoo

After the initial surprise of the hot Jupiters, planet hunters settled down for the long haul, lowering their detection thresholds and accumulating statistics. After nearly 20 years it's still too early to measure the abundance of normal gas giants on orbits like those in the Solar System, although some proxies have finally been detected. Observational biases still strongly favor the more massive and rapidly-orbiting exoplanets, but we're gradually getting a better sense of the exoplanet "zoo." The range of physical properties makes is challenging to decide what is normal or typical in the underlying population. Pulsar planets were an early oddity, but the following sampling may give a sense of the bestiary.

The Methuselah planet, or PSR B1620-26b, is 12,400 light years away and is the oldest known exoplanet, with an age of 12.7 billion years. It orbits a pulsar and a white dwarf. It most likely formed around a Sun-like star but when they entered the dense environment of the M4 globular cluster, the planet was captured by a neutron star and its companion while its original host was ejected from the system. The Jupiter-sized planet settled into a distant orbit with a good view of a binary where material from a red giant turned a neutron star into a pulsar spinning 100 times a second. Some exoplanets have extreme eccentricities. HD 80606b goes from distance like the Earth's from its star to a distance less than Mercury's, getting blasted by a blowtorch every 4 months. Other exoplanets are scorched all the time. Corot-7b is five times Earth's mass and is in a tight orbit with its star-facing side at 2,330 °C (4,220 °F) and its outward-facing side at −220 °C (−370 °F). With an atmosphere of sodium and oxygen, the hot side probably has molten pebbles raining down from the sky. SWEEPS-10 is even closer to its parent star, which is a red dwarf. It whips around in 10 h, 200 times faster than Mercury. The "Tatooine" planet, or Kepler 16b, orbits twin red dwarfs and is near the edge of the habitable zone. From Tatooine, double sunsets would be visible as from the fictional planet in the Star Wars movies. In addition to these extremes there are dozens of hot and icy giants, water worlds, rocky super-Earths, and even free-floating planets. The Sun-like star HD 10180 has at least seven and possibly as many as nine planets, rivaling the Solar System in richness.

10.8 The Hunt for Earths

The bulk of the heavy lifting in extrasolar planet research has used, and continues to use, the indirect Doppler method. In the past decades, eclipses have given the extra information on size, and so a constraint on mean density, while direct imaging has become effective with space-based observations and nulling interferometry on the ground. About 10 % of Sun-like stars have planets, with indications that the true fraction might be much higher and that rocky terrestrial planets may outnumber gas giants (Marcy et al. 2005). Over a hundred multiple planet systems are known. Simulations do what NASA does by "following the water" as a nebula forms and planets grow by accretion, and they suggest that the Earth has a typical inventory of water so terrestrial planets with all the ingredients needed for life should not be rare (Raymond et al. 2004). Even giant planet migration does not preclude habitable planets because it happens so rapidly that rocky planets can grow after the gas giant has moved in and parked. The Doppler method has detected several dozen super-Earths, rocky planets with three to ten Earth's mass, however, a true Earth clone is just beyond reach (Mayor et al. 2013).

The transit method requires a precision that depends on exoplanet size: 1 % for a Jupiter and 0.01 % for an Earth. Atmospheric turbulence and transmission variations make it impossible to measure variations much less than 0.1 % from the ground, putting Earth's beyond reach. But within the last few years, the European CoRoT satellite and NASA's Kepler satellite have been launched and the stability of the space environment gives much better photometric precision. Very few exoplanet systems will happen to be aligned suitably for a transit, so the strategy is to "stare" at a large patch of sky containing a large number of stars. Kepler uses a one-meter mirror to measure the brightness of 170,000 stars in the direction of the Cygnus constellation every 7 min; after a recent mission extension it will do this for a total of 7 years. Three transits have to be observed to confirm a planet. Once the size is measures by a transit, the Doppler method can be used to measure mass and characterize the exoplanet.

NASA's Kepler mission has blown the lid off the search for low mass planets. The team announced over 1,200 candidates in early 2011, over fifty of which were in their habitable zones, among which five are probably less than twice the Earth's size (Borucki et al. 2011). By early 2012, the number of candidates had grown to over 2,300, nearly 250 of which are less than 1.25 times Earth's size (Batalha et al. 2013). It's just a matter of time before Earth-like planets are found in Earth-like orbits. Mission leader Bill Borucki and his team pitched the project to NASA Headquarters in 1992, but it was rejected as being technically too difficult. In 1994 they tried again, but this time it was rejected as being too expensive. In 1996, and then again in 1998, the proposal was rejected on technical grounds, even though lab work had proved the concept and exoplanets had recently been discovered. By the time the project was finally given the go ahead as a NASA Discovery class mission in 2001, the first transits had been detected from the ground. Kepler launched in 2009 and it promises to rewrite the book on exoplanets. Persistence paid off.

10.9 Habitable Real Estate

Astronomers adhere to a conventional and conservative definition of habitability: the zone around a star within which water can be in stable liquid form on the surface of a rocky planet. This calculation is strongly affected by atmospheric thickness and composition; Venus and the Earth are similar in mass and size and would be equally detectable by the Doppler or transits methods, yet Venus is almost certainly uninhabitable due to a strong Greenhouse effect. Another complication is the fact that habitable zones evolve as stars age and the amount of radiation they deliver changes, and planetary atmospheres also evolve due to geological activity (and of course, life). The definition is conservative because it supposes that stellar radiation is the only energy source that can power biology. On Earth, life can exist above the boiling point of water and below its freezing point, and in total darkness on the sea floor or deep inside rock. In the Solar System, there may be a dozen habitable "spots," many of which are in a cryogenic habitable zone where icy and rocky surfaces conceal water kept liquid by pressure, geological heating and, in the case of moons around giant planets, tidal heating. Enceladus provides the perfect example of a Solar System body that may harbor microbial life and yet is completely unnoticeable in a distant solar system.

Planet hunters have concentrated on Sun-like stars for obvious reasons, but simple arguments suggest that the habitable "real estate" around dwarf stars far exceeds that around Sun-like stars, motivating new wide-field surveys for transits associated with stars much nearer and brighter than Kepler's faint targets. Observational selection effects favor the detection of Earths around M stars rather than G stars in almost every way. In fact the two worlds closest to habitability discovered so far are Gliese 581 c and d, in orbit around an M dwarf (Mayor et al. 2009). Exoplanet research is a burgeoning but still young field, with many observational and theoretical puzzles to solve before we can confidently project a number of habitable worlds (Baraffe et al. 2010). However, rough estimates based on the relatively unbiased method of microlensing suggest at least one planet per star in the Milky Way, or a total of 100 billion (Cassan et al. 2012). That conservatively (but uncertainly) projects to 100 million terrestrial planets around Sun-like stars in the Milky Way, several million of which are probably both Earth-like and habitable.

10.10 Biomarkers and Life

Biomarkers are required to take the huge step forward from demonstrating habitability to the first detection of life beyond Earth. That detection—keenly anticipated by all astrobiologists and by members of the general public with an interest in science—might come in the form of a shadow biosphere on our planet, from trace fossils in a Mars rock, from future exploration of targets in the outer Solar System, from a spectral signature in the atmosphere of an extrasolar planet, or even from success in the campaign to detect signals from remote civilizations. Each possibility implies a different type of evidence, which must be matched against very uncertain criteria for the definition of success.

Mars gives an indication of the challenges in life detection. It is in our cosmic back yard and we have landed over a dozen probes on it and mapped the entire surface with a resolution of a couple of meters. Geochemical traces in the Martian meteorite ALH 84001 and the more recent remote sensing of methane seemed to implicate biological activity, but in both cases we're left with the Scottish verdict "not proven" (McKay et al. 1996; Mumma 2009). If there is extant life on Mars, it is likely to be in a subsurface aquifer that is beyond the reach of any lander that has yet been contemplated. Titan presents a different conundrum. We simply do not have a basis in lab experiments or in a general theory of biochemistry to predict what to look for (e.g. Bains 2004). All astrobiology is based on life as we know it rather than life as it could be.

Extraolar planets simplify the problem because the bar is set at the global alteration of atmospheric composition by metabolic processes. An important observational advance in the early 2000s was taking spectra of stars during exoplanet transits; the exoplanet atmosphere is backlit by the star, which imprints extra absorption due to constituents in the atmosphere of the exoplanet (Charbonneau et al. 2002). Alternatively, the star can be used as a natural coronagraph to enable an emission or a reflection spectrum to be taken of the exoplanet at different phases (e.g. Charbonneau et al. 2005; Knutson et al. 2007). These difficult observations require the stability of the Hubble and Spitzer Space Telescopes and they have been done for less than a dozen objects, all gas giants. But new information can be derived with transit spectroscopy, including albedo, "weather," and hints of atmospheric composition (Seager and Deming 2010). At infrared wavelengths, H_2O, CO, CO_2 and CH_4 have been detected (Tinetti et al. 2010).

This work is "proof of concept" for spectroscopy of rocky exoplanets that will require upcoming facilities like the James Webb Space Telescope and as-yet-unfunded NASA and ESA missions. Oxygen, and its photolytic product, ozone, are the "gold standards" of biomarkers because their reactivity means they are rapidly depleted on any Earth-like planet without continual replenishment by biogenic photosynthesis. Methane and nitrous oxide are also good biomarkers. Even with the 6.5 m JWST, these observations are extremely challenging. The stars that Kepler is studying in one small patch of sky are thousands of times fainter than the bright stars that will yield the most sensitive Doppler measurements, so wide but shallow surveys are needed to identify the closest Earth-like planets as biomarker targets. In practice, a suite of biomarkers will be needed to confidently assert microbial life on another planet, bolstered by simulations and lab experiments (Kaltenegger et al. 2010). Yet this may be the approach that yields the first detection of life beyond Earth.

10.11 Conclusion

The Copernican Principle has been robust enough to bear our weight at every turn in the long history of astronomy. Our situation on a rocky planet that orbits a middle-weight star on the outskirts of an unexceptional spiral galaxy appears not be unusual or unique. In just two decades astronomers have come close to measuring two terms in the Drake equation: the fraction of stars in the Milky Way that have planets, and the number of planets per system that can potentially support life. A conservative estimate might be a billion habitable "spots"—terrestrial planets in conventionally defined habitable zones, plus moons of giant planet harboring liquid water—in the Milky Way alone (Impey 2011). That number must be multiplied by 10^{11} for the number of "Petri dishes" in the observable cosmos. Do we imagine that they are all stillborn and inert? Or do we think a significant fraction of them host biological experiments, either like or unlike the experiment that took place on Earth? That is the central question of astrobiology, and it feels like we're finally getting much closer to the answer.

Acknowledgments   Astrobiology is not my research specialty so I gain knowledge and stay current thanks to the good graces and patience of many of my colleagues. In particular, this chapter has benefitted from conversations with John Baross, Alan Boss, Adam Burrows, David Charbonneau, Debra Fischer, Lisa Kaltenegger, Jonathan Lunine, Geoff Marcy, Sara Seager, and Giovanna Tinetti, Any errors or misrepresentations that remain are, however, purely my own.

References

Bains, William. 2004. "Many Chemistries Could Be Used to Build Living Systems." *Astrobiology* 4: 137–167.

Baraffe, Isabelle, Gilles Chabrier, and Travis Barman. 2010. "The Physical Properties of Exoplanets." *Reports on Progress in Physics* 73: 16901–16940.

Batalha, Natalie M., et al. 2013. "Planetary Candidates Observed by Kepler. III. Analysis of the First 16 Months of Data." *The Astrophysical Journal Supplements* (in press).

Bond, Ian, et al. 2004. "OGLE 2003-BLG-235/MOA 2003-BLG-53: A Planetary Microlensing Event." *The Astrophysical Journal Letters* 606: 155–158.

Borucki, William J., et al. 2011. "Characteristics of Planetary Candidates Observed by Kepler, II. Analysis of the First Four Months of Data." *The Astrophysical Journal* 736: 19–40.

Boss, Alan P. 2009. *The Crowded Universe: The Search for Living Planets.* New York, NY: Basic Books.

Campbell, Bruce, Gordon Walker, and Stevenson Yang. 1988. "A Search for Sub-stellar Companions to Solar-type Stars." *Astronomical Journal* 331: 902–921.

Cassan, Arnaud, et al. 2012. "One or More Bound Planets per Milky Way Star from Microlensing Observations." *Nature* 481: 167–169.

Charbonneau, David, et al. 2000. "Detection of Planetary Transits across a Sun-like Star." *The Astrophysical Journal Letters* 529: 45–48.

Charbonneau, David, et al. 2002. "Detection of an Extrasolar Planet Atmosphere." *The Astrophysical Journal* 568: 377–384.

Charbonneau, David, et al. 2005. "Detection of Thermal Emission from an Extrasolar Planet." *The Astrophysical Journal* 626: 523–529.

Chauvin, Gael, et al. 2004. "A Giant Candidate Planet around a Young Brown Dwarf. Direct VLT/NACO Observations using IR Wave-front Sensing." *Astronomy and Astrophysics* 425: L29–L32.

Dick, Steven J. 2013. "Lessons from the History of Astrobiology." In *Astrobiology, History, and Society: Life Beyond Earth and the Impact of Discovery*, ed. Douglas A. Vakoch. Heidelberg: Springer.

Gaudi, B.Scott, et al. 2008. "Discovery of a Very Bright, Nearby Gravitational Microlensing Event." *The Astrophysical Journal* 677: 1268–1277.

Hatzes, Artie P., et al. 2003. "A Planetary Companion to γ Cephei A." *The Astrophysical Journal* 599: 1383–1394.

Impey, Chris D, ed. 2010. *Talking About Life: Conversations on Astrobiology.* Cambridge: Cambridge Univ. Press.

Impey, Chris D. 2011. *The Living Cosmos: Our Search for Life in the Universe*. Cambridge: Cambridge Univ. Press.

Jayawardhana, Ray. 2011. *Strange New Worlds*. Princeton, NJ: Princeton Univ. Press.

Kaltenegger, Lisa, et al. 2010. "Deciphering Spectral Fingerprints of Habitable Exoplanets." *Astrobiology* 10: 89–102.

Knutson, Heather A., et al. 2007. "Using Stellar Limb-darkening to Refine the Properties of HD 209458b." *The Astrophysical Journal* 655: 564–575.

Latham, David W. 1989. "The Unseen Companion of HD 114762—a Probable Brown Dwarf." *Nature* 339: 38–40.

Lin, Douglas N.C., Peter Bodenheimer, and Derek C. Richardson. 1996. "Orbital Migration of the Planetary Companion of 51 Pegasi to its Present Location." *Nature* 380: 606–607.

Marcy, Geoff W., and R. Paul Butler. 1998. "Detection of Extrasolar Giant Planets." *Annual Reviews of Astronomy and Astrophysics* 36: 57–97.

Marcy, Geoff W., et al. 2005. "Observed Properties of Exoplanets: Masses, Orbits, and Metallicities." *Progress of Theoretical Physics Supplement* 158: 24–42.

Marois, Christian, Bruce Macintosh, and Travis Barman, et al. 2008. "Direct Imaging of Multiple Planets Orbiting the Star HR 8799." *Science* 322: 1348–1352.

Mayor, Michel, and Didier Queloz. 1995. "A Jupiter-Mass Companion to a Solar-Type Star." *Nature* 378: 355–359.

Mayor, Michel, et al. 2009. "The HARPS Search for Southern Extrasolar Planets. XVIII. An Earth-mass Planet in the GJ 581 Planetary System." *Astronomy and Astrophysics* 507: 487–494.

Mayor, Michel, et al. 2013. "The HARPS Search for Southern Extrasolar Planets XXXIV. Occurrence, Mass Distribution and Orbital Properties of Super-Earths and Neptune-mass Planets." *Astronomy and Astrophysics* (in press).

McKay, David, et al. 1996. "Search for Past Life on Mars: Possible Relic Biogenic Activity in Martian Meteorite ALH84001." *Science* 273: 924–930.

Mumma, Michael J. 2009. "Strong Release of Methane on Mars." *Science* 323: 1041–1045.

Perryman, Michael A.C. 2011. *The Exoplanet Handbook*. Cambridge: Cambridge Univ. Press.

Raulin Cerceau, Florence. 2013. "Planetary Habitability in Historical Context." In *Astrobiology, History, and Society: Life Beyond Earth and the Impact of Discovery*, ed. Douglas A. Vakoch. Heidelberg: Springer.

Raymond, Sean N., Tom Quinn, and Jonathan I. Lunine. 2004. "Making Other Earths: Dynamical Simulations of Terrestrial Planet Formation and Water Delivery." *Icarus* 168: 1–17.

Seager, Sara, and Drake Deming. 2010. "Exoplanet Atmospheres." *Annual Reviews of Astronomy and Astrophysics* 48: 631–672.

Tinetti, Giovanna, et al. 2010. "Probing the Terminator Region Atmosphere of the Hot-Jupiter XO-1b with Transmission Spectroscopy." *The Astrophysical Journal Letters* 712: 139–142.

Udry, Stephane, and Nuno C. Santos. 2007. "Statistical Properties of Exoplanets." *Annual Reviews of Astronomy and Astrophysics* 45: 397–439.

Van de Kamp, Peter. 1969. "Alternate Dynamical Analysis of Barnard's Stars." *Astronomical Journal* 74: 757–759.

Wolszczan, Alexander, and Dail A. Frail. 1992. "A Planetary System around the Millisecond Pulsar PSR1257+12." *Nature* 355: 145–147.

Chapter 11
Extraterrestrial Life in the Microbial Age

Aaron L. Gronstal

Abstract Humankind has long been fascinated with the potential for alien civilizations within the Solar System and beyond (e.g., Crowe and Dowd 2013; Sullivan 2013). Despite the early optimism for life beyond Earth, humankind has yet to make *first contact* with an alien race. Historical discourse on the topic of alien life can provide some useful input into questions about how the people of Earth today might respond to contact with alien life (e.g., Dick 2013). However, this discourse is primarily devoted to understanding humankind's response to intelligent life. We must recognize that the search for life's potential beyond Earth has dramatically changed since the dawn of the Space Age. We now know that advanced civilizations are not common on planets in our solar system. The search for life on nearby worlds is now limited to non-intelligent, microbial life. Any chance we have of contacting intelligent life lies in receiving transmissions from distant worlds, and contact with such cultures would be greatly limited by the vast expanse of space. This chapter discusses the need for more attention paid to the possible social, economic, and legal ramifications that the discovery of non-intelligent, alien microbial life might bring.

11.1 Changing Views of Extraterrestrial

The underlying theme of research encompassed by the many disciplines of *astrobiology* is the search for and understanding of life's *potential* in the Universe. Today, research concerning life beyond planet Earth generally falls into three categories: the search for advanced civilizations beyond our solar system, the search for planets that are habitable for life as we know it, and the search for life's

A. L. Gronstal (✉)
Astrobiology Magazine, 470 Park Avenue South, New York, NY, USA
e-mail: algronstal@gmail.com

D. A. Vakoch (ed.), *Astrobiology, History, and Society*, Advances in Astrobiology and Biogeophysics, DOI: 10.1007/978-3-642-35983-5_11,
© Springer-Verlag Berlin Heidelberg 2013

potential in our own solar system (through robotic missions, the study of biosignatures and chemicals in meteorites, etc.) (Race 2008). A fourth category of study that could have implications similar to the discovery of alien life is the search for 'weird life' or a 'shadow biosphere' on Earth itself (e.g., Wolfe-Simon, Davies and Anbar 2009; Davies et al. 2009; Benner et al. 2010). This forth possibility refers to life forms that operate using a different biology than the life we are familiar with [for instance, life that uses molecules other than deoxyribonucleic acid (DNA) to store information]. In some ways, a shadow biosphere on our planet could be considered 'alien' as it rests outside of our current definition of Earth biology.

Public discourse on the topic of alien life is often dominated by our search for intelligent civilizations; but much of the research being undertaken today is focused on more fundamental examples of living organisms—namely single-cellular life (e.g., Randolph et al. 1997). Intelligent life, and our ability to identify and interact with intelligent life, denotes a level of complexity that we are yet unable to define, much less identify on distant worlds.

Humankind has been posing questions about life beyond our planet ever since our eyes first turned toward the heavens (e.g., Crowe and Dowd 2013; Danielson 2013; Peters 1994; Bonting 2004). Today, as with generations past, many people have an inherent belief that space is populated with a wide array of alien life forms (e.g., Chequers et al. 1996; Oliveira 2008).[1] Science, however, has yet to lend any validity to such beliefs. Alongside ever-changing technologies, our perception of life's potential in the Solar System has undergone dramatic changes over the past centuries (e.g., Crowe and Dowd 2013).

When telescopes were first applied to the field of astronomy in the early 17th century, direct scientific observation of the stars became more accessible (King 1955). The theoretical, philosophical and spiritual interpretations of the heavens made way to include scientific observation, and the contributions of astronomers dramatically altered our understanding of the Solar System. The knowledge we held of life on our own planet Earth shaped the conclusions that early astronomers drew from their observations (e.g., Crowe and Dowd 2013). A few examples include Christiaan Huygens' theory that life on Earth signified a potential for unique inhabitants on the rest of the Solar System's planets that receive light from the Sun (Danielson 2013), and the Russian astronomer Gavriil Adrianovich Tikhov theories of astrobotany on Mars based on his study of Earthshine (Briot 2013).

As the Space Age dawned, our ability to observe the Earth's celestial neighbors in great detail improved—and our hopes of contacting neighboring civilizations on planets and moons of the Solar System came to an abrupt end. The Moon was a barren wasteland of ancient impact craters. Venus was a boiling cauldron of molten rock shrouded in a dense and poisonous atmosphere. And Mars, once the purported home of fantastical canals built by intelligent hands and canyons dense with vegetation, was a barren wasteland of desiccated sand and rock.

---

[1] It should be noted that documentation of this belief in life beyond the Earth is mostly limited to North America and Europe (e.g., Weigel and Coe 2013).

11.2 Microbiology: A New Understanding of Life's Potential

For obvious reasons, the science of microbiology does not predate the invention of the microscope. Early microscopes sprung from the same series of discoveries that led to the telescope. It was in 1664 that Englishman Robert Hooke coined the term *cell* in his seminal book *Micrographia* (Hooke 1664). In the Netherlands, during September of the same year, Antony van Leeuwenhoek observed bacteria in a drop of lake water, and dubbed the organisms *animalicules* (Dobell 1932). However, it wasn't until the mid-1800s that scientists understood all living organisms were composed of cells and that a clear description of germ theory was introduced (Janes et al. 1986). Only by 1875 was the first classification of a bacterium made under the genus *Bacillus* (Drews 2000). Through the end of the 1800s and into the 20th century, discoveries in microbiology came in quick succession. Scientists learned how to culture and identify bacteria in the laboratory, and we began to see the integral role that the microbial biosphere played in exploiting and maintaining the habitability of Earth (e.g., Waksman 1927; Van Niel 1931; Werkman and Wood 1942).

In 1958, Joshua Lederberg, Edward Lawrie Tatum, and George Wells Beadle received the Nobel Prize for their work demonstrating the transfer of DNA from one bacteria to another. Lederberg's work would provide a foundation for the field of bacterial genetics. With his knowledge, Lederberg would also play a central role in NASA's interests in biology and microbiology.

As the Space Age dawned, Lederberg witnessed the flight of Sputnik through the night sky. At the time, he was paying a visit to the English scientist J.B.S. Haldane, who performed some of the early work in origin of life research in the 1920s (Dick and Strick 2005). Lederberg and Haldane both saw a danger in humankind's 'reckless' exploration of the Solar System (Dick and Strick 2005). With his knowledge of microbiology, Lederberg was fearful that space missions could contaminate alien biospheres with Earth bacteria, causing irreversible damage to ecosystems on other worlds. He also feared for the safety of Earth if alien contaminants were returned to our own planet (Morange 2007).

In the same year that Lederberg received the Nobel prize, US President Dwight D. Eisenhower signed the National Aeronautics and Space Act. Lederberg was quick to contact NASA's first administrator, Hugh Latimer Dryden, and was also appointed head of the National Research Council's Space Science Board's panel on extraterrestrial life. Lederberg coined the term 'exobiology,' and played an important role in guiding the early days of NASA research concerning the origin, evolution and distribution of life in the Universe.

As humankind began to expand our influence beyond Earth's atmosphere with observations and robotic missions, our dream of finding a companion among our neighboring planets was dashed; but with our knowledge of the microbial world, scientists still hoped that microbial life could be found in the soils and atmospheres of some planets. Theories of microorganisms in the clouds of Venus or the soils of Mars pervaded the early Space Age, and in many cases still hold weight today (e.g., Cockell 1999; Ivarsson and Lindgren 2010; Morowitz and Sagan 1967; Schulze-Makuch et al. 2004).

In 1975, NASA launched the first missions dedicated to studying life's potential on the surface of Mars. The Viking 1 and 2 missions each had a lander and an orbiter, and each lander carried 14 experiments to the martian surface. This included a set of experiments specifically designed to search for evidence of martian life. The results of the Viking experiments—and the controversy surrounding them—are well documented (Dick and Strick 2005). Ultimately, the scientific community came to the conclusion that no *definitive* evidence of the existence of extant life on Mars had been identified by Viking.

Following Viking, the astrobiology community went through something of an identity crisis (Dick and Strick 2005). Humankind began with the belief that the Solar System would harbor a multitude of inhabited planets filled, like Earth, with an immense diversity of life and culture. As scientific knowledge of life on Earth improved, we became increasingly aware of life's complexity and it's interconnections with the Earth system as a whole. When technology allowed us to make more in-depth observations of our neighboring celestial bodies and we revealed that complex, intelligent life in Solar System was an impossibility, we clung to the idea that simple forms of life could still persist in the harsh environments of the rocky bodies in our solar system. Now, we were faced with the prospect that Earth-like life was restricted to Earth itself.

Today, the prospects for life in our Solar System have undergone somewhat of a revival (Dick and Strick 2005; Olson and Tobin 2008). As technologies have continued to improve, our ability to identify and study life in some of the most extreme environments on Earth has increased by leaps and bounds. With this terrestrial exploration, we have found life forms that thrive in environments that were previously thought to be thoroughly uninhabitable. There was a time when scientists assumed that all energy for life was derived from the Sun. However, science has now allowed us to unravel the cellular methods by which life can survive independent of the Sun—deep in ocean sediments, or kilometers beneath the ground (Cockell et al. 2012; Gronstal et al. 2009; Satyanarayana 2005; Amalie et al. 2006; Li-Hung et al. 2006; Olson and Tobin 2008).

Regions of Earth that were once thought to be void of life actually support diverse microbial ecosystems. This incredible adaptability of life at the microscopic scale has renewed interest in the search for life in our solar system. We again imagine scenarios in which microbial life could gain a foothold in select environments on planets like Mars, particularly early in the planet's history when temperatures were warmer and water is thought to have persisted at the surface (e.g., Pollack and Kasting 1987; Squyres and Kasting 1994). Could microbial ecosystems exist deep below the surface of Mars (e.g., Ivarsson and Lindgren 2010)? Could hydrothermal vents provide the energy for life's origin and evolution beneath the icy crust of Jupiter's moon Europa (e.g., Prieto-Ballesteros et al. 2010)?

11.3 Social Implications

There has been a great deal of discourse concerning the effects of discovering intelligent alien life beyond the Earth (e.g., Dick 2013; Peters 2013). There has also been a fair amount of interest in the potential implications of discovering alien microbial life, particularly following claims in 1996 that alien microfossils had been discovered in the Mars meteorite Allan Hills 84001 (ALH 84001) (e.g., Bertka 2009; Jones 2013; McKay et al. 1996; Olson and Tobin 2008). However, considering that the search for life in our solar system is now limited to microorganisms, and current missions and scientific investigations could theoretically yield positive results in the very near future, there is still insufficient data concerning what our response could or should be (Race 2008). The SETI[2] community has led international discussion on how humankind should respond to a signal from an alien civilization, which has produced the 'SETI principles' that could act as some kind of guideline for first contact (Race 2008). A similar framework does not exist for the discovery of alien microbial life (Race and Randolph 2002). There is no guideline for how humankind should or would respond to the discovery of 'non-intelligent' life, even though such a discovery could have profound scientific, governmental, legal and societal implications (Race 2008).

There are completely different issues in terms of the legal, ethical and societal implications of finding microbial life beyond the Earth when compared to contacting intelligent life (Race 2008). For instance, the discovery of life on a planet like Mars, and the issues surrounding forward and backward contamination,[3] have already been addressed in a legal framework in the form of the Outer Space Treaty of 1967. What is not entirely clear are the social and legal implications of scientific discoveries; including issues like patent rights, commercialization, extraterrestrial property and resource rights, and environmental ethics (Race 2008; Olson and Tobin 2008).

The general public now has a working knowledge of microorganisms, yet we obviously cannot interact with bacteria and archaea on the same level as with complex organisms. Because the public in general now understands and accepts that microorganisms exist all around us (and inside of us) in nearly unfathomable numbers, it may not come as much of a surprise if we find them on other worlds in our solar system. Even so, the discovery of a native microbiology on planets such as Mars or moons like Saturn's Enceladus could still have an impact on societal and theological perceptions of the existence of life on Earth (Lowrie 2013).

The discovery of a second *origin* of life beyond Earth could directly challenge interpretations of creation stories that exist in many ancient and modern religions. In particular, if alien microbes had a distinct and independent origin from those of

---

[2] Search for Extraterrestrial Intelligence (SETI).

[3] For example, the contamination of Mars by organisms originating from Earth, or the contamination of Earth by organisms originating from Mars.

Earth, this "Second Genesis" of life could have profound meaning to many theological doctrines concerning the origin and meaning of life.

The history of humankind's discourse on the potential for alien life holds value in that it has provided a framework for how we can analyze first contact with a non-intelligent alien life form (Olson and Tobin 2008; Peters 1994). In fact, many major religions have broached the subject (e.g., Bertka 2013). It has been suggested by Olson and Tobin (2008) that such a discovery would only pose a problem for more fundamentalist religious traditions, whereby literal interpretations of scripture tend to be Earth-centric. Ultimately, many major religions have reconciled the possibility of non-intelligent life and have come to the conclusion that the discovery of microbial life beyond Earth would have no detrimental effect on the interpretation of current beliefs (Olson and Tobin 2008). In fact, it may present a 'mandate' for humankind's stewardship over the safety and cultivation of such life (Olson and Tobin 2008).

Contact with alien microbes would bring with it discussions of how theology's more Earth-centric view of creation and existence translate to a broader reality. However, we cannot disregard the fact that microorganisms have no discernible 'consciousness' that we can yet identify. The discovery of microorganisms on Mars would not challenge concepts that we are alone and unique as 'intelligent' life in the Universe. Microbial life forms would also not present their own native theologies or beliefs that could challenge or contradict religious views on Earth in the same way that an intelligent alien culture might. In terms of the general perception of alien microbial life, it is possible that the public will recognize the importance the discovery in terms of our understanding of life's existence in the Universe, but our Earth-centric view of the *value* of our own existence would not be challenged. The differences between extraterrestrial microorganisms and those native to Earth may be dramatic at the cellular or molecular level when viewed under the microscope, but these differences could easily be dismissed by a lay person who takes the existence of microorganisms, which are of course invisible to the naked eye, on faith alone.

The most likely sentiment that would resonate in the public is *not* concern over the religious implications of the discovery, but of the potential for contamination of our home planet. Society is most familiar with microbial life in the form of bacteria or 'germs' that cause illness. The existence of an alien microbe would likely generate a fear of alien disease, ala numerous stories in popular science fiction and fantasy. This is perhaps why issues of contamination have been a major thread through scientific exploration of the Solar System since the days of Joshua Lederberg—and why these concerns were addressed as early as 1967 with the Outer Space Treaty (Dembling and Arons 1967).

We may not be able to interact with microbial life on a 'social' level, but the presence of alien microbiology within our solar system would, in a sense, be more immediately interactive than an intelligent civilization on a distant, extra solar world. We now know that there is no intelligent life that we can identify in our solar system beyond the Earth. Our most likely identification of intelligent life will be in the form of a SETI-type signal from a distant world, with a significant gap in time between the signal's origination and when we receive it. Over the vast distances of space, actual *interaction* with an alien civilization may be extremely difficult and communication would incur significant delays in call and response times.

The discovery of a microorganism on Mars, on the other hand, would allow relatively immediate opportunities to interact with the alien life (of course, with consideration for the relevant legal and safety implications and the time it takes to develop a mission). Missions can be sent to Mars, and other locations in our solar system, to sample, retrieve and perform scientific investigations. The amount of data that could be gathered from 'first contact' with an alien microbe as opposed to an alien civilization could be, in many way, vastly greater and more complete than the data incurred by contacting an alien civilization (Race 2008).

11.4 Economic Implications

The advent of microbiology and molecular biology has also revealed that microbial ecosystems hold immense economic value. Microbes that inhabit some of Earth's most 'extreme' environments have provided chemical and molecular products that have completely changed areas of medicine and industry (e.g., Aguilar 2006; Dijkshoorna 2010; Gomes and Steimer 2004; van den Burg 2003; Kumar 2011). The unique conditions under which these microbes have evolved have allowed them to produce novel responses to environmental pressures. Microorganisms are now known to survive in a wide array of habitats previously assumed to be uninhabitable. They grow and reproduce in environments that are highly acidic, desiccated, high in radiation, low in nutrients and at extreme temperatures both hot and cold.

Microbial ecosystems on other worlds may or may not be biologically similar to life on Earth; but if they are, their adaptation to unique environmental stresses would likely produce a unique set of biomolecules that could have incredible value in areas of biotechnology and medicine. The economic value of extremophile research on Earth has been widely recognized by private and governmental institutions around the world (e.g., Aguilar 2006; Schiraldi 2002), and alien life on planets and moons that are accessible in our solar system would potentially provide a new set of laboratories in which this research could continue. In the same way that extraterrestrial civilizations could have been a unique source of trade and industry, providing humankind with new knowledge and technologies; the existence of microbial biospheres on other planets could hold a wealth of unimagined opportunities in economically relevant fields. Discovering unique proteins and enzymes produced by 'martian' cells may not be quite the same as opening trade relations with a martian civilization—but the economic and cultural implications for humankind would be similarly profound.

Some scientists have also posited theories of 'weird' life on Saturn's moon Titan, where organisms might rely on liquid methane as a solvent for cellular functions rather than the familiar liquid water of Earth (Benner, Ricardo and Carrigan

2004). Although possibly far-fetched, the discovery of life that operates under completely different conditions, such as a genetic system other than deoxyribonucleic acid (DNA), would also have profound implications for life on Earth. A 'second origin' for life may or may not provide valuable biomolecules that could function in an Earth-life system. Yet, having a second example of life would open an entirely new field of comparative biology. Much like comparative planetology, where the study of other planets like Venus can be used to understand planetary and climate processes on Earth, comparative biology could provide new insights into how biology functions—such as alternative routes for origins of life and evolutionary processes.

We currently know very little about the likelihood of life originating on Earth-like worlds in the Universe. Our ability to improve our knowledge is hindered by one simple fact—we have only one example of a habitable world thus far. Clever mathematicians have previously attempted to draw some sort of estimate on the likelihood of alien life based on this 'limited' data set—many concluding that life should be rare (e.g., Spiegel et al. 2011) and others that life has the potential to be plentiful (e.g., Lineweaver and Davis 2002; Michaud 2007).

We do not know the probability of life arising from pre-biotic environments, and we do not know how common such environments are in the Universe. We only know that life arose at least once, and can therefore not make a statistical estimate of life's prevalence beyond the Earth (Spiegel and Turner 2011). The 'optimistic' approach, as cited by Spiegel and Turner (2011), assumes that because life arose so quickly on the Earth after the conditions and climate were right, the origin of life must surely be a common process in the Universe. Rather than attempting to estimate life's prevalence in the Universe, Spiegel and Turner (2011) instead used a Bayesian statistical framework to estimate the frequency of life's origin on Earthlike worlds. Their study focuses on the elapsed time between when the conditions for life's origins arise on a planet, and when life actually arises (i.e., the time in which it took Earth to go from 'habitable' to 'inhabited').

Of course, the lack of data concerning life's abundance in the Universe (and the abundance of habitable worlds) means that there are numerous problems in such statistical estimates. Spiegel and Turner (2011, 395) readily admit this, and they sight a number of assumptions that need to be made in order to perform their calculations. However, one important conclusion of their work is this:

> Finding a single case of life arising independently of our lineage (on Earth, elsewhere in the Solar System, or on an extra solar planet) would provide much stronger evidence that abiogenesis is not extremely rare in the Universe.

We cannot begin to estimate or truly understand the abundance of life in the Universe until we identify at least one other instance of life's occurrence. If the most likely scenario for discovering alien life is to identify a microorganism on a world like Mars or Europa (or a second and independent example of life's origin on Earth)—then this discovery could help lay the foundation for a more accurate understanding of whether or not multiple origins of life have occurred in the Universe. While an alien microbe may not bring the same shock and awe as a message from an alien civilization—it's discovery would go a long way in letting us know whether or not we should be preparing for a call from somewhere deep among the stars.

11.5 Conclusion

From academia to religion and literature to film, there has been a wide range of discussion concerning our response to *first contact* with an intelligent alien culture. Academics, theologians and scientists have constructed various responses to the potential discovery of intelligent life and guidelines for how humankind should or possibly would respond. However, in light of our current efforts of exploration in the Solar System and beyond, it seems more likely that the first evidence of alien life will come in the form of a microorganism—and possibly the fossil remnants of a long-extinct alien microbe. Even if the discovery of martian or europan microorganisms would not necessarily cause any dramatic changes in societal or theological *perceptions* of life on Earth, it could cause dramatic and far-reaching waves in the technological and economical systems of humankind.

If we are to learn from historic conceptions of life beyond our planet, and provide useful criteria for how we might respond to *first contact* if and when it happens, more effort must be made to understand the potential social, economic, and legal ramifications of discovering non-intelligent life.

References

Aguilar, Alfredo. 2006. "Extremophile Research in the European Union: From Fundamental Aspects to Industrial Expectations." *FEMS Microbiology Reviews* 18 (2–3): 89–92.

Benner, Steven A., Hyo-Joong Kim, and Zunyi Yang. 2010. "Setting the Stage: The History, Chemistry, and Geobiology behind RNA." *Cold Spring Harbor Perspectives in Biology* 4 (1).

Benner, Steven A., Alonso Ricardo, and Matthew A. Carrigan. 2004. "Is There a Common Chemical Model for Life in the Universe?" *Current Opinion in Chemical Biology* 8 (6): 672–689.

Bertka, Constance M. (ed.) 2009. *Exploring the Origin, Extent, and Future of Life: Philosophical, Ethical, and Theological Perspectives*, Cambridge: Cambridge University Press.

Bertka, Constance M. 2013. "Christianity's Response to the Discovery of Extraterrestrial Intelligent Life: Insights from Science and Religion and the Sociology of Religion." In *Astrobiology, History, and Society: Life Beyond Earth and the Impact of Discovery*, ed. Douglas A. Vakoch. Heidelberg: Springer.

Briot, Danielle. 2013. "The Creator of Astrobotany, G. A. Tikhov." In *Astrobiology, History, and Society: Life Beyond Earth and the Impact of Discovery*, ed. Douglas A. Vakoch. Heidelberg: Springer.

Bonting, Sjoerd L. 2004. "Theological Implications of Possible Extraterrestrial Life." *Sewanee Theological Review* 4: 420–435.

Carter, Brandon, and William H. McCrea. 1983. "The Anthropic Principle and Its Implications for Biological Evolution." *Royal Society of London Philosophical Transactions Series A* 310 (1512): 347–363.

Chequers, James, Stephen Joseph, and Debbie Diduca. 1996. "Belief in Extraterrestrial Life, UFO-related Beliefs, and Schizotypal Personality." *Personality and Individual Differences* 23 (3): 519–521.

Cockell, Charles S. 1999. "Life on Venus." *Planetary and Space Science* 47 (12): 1487–1501.

Cockell, Charles S., Mary A. Voytek, Aaron L. Gronstal, Kai Finster, Julie D. Kirshtein, Kieren Howard, Joachim Reitner, Gregory S. Gohn, Ward E. Sanford, J. Wright Horton Jr., Jens Kallmeyer, Laura Kelly, and David S. Powars. 2012. "Impact Disruption and Recovery of the Deep Subsurface Biosphere." *Astrobiology* 12 (3): 231–246.

Crowe, J.Michael, and Matthew F. Dowd. 2013. "The Extraterrestrial Life Debate from Antiquity to 1900." In *Astrobiology, History, and Society: Life Beyond Earth and the Impact of Discovery*, ed. Douglas A. Vakoch. Heidelberg: Springer.

Danielson, Dennis. 2013. "Seventeenth-century ET, Reflexive Telescopics, and Their Relevance Today." In *Astrobiology, History, and Society: Life Beyond Earth and the Impact of Discovery*, ed. Douglas A. Vakoch. Heidelberg: Springer.

Davies, Paul C.W., Steven A. Benner, Carol E. Cleland, Charles H. Lineweaver, Christopher P. McKay, and Felisa Wolfe-Simon. 2009. "Signatures of a Shadow Biosphere." *Astrobiology* 9 (2): 241–249.

Dembling, Paul G., and Daniel M. Arons. 1967. "The Evolution of the Outer Space Treaty." *Journal of Air Law and Commerce* 33: 419–456.

Dick, Steven J. 2013. "Lessons from the History of Astrobiology." In *Astrobiology, History, and Society: Life Beyond Earth and the Impact of Discovery*, ed. Douglas A. Vakoch. Heidelberg: Springer.

Dick, Steven J., and James E. Strick. 2005. *The Living Universe: NASA and the Development of Astrobiology*. New Brunswick: Rutgers Univ. Press.

Dijkshoorna, Lenie, Paul De Vosb, and Tom Dedeurwaerderec. 2010. "Understanding Patterns of Use and Scientific Opportunities in the Emerging Global Microbial Commons." *Research in Microbiology* 161 (6): 407–413.

Dobell, Clifford. 1932. *Antony Van Leeuwenhoek and his "Little Animals."* New York: Dover Publications, Inc.

Drews, Gerhart. 2000. "The Roots of Microbiology and the Influence of Ferdinand Cohn on Microbiology of the 19th Century." *FEMS Microbiology Reviews* 24 (3): 225–249.

Gomes, Joseph, and Walter Steiner. 2004. "The Biocatalytic Potential of Extremophiles and Extremozymes." *Extremophiles* 42 (4): 223–235.

Gronstal, Aaron L., Mary A. Voytek, Julie D. Kirshtein, Nicole M. von der Heyde, Michael D. Lowit, and Charles S. Cockell. 2009. "Contamination Assessment in Microbiological Sampling of the Eyreville Core, Chesapeake Bay Impact Structure." *GSA Special Papers* 458: 951–964.

Hooke, Robert. 1664. *Micrographia: Or Some Physiological Descriptions of Minute Bodies Made by Magnifying Glasses with Observations and Inquiries thereupon*. London: Martyn and Allestry.

Ivarsson, Magnus, and Paula Lindgren. 2010. "The Search for Sustainable Subsurface Habitats on Mars, and the Sampling of Impact Ejecta." *Sustainability* 2 (7): 1969–1990.

Janes, Craig Robert, Ron Stall, and Sandra M. Gifford. 1986. *Anthropology and Epidemiology: Interdisciplinary Approaches to the Study of Health and Disease*. Boston: D. Reidel Publishing Company.

Jones, Morris. 2013. "Mainstream Media and Social Media Reactions to Extraterrestrials." In *Astrobiology, History, and Society: Life Beyond Earth and the Impact of Discovery*, ed. Douglas A. Vakoch. Heidelberg: Springer.

King, Henry. 1955. *The History of the Telescope*. High Wycombe: Charles Griffin & Company Ltd.

Kumar, Lokendra, Gyanendra Awasthi, and Balvinder Singh. 2011. "Extremophiles: A Novel Source of Industrially Important Enzymes." *Biotechnology* 10: 121–135.

Lin, Li-Hung, Pei-Ling Wang, Douglas Rumble, Johanna Lippmann-Pipke, Erick Boice, Lisa M. Pratt, Barbara Sherwood Lollar, Eoin L. Brodie, Terry C. Hazen, Gary L. Anderson,

Todd Z. DeSantis, Duane Moser, Dave Kershaw, and Tullis C. Onstott. 2006. "Longterm Sustainability of a High-Energy, Low-Diversity Crustal Biome." *Science* 314: 479–482.

Lineweaver, Charles H., and Tamara M. Davis. 2002. "Does the Rapid Appearance of Life on Earth Suggest that Life is Common in the Universe?" *Astrobiology* 2 (2): 293–304.

Lowrie, Ian. 2013. "Cultural Resources and Cognitive Frames: Keys to an Anthropological Approach to Prediction." In *Astrobiology, History, and Society: Life Beyond Earth and the Impact of Discovery*, ed. Douglas A. Vakoch. Heidelberg: Springer.

Michaud, Michael A.G. 2007. *Contact with Alien Civilizations: Our Hopes and Fears About Encountering Extraterrestrials*. New York: Copernicus Books, Springer Science + Business Media.

McKay, David S., Everett K. Gibson Jr., Kathie L. Thomas-Keprta, Hojatollah Vali, Christopher S. Romanek, Simon J. Clemett, Xavier D.F. Chillier, Claude R. Maechling, and Richard N. Zare. 1996. "Search for Past Life on Mars: Possible Relic Biogenic Activity in Martian Meteorite ALH84001." Science 273 (5277): 924–930.

Morange, Michel. 2007. "What History Tells Us: X. Fifty Years Ago: The Beginnings of Exobiology." *Journal of Biosciences* 32 (6): 1083–1087.

Morowitz, Harold, and Carl Sagan. 1967. "Life in the Clouds of Venus?" *Nature* 215: 1259–1260.

Oliveira, Carlos. 2008. "Astrobiology for the 21st century." *CAP Journal: Communicating Astronomy with the Public* 2: 24–25.

Olson, A. Randall, and Vladimir V.M. Tobin. 2008. "An Eastern Orthodox Perspective on Microbial Life on Mars." *Theology and Science* 6 (4): 421–437.

Pakchung, Amalie A.H., Philippa J.L. Simpson, and Rachel Codd. 2006. "Life on Earth: Extremophiles Continue to Move the Goal Posts." *Environmental Chemistry* 3 (2): 77–93.

Peters, Ted. 1994. "Exo-Theology: Speculations on Extra-Terrestrial Life." *CTNS Bulletin* 14: 1–9.

Peters, Ted. 2013. "Would the Discovery of ETI Provoke a Religious Crisis?" In *Astrobiology, History, and Society: Life Beyond Earth and the Impact of Discovery*, ed. Douglas A. Vakoch. Heidelberg: Springer.

Pollack, James B., James F. Kasting, Steven M. Richardson, and K. Poliakoff. 1987. "The Case for a Wet, Warm Climate on Early Mars." *Icarus* 71 (2): 203–224.

Prieto-Ballesteros, Olga, Elena Vorobyova, Victor Parro, Jose A. Rodriguez Manfredi, and Felipe Gómez. 2010. "Strategies for Detection of Putative Life on Europa." *Advances in Space Research* 48 (4): 678–688.

Race, Margaret S., and Richard O. Randolph. 2002. "The Need for Operating Guidelines and a Decision-Making Framework Applicable to the Discovery of Non-Intelligent Extraterrestrial Life." *Advances in Space Research* 30: 1583–1591.

Race, Margaret S. 2008. "Communicating About the Discovery of Extraterrestrial Life: Different Searches, Different Issues." *Acta Astronautica* 62: 71–78.

Randolph, Richard O., Margaret S. Race, and Christopher P. McKay. 1997. "Reconsidering the Theological and Ethical Implications of Extraterrestrial Life." *CTNS Bulletin* 17: 1–8.

Satyanarayana, Tulsi, Chandralata Raghukumar, and S. Shivaji. 2005. "Extremophilic Microbes: Diversity and Perspectives." *Current Science* 89 (1): 78–90.

Schiraldi, Chiara, and Mario De Rosa. 2002. "The Production of Biocatalysts and Biomolecules from Extremophiles." *Trends in Biotechnology* 20 (12): 515–521.

Schulze-Makuch, Dirk, David H. Grinspoon, Ousama Abbas, Louis N. Irwin, and Mark A. Bullock. 2004. "A Sulfur-Based Survival Strategy for Putative Phototrophic Life in the Venusian Atmosphere." *Astrobiology* 4 (1): 11–18.

Spiegel, David S., and Edwin L. Turner. 2011. "Bayesian Analysis of the Astrobiological Implications of Life's Early Emergence on Earth." *Proceedings of the National Academy of Sciences of the United States of America (PNAS).* 109 (2): 395–400.

Squyres, Seven W. and James F. Kasting. 1994. "Early Mars: How Warm and How Wet?" *Science* 265 (5173): 744–749.

Sullivan, Woodruff T., III. 2013. "Extraterrestrial Life as the Great Analogy, Two Centuries Ago and in Modern Astrobiology." In *Astrobiology, History, and Society: Life Beyond Earth and the Impact of Discovery*, ed. Douglas A. Vakoch. Heidelberg: Springer.

van den Burg, Bertus. 2003. "Extremophiles as a Source for Novel Enzymes." *Current Opinion in Microbiology* 6 (3): 213–218.

Van Niel, B. Cornelius, and F.M. Muller. 1931. "On the Purple Bacteria and Their Significance for the Study of Photosynthesis." *Rec. Trav. Bot Neer* 28: 245–274.

Waksman, Selman A. 1927. *Principles of Soil Microbiology*. Baltimore: Williams and Wilkins Company.

Weigel, M. Margaret, and Kathryn Coe. 2013. "Impact of Extraterrestrial Life Discovery for Third World Societies: Anthropological and Public Health Considerations." In *Astrobiology, History, and Society: Life Beyond Earth and the Impact of Discovery*, ed. Douglas A. Vakoch. Heidelberg: Springer.

Werkman, Chester Hamlin, and Harland Goff Wood. 1942. "On the Metabolism of Bacteria." *The Botanical Review* 8 (1): 1–68.

Wolfe-Simon, Felisa, Paul C.W. Davies, and Ariel D. Anbar. 2009. "Did Nature Also Choose Arsenic?" *International Journal of Astrobiology* 8: 69–74.

Part III
Societal Impact of Discovering Extraterrestrial Life

Chapter 12
The Societal Impact of Extraterrestrial Life: The Relevance of History and the Social Sciences

Steven J. Dick

Abstract This chapter reviews past studies on the societal impact of extraterrestrial life and offers four related ways in which history is relevant to the subject: the history of impact thus far, analogical reasoning, impact studies in other areas of science and technology, and studies on the nature of discovery and exploration. We focus particularly on the promise and peril of analogical arguments, since they are by necessity widespread in the field. This chapter also summarizes the relevance of the social sciences, particularly anthropology and sociology, and concludes by taking a closer look at the possible impact of the discovery of extraterrestrial life on theology and philosophy. In undertaking this study we emphasize three bedrock principles: (1) we cannot predict the future; (2) society is not monolithic, implying many impacts depending on religion, culture and worldview; (3) the impact of any discovery of extraterrestrial life is scenario-dependent.

12.1 Introduction: Past Studies and General Principles

The question of the societal impact of extraterrestrial life has received increasing attention in the last two decades, since John Billingham, the head of NASA's SETI program at the time, convened a series of workshops on "The Cultural Aspects of SETI" (CASETI) on the eve of the inauguration of NASA SETI observations in 1992 (Billingham et al. 1999). During the formulation and initiation of the first Astrobiology Roadmap in 1998 (Des Marais et al. 2008), calls were made for the study of cultural impacts of astrobiology (Dick 2000b), and in 1999 NASA Ames Research Center organized a workshop on the societal implications of astrobiology (Harrison and Connell 2001). Other organizations, including the John Templeton Foundation and the Foundation For the Future, organized meetings

S. J. Dick (✉)
National Air and Space Museum, Washington, DC, USA
e-mail: stevedick1@comcast.net

D. A. Vakoch (ed.), *Astrobiology, History, and Society*, Advances in Astrobiology and Biogeophysics, DOI: 10.1007/978-3-642-35983-5_12,
© Springer-Verlag Berlin Heidelberg 2013

on the subject at about the same time (Dick 2000c; Harrison and Dick 2000; Tough 2000). Interest has increased in the last decade, notably with the American Association for the Advancement of Science (AAAS) series of workshops sponsored by its program on Dialogue on Science, Ethics and Religion (Bertka 2010); several meetings at the Royal Society of London (Dominik and Zarnecki 2011); and a series of sessions at the American Anthropological Association (Vakoch 2009).

While most of the attention has focused on the impact of the discovery of extraterrestrial intelligence, the AAAS volume also addresses the quite different scenario of the impact of discovery of microbial life. Most recently Race et al. (2012) has taken the lead in marshalling the astrobiology, social sciences and humanities communities to address these issues in the context of the latest Astrobiology Roadmap, with the support of the NASA Astrobiology Institute. Individual efforts have also concentrated on different aspects of the problem (White 1990; Almar 1995; Davies 1995; Randolph, Race, and McKay 1997; Achenbach 1999; Vakoch 2000; Harrison et al. 2000; Michaud 2007; Arnould 2008; Denning 2009), including a comparison to the impact of other scientific endeavors such as biotechnology (Race 2007).

The results of these studies have been to demonstrate the serious impact that the discovery of extraterrestrial life could have on society, especially in the case of extraterrestrial intelligence. The report on the workshops of the Cultural Aspects of SETI, the first high-level conference on the subject, makes this clear from a variety of perspectives, including history, the social sciences, theology, policy, and education. It provides numerous recommendations for studying possible impacts, some of which have been elaborated in subsequent studies, but most of which remain to be examined in detail. As with subsequent studies, it emphasizes that the discovery of extraterrestrial life will be scenario-dependent. For example, any serious study of the impact of a discovery of extraterrestrial intelligence must categorize the types of contact. A general matrix of scenarios as terrestrial or extraterrestrial, direct or remote, is given in Table 12.1, together with examples from science fiction. Even a brief consideration of the societal implications of SETI demonstrates that the subject is complex, involving matrices embedded within matrices. Nevertheless, these complexities may be approached systematically in discrete parts. A similar kind of matrix, with different parameters, would apply to the discovery of microbial life.

In this chapter we focus first on the relevance of history to the societal impact of extraterrestrial life. We then address other possible approaches, including the social sciences, and especially anthropology. We end by summarizing what has emerged as one of the hottest topics in the field: the debate over the theological implications of extraterrestrial life. In doing so we insist on three bedrock principles: (1) we cannot predict the future; (2) society is not monolithic, implying many impacts depending on the religious, cultural and worldview aspects of each society; (3) the impact of any discovery of extraterrestrial life is scenario-dependent. The impact of the discovery of fossil or living microbial life will presumably be different from the impact of the discovery of extraterrestrial intelligence, and

Table 12.1 Modes of contact with extraterrestrial intelligence (and some representative science fiction scenarios)

	Terrestrial	Extraterrestrial
Direct	Wells	Clarke
	War of the Worlds	*Rendezvous with Rama*[a]
	Clarke	Bradbury
	Childhood's End	*Martian Chronicles*
	ET: The Extraterrestrial	*Alien* (and its sequels)
Indirect	Clarke	Gunn
	2001: A Space Odyssey	*The Listeners*
	McCollum	
	Lifeprobe	
	Hoyle	Sagan
	The Black Cloud	*Contact*[a]

[a] More than one mode of contact takes place

the latter will be different again if we receive a message, different yet again if we decipher a message, and very, very different depending on what the message says. Direct contact is another matter altogether.

We also distinguish the impact of the *idea* of extraterrestrial life from the impact of its *actual discovery*. Some argue that since the majority of the population (at least in the West) already believes life exists beyond Earth, we have already seen the impact. I think not. Others argue that even when the discovery is made the impact will not be very great. That may be true in an immediate sense, as the Copernican worldview had little immediate impact. But in the long run, just as Copernicus changed everything, so, I believe, will the actual discovery of extraterrestrial life. And we should remember that communications are much more rapid than in the 16th century, not only allowing for the quick spread of reliable information, but also rumors, as the internet era has amply demonstrated.

As a final introductory caution we note the often changing and confusing uses of the terms "society" and "culture." Anthropologist Clifford Geertz defined culture as "an historically transmitted pattern of meanings embedded in symbolic forms by means of which men communicate, perpetuate and develop their knowledge about and attitudes toward life" (Geertz 1973, 289). According to Harvard biologist E. O. Wilson—famed for his work on sociobiology—each society creates culture and is created by it (Wilson 1998). One need not read too much literature in the social sciences to realize that the idea of "culture" is a moving target, evolving with time and in space (Denning 2009, 65). A recent book on the key concepts in social and cultural anthropology put it this way: "Throughout the modernist period, a concept of society has underpinned the construction of all social theory, whatever its hue or denomination. If the concept of culture has played the role of queen to all analytic categories of the human sciences, the notion of society has been king. It is the master trope of high modern social thought" (Rapport and Overing 2000, 333). It is, the authors emphasized, a treacherous friend, a necessary term, but a term to be used at one's risk. In the interdisciplinary field of astrobiology, it is not surprising that the terms have been used interchangeably, and for practical purposes we do not distinguish them here.

12.2 The Relevance of History

The first thing that must be said about the societal impact of extraterrestrial life is that we cannot predict the future. As Yale historian John Lewis Gaddis says in his book *The Landscape of History*, as historians "we pride ourselves on *not* trying to predict the future, as our colleagues in economics, sociology, and political science attempt to do. We resist letting contemporary concerns influence us—the term 'presentism,' among historians, is no compliment. We advance bravely into the future with our eyes fixed firmly on the past: the image we present to the world is, to put it bluntly, that of a rear end" (Gaddis 2002, 2).

How, then can history be relevant to the future, in particular, how can history be relevant to the societal impact of extraterrestrial life? That is the primary question of this section. In this section we distinguish four ways in which history may be illuminate this enterprise: (1) documenting past impact, (2) investigating the validity and role of analogues, (3) utilizing past impact studies in science and technology, (4) analyzing the structure and patterns of discovery and placing the discovery of life beyond Earth in the context of exploration. Many of these involve, in one way or another, the use of analogy or meta-analogy. Therefore a primary focus of this section will be the question of how valid analogy is as an analytical tool. In any case, history arguably may be seen as a long-running set of natural experiments. The importance of history is that it provides real data potentially applicable to our problem; the question—as in politics and other areas of human endeavor—is how to use this data, or at least how to avoid mis-using it.

12.2.1 Past Impact

History can document what impact astrobiology has already had on society. And here I use "astrobiology" in its broadest sense to mean the very idea of that extraterrestrial life might exist. This is a clearly defined historical problem amenable to standard historical methods, as long as we keep in mind the second principle stated above, namely, that "society" is not monolithic and that impacts will likely vary across society. If anything is clear from the histories of the extraterrestrial life debate (Dick 1982, 1996, 1998; Crowe 1986; Guthke 1983), it is that a broad swath of Western society has been "captured by aliens" in the felicitous title of *Washington Post* writer Joel Achenbach (1999). From the popular culture of UFOs, alien science fiction literature and film, to philosophy, religion, and the scientific and scholarly study of microbes and extraterrestrial intelligence, what has happened in the past is open to historical scrutiny, and possible lessons learned—with all the cautions enunciated earlier in this volume (Dick 2013b).

Historians can study the public reaction when scientists or the public *thought* we had found life, or that it had found us, ranging from the Moon Hoax of 1835 (Crowe 1986; Goodman 2008) to the Lowellian canals of Mars in the late 19th century (Crowe 1986; Dick 1996), the famous Orson Welles 1938 radio broadcast of "the War of the Worlds" (Cantril 1940) and the reaction to the claim in 1996 that nanofossils had been found in the Mars rock ALH 84001. For those who think astrobiology will have no impact, these case studies indicate otherwise; even when the subject was Martian nanofossils, let alone intelligent extraterrestrials, the media was full of speculation, theologians were commenting on the impact to religion, and (in the case of the Martian fossils) scientists were at odds over the scientific veracity of the claims. These are all likely to be elements of the reaction to any such discovery in the future.

In addition to such case studies, historians can, and already have to some extent, documented the 500 year-old theological discussion over the impact of extraterrestrial life on religion and philosophy, as we shall see below (Peters 1995; Dick 1996, 2000c; Crowe 1997; Vakoch 2000; Peters 2009; Bertka 2010; Peters 2011), with some concluding that Abrahamic religions with a godhead might be more affected by contact with extraterrestrials than Eastern religions. Surely, at the very least, scholars today should be aware of historical discussions of the societal impact of extraterrestrial life, not only for theology but also for other aspects of culture. In a broader sense the history of the extraterrestrial life debate provides context for the modern problem, as history at its best always does for any problem. Those ignorant of history may not necessarily be condemned to repeat it, but they will be far less enlightened than those who are aware of history.

12.2.2   The Validity and Role of Analogues

This leads to my second point: the validity and role of analogues. As already shown in Chaps. 3 and 4 in the first section of this volume, analogues have been used throughout history, both inside and outside the extraterrestrial life debate (Sullivan 2013; Ross 2013). History offers the opportunity to study analogues in a variety of forms. For those skeptical of analogues, or who consider them a fuzzy form of reasoning, two pioneering studies by philosophers should ease the mind. Almost 50 years ago the philosopher of science Mary Hesse penned her classic *Models and Analogies in Science* (Hesse 1966). Here she argued that models and analogies are integral to scientific practice and advancement, and she distinguished positive analogies, negative analogies, and neutral analogies. She pointed to exemplars of analogy in science ranging from the German organic chemist Friedrich Kekulé's dream of a snake with its tail in its mouth that (by Kekulé's own account) helped him arrive at the structure of benzene, to wave models for sound and light, and the use of billiard balls in random motion as a model for the behavior of gases. She pointed out that, while no account of the snake appears in textbooks of organic chemistry, models have been essential to the logic of scientific theories. Her book is largely a debate about just how essential analogies and models are to science, and she comes down strongly on the side of their overwhelming importance.

The more recent study is philosopher Paul Bartha's volume *By Parallel Reasoning*: *The Construction and Evaluation of Analogical Arguments*. Bartha points out that analogy is a hot topic in artificial intelligence research, psychology and cognitive science, and he lays out a widely accepted process of analogical thinking as follows (Bartha 2010):

(a) Retrieval or access of a relevant "source" analogue
(b) Mapping that sets up systematic correspondences between the elements of the source and "target" analogues
(c) Analogical inference or transfer of information from source to target
(d) Learning of new categories or schemas in the aftermath of analogical reasoning.

This process, I would suggest, could and should be applied to the problem of the impact of astrobiology. Certainly astrobiologists dealing with the microbiological parts of the subject make heavy use of analogy in the most general sense: since we have no extraterrestrial microbes, we use terrestrial microbes as analogues. As anthropologist Stefan Helmreich has emphasized "The shift in attention from alien intelligence to alien nature has suggested a novel methodological strategy to those who would scout for extraterrestrial life. Astrobiologists treat unusual environments on Earth, such as methane seeps and hydrothermal vents, as models for extraterrestrial ecologies. Framing these environments as surrogates for alternative worlds has made marine microbes like hyperthermophiles attractive understudies—what scientists call analogs—for aliens" (Helmreich 2009, 255). While Helmreich is incorrect about this being "a novel methodological strategy," arguably astrobiology would not exist without this most general use of analogy. Put another way, withdrawal of analogy as a form of argument would be an existential threat to the survival of astrobiology as a discipline.

In addition to microbes themselves, astrobiologists employ geographic conditions as analogues to other planetary conditions: to use the language of Bartha, Lake Vostok in the Antarctic is deployed as the source analogue to the target analogues Europa and Ganymede, and the Atacama desert is the source analogue to the target analogue Mars, to name only two specific cases (for others see Pyle 2012, 271 ff). Analogues may also be two-dimensional, combining both microbes and conditions: biogeochemists have recently used observations of microbial mats in sinkholes in Lake Huron (which thrive and persist today under low oxygen concentrations) as models for the low-oxygen early Earth, prior to the Great Oxidation Event 2.4 billion years ago, as well as for the pre-Cambrian, when oxygen concentration was still relatively low (Biddanda, Nold, Dick, et al., 2012). That is an example of analogy working in the backward direction—a current condition (the source analogue) is used to illuminate the past (target). Analogues can also work in the forward direction: climate scientists use past climate change records to predict, controversially to be sure, what may happen in the future. Without analogical reasoning much of science would come to a standstill.

History offers the opportunity to study analogues to the societal impact of astrobiology in a variety of forms that apply to different scenarios. Analogues have most often been applied to contact scenarios with extraterrestrial intelligence (Table 12.1), but some of them, as well as others, may also be used for the discovery of microbial life. Examples of general analogues include: (a) the impact of new worldviews such as the Copernican and Darwinian, as applied to the discovery of both microbial and intelligent life; (b) historical culture contacts on Earth, and unforeseen consequences such as the so-called "Columbian exchange," applied to both microbial and intelligent life scenarios; (c) the transmission of new ideas among cultures, as applied to information deciphered from a SETI signal; (d) cases where extraterrestrial life has been briefly considered a plausible scientific hypothesis, as in the discovery of pulsars, also apply to the intelligent life scenario.

A look at two of these general cases will suffice to illustrate the promise and perils of analogy: information derived from a SETI signal (case c) and the impact of new worldviews (case a), which might follow such a signal, but which could also apply to the discovery of microbial life. Assuming that a SETI signal is deciphered and significant information is transmitted, the flow if information between terrestrial civilizations across time finds a tantalizing analogue in the transmission of Greek and Arabic knowledge by way of the Arabs to the Latin West in the 12th and 13th centuries (Dick 1995). This is an example of what historian Arnold Toynbee, in his massive *Study of History*, called "encounters between civilizations in time" (Toynbee 1957, 241–260). "First a trickle and eventually a flood," one historian of science wrote about this endeavor, "the new material radically altered the intellectual life of the West" (Lindberg 1992, 215). Western Europe, which had been struggling to keep the intellectual flame from being extinguished, now had to assimilate a torrent of new ideas (Grant 1971; Lindberg 1978, 1992). While we do not fancy our civilization analogous to the Middle Ages, the torrent of new ideas would be analogous to a significant flow of information from an extraterrestrial civilization to one probably less knowledgeable but eager to learn. The army of translators involved in the recovery of lost learning in the Middle Ages may find its analogy in the legions of scientists, cryptographers, linguists and others sure to participate in any attempt to decipher an extraterrestrial signal. The result of the newly recovered knowledge is a matter of record. Thomas Aquinas and other scholars, often with agendas of their own, attempted to reconcile the new Greek and Arabic knowledge with Christianity, and with current knowledge—such as it was. The result was the European Renaissance, which spread gradually through the continent. While one cannot guarantee a global terrestrial renaissance based on extraterrestrial knowledge (it might have an opposite and depressing effect), one can project with some certainty that personal and institutional agendas would play a role in deciphering and spreading the information.

Even assuming a message was not deciphered, and perhaps even in the case that microbial life was discovered constituting a "second Genesis," a change in worldview would likely gradually take place. Such changes might be analogous to changes in cosmological worldview, exemplified in the Copernican worldview originated in the 16th century, or the "galactocentric" worldview of the 20th century, in which our solar system was demonstrated to be at the periphery of our Milky Way Galaxy, itself only one of billions of galaxies. The gradual acceptance of the Copernican theory, followed by its triggering of the Scientific Revolution and indeed its impact in all areas of human thought, has now been studied extensively (Kuhn 1957; Blumenberg 1987; Stimson 1972; Westman 1975, 2011). The Copernican theory eventually gave birth to a new physics, caused wrenching controversy in theology, and made the Earth a planet and the planets potential Earths. Gradually, and more broadly, it changed the way humans viewed themselves and their place in the universe. The galactocentric revolution, on the other hand, was more silent in nature. Astronomers celebrated the discovery, the press routinely reported it, and the general population went about its business as usual despite humanity's slide from the center to the edge of the Galaxy (Bok 1974; Berenzden et al. 1976; Smith 1982). The long-term implications of both discoveries, however, continue to reverberate today in the form of the anthropocentric versus the de-centered worldviews (Danielson 2001, 2013).

Yet another relevant change in worldview was the Darwinian revolution, still very much with us as a controversy, especially among a minority segment of the American public. Like the Darwinian theory, the interpretation of an extraterrestrial signal is likely to be ambiguous and debatable, and the diverse reaction to such a signal may therefore be comparable. The details of that revolution are well known, thanks to the "Darwin industry" of scholars who have studied it. From the early general historical treatments of Darwinism to recent historical, philosophical and scientific analyses, the Darwin industry itself provides a model of scholarship likely to be precipitated by a discovery of extraterrestrial intelligence. The debates over Darwinism raged over Europe and the Western world, and eventually over the entire world. Studies have shown how Darwin's theory had distinctive impacts over the short term (Vorzimmer 1970) and the long term (Bowler 1989), and among scientists (Hull 1973), theologians and other segments of the population. The title of one of the studies, *Science, Ideology and World View* (Greene 1981) is likely to express succinctly the general tenor of the debate in the aftermath of the discovery of extraterrestrial life. The construction of worldviews and their influence on our thinking is a deep philosophical problem (Vidal 2007, 2012) that can likely be applied to this issue.

Analogues are also possible from fields outside history; one of the immediate analogues suggested for first contact was the meeting of Neanderthals and *Homo sapiens*. Ian Tattersall has argued that we are not justified in using modern humans as ethnographic analogues to make sense of Neanderthals: "When we look at homo neanderthalensis," he observed "we are looking at a creature possessed of another sensibility entirely" (Tatersall 1995, 153). As Paul K. Wason observed, "Altogether, we might well expect any encounter between Neanderthals and Cro-Magnons to have been a difficult and ineffective affair, fraught with misunderstanding" (Wason 2011, 44). While they may be right in terms of ethnographic analogues, as an analogue for physical first contact their description seems likely to be quite accurate—if only we had any data for the actual meeting of Neanderthal and *Homo sapiens*! Poorly understood or misused source analogues do not inspire confidence in illuminating target analogues.

Many other authors have written on the importance of analogy in both general and specific thinking; cognitive scientist Douglas Hofstadter, for example, argues that analogy is the core of cognition: "One should not think of analogy-making as a special variety of *reasoning* (as in the dull and uninspiring phrase 'analogical reasoning and problem-solving,' a long-standing cliché in the cognitive-science world), for that is to do analogy a terrible disservice. After all, reasoning and problem-solving have (at least I dearly hope!) been at long last recognized as lying far indeed from the core of human thought. If analogy were merely a special variety of something that in itself lies way out on the peripheries, then it would be but an itty-bitty blip in the broad blue sky of cognition. To me, however, analogy is anything but a bitty blip—rather, it's the very blue that fills the whole sky of cognition—analogy is *everything*, or very nearly so, in my view" (Hofstadter 2001, 499).

Analogies have been extensively analyzed as arguments and applied in other disciplines. A specific study 50 years ago, *The Railroad and the Space Program: An Exploration in Historical Analogy*, concluded that analogy is not predictive, but can be suggestive of the topology of the future (Mazlish 1965). Analogies abound not only in science (the Bohr atom and the solar system), but also in history, as in Cullen Murphy's volume *Are We Rome?* (Murphy 2007), which by its very title indicates the inherent difficulties. Nevertheless, as Gaddis says "It's here, I think, that science, history, and art have something in common: they all depend on metaphor, on the recognition of patterns, on the realization that something is 'like' something else" (Gaddis 2002, 2).

These ruminations indicate that the systematic application of analogy to the problem of the societal impact of extraterrestrial life is a field that may hold much promise. At the same time serious precautions are in order. First, analogical reasoning can be misleading. Examples are attempts to show that religion is analogous to science, or to spaceflight (Harrison 2007), or to SETI (Basalla 2006; Harrison 2007, 95 ff), or to Eastern mysticism and quantum mechanics (Capra 1975; Stenger 2011, 258). The often-heard analogy of the Book of Genesis compared to the details of the Big Bang theory has been thoroughly debunked (Stenger 2011, 122). These are often what I would call "polemical analogies," and the goal of analogical argument should not be to polemicize but to illuminate. Secondly, as Denning (2013) warns in this volume, it is easy to get carried away with analogy, descending to a level of detail unlikely to be useful and maybe even irrelevant, if not downright harmful. We conclude *analogy must not be so general as to be meaningless, nor so specific as to be misleading. The middle "Goldilocks" ground is where analogies may serve as useful guideposts.* On the one hand it does little good to argue that science and religion are both searching for our place in the universe, when one addresses the natural world and the other invokes the supernatural—differences so great as to swamp any comparison whatsoever. On the other hand it is hopelessly naïve to expect that contact with extraterrestrial intelligence will change our worldview in ways precisely mirroring past revolutions in thought, leading us to reiterate our first principle above: under no circumstances will analogy predict the future.

In the end the skeptic may be left with a nagging doubt: just because analogy has proven useful in some areas of science and scholarship, how do we know if it is useful or valid in our particular problem of the societal impact of extraterrestrials? This is a meta-analogical problem, also discussed in considerable detail by philosophers (Bartha 2012). And again their results are encouraging for this particular endeavor.

12.2.3 The Utility of Past Impact Studies in Science and Technology

Substantial studies have been undertaken on the societal impact of other scientific endeavors such as the Human Genome Project, biotechnology, spaceflight, and cosmic evolution. Such studies should prove useful for the current problem, avoiding the reinvention of the wheel. The Human Genome Project literature notes that "The U. S. Department of Energy (DOE) and the National Institutes of Health (NIH) devoted 3–5 % of their annual Human Genome Project (HGP) budgets toward studying the ethical, legal, and social issues (ELSI) surrounding availability of genetic information. This represents the world's largest bioethics program, which has become a model for ELSI programs around the world" (Human Genome Project 2012). Among the societal concerns embraced for study were (1) conceptual and philosophical implications regarding human responsibility, free will versus genetic determinism, and concepts of health and disease; (2) fairness in the use of genetic information; (3) privacy and confidentiality of genetic information; (4) psychological impact and stigmatization due to an individual's genetic differences; and (5) commercialization of products, including property rights and accessibility of data and materials. Huge amounts of money were spent analyzing these subjects. Surely a great deal may be learned from these studies and their approaches.

Less sweeping (and less well funded) studies have been undertaken in other areas. NASA has funded several of these, including on the societal impact of spaceflight (Dick and Launius 2007) and cosmic evolution (Dick and Lupisella 2009). Even closer to astrobiology's core interests are planetary protection protocols, which are certainly studies of potential impact (Race 2007). In addition to these studies, which are relevant both for methodology and substance, "Biology and Society" program exist at several universities, and their approaches might well be applied to the present problem. We should be under no illusion that millions of dollars are going to be spent on the implications of extraterrestrial life—not, that is, until it is discovered, in which case the floodgates may open as they did with the Human Genome Project, now in the form of a practical problem rather than a theoretical one.

12.2.4 *Extraterrestrial Life in the Context of Discovery and Exploration*

The discovery of life beyond Earth would be one of the signal events in the history of science. History can help illuminate the nature of discovery, and one of the primary conclusions of historians of science working in this area is that discovery is an extended process, consisting of detection, interpretation and understanding, each with its own technological, conceptual and social roles (Kuhn 1962a, b; Caneva 2005; Dick 2013a). This extended process of discovery (Fig. 12.1) can take place over periods ranging from days to centuries. The discovery of extraterrestrial life, whether microbial or intelligent, is likely to follow a similar pattern. This "natural history" of discovery will help us understand possible scenarios.

Consider the discovery of microbial life: as we have seen with the Viking experiments (Dick 2013b), detection was only the first stage, followed by interpretation, which is still ongoing almost 40 years later, particularly with the Phoenix spacecraft discovery of perchlorates on Mars. We are likely a long way from understanding. Consider again the discovery of a signal from extraterrestrial intelligence. This is likely to entail the detection of an unusual narrow-band signal, followed by a more-or-less extended period of interpretation before understanding gels, perhaps many years later. Even then, and especially in the case of a signal with information content that can be deciphered, an even more extended period of study is likely to follow. As Philip Morrison has emphasized, the complex signal arriving at our radio telescope "is the object of intense socially required study for a long period of time. I regard it as a much more like the enterprise of history of science than like the enterprise of reading an ordinary message… The data rate will for a long time exceed our ability to interpret it." He went on to say that "the recognition of the signal is the great event, but the interpretation of the signal will be a social task comparable to that of a very large discipline, or branch of learning…

The Anatomy of Discovery:
An Extended Process

Discovery

| Detection | Interpretation | Understanding |

Technological, Conceptual and Social Roles at Each Stage

Pre-Discovery	Post-Discovery
• Theory	• Issues of credit & reward
• Casual or Accidental observations	• How do discoveries end?
• Classification of Phenomena	• Classification of "The Thing
(Harvard spectral types)	Itself" (MK spectral types)

Fig. 12.1 Studies have shown that discovery in science is an extended process. For example, the discovery of any new class of astronomical object, whether in the realm of the planets, stars or galaxies, consists of detection, interpretation and understanding. Pre-discovery and post-discovery phases help to delimit discovery, and classification is common in both phases, based on phenomena in the first case, and on a real understanding of "the thing itself" in post-discovery, as illustrated here for stellar classification. The discovery of extraterrestrial microbial or intelligent life will likely follow a similar extended pattern. From Dick (2013a)

We could imagine the signal to have great impact—but slowly and soberly mediated" (Morrison 1973, 336–337). In this process, decipherments of past languages on Earth is likely to play a role, again by way of analogy.

Finally astrobiology pioneers like Morrison and Baruch Blumberg like to place astrobiology in the context of exploration. As Morrison said, "unlike most of science, this topic extends beyond the test of a well-framed hypothesis; here we try to test an entire view of the world, incomplete and vulnerable in a thousand ways. That has a proud name in the history of thought as well; it is called exploration. We are joined in the early ingenuous stages of a daring exploration, become real only during recent years. It is a voyage whose end we do not know, like that of science itself" (Morrison 1995 211). Similarly, Blumberg (2003) specifically compared astrobiology to the Lewis and Clark expeditions in the American tradition. This comparison should be analyzed in more detail, with an eye toward its usefulness in understanding the societal impact of astrobiology.

In summary, history can be useful in multiple ways in analyzing the societal impact of astrobiology. History grounds this study in what otherwise might be pure speculation. As problematic as analogy based on terrestrial history may be, like astrobiology science it gives substance to studies of the societal impact of what would be one of the greatest discoveries in the history of science, while at the same time adding to terrestrial history an element never before present—the possibility of extraterrestrial minds. Whether or not they exist, the possibility of such minds raises discussion of terrestrial issues in history, philosophy religion and the social sciences to a new level of generality, providing a perspective, and expanded conceptual spaces, otherwise lacking.

12.3 The Relevance of the Social Sciences

Aside from history, many other approaches may be taken to the problem of the impact of the discovery of extraterrestrial life. In particular, it would seem that the broader social sciences have the potential to illuminate a subject whose central concerns are, at least in the extraterrestrial intelligence mode, societies and cultural evolution, even if the setting happens to be extraterrestrial (Harrison et al. 2000). Even in the microbial life aspects of astrobiology, social scientists have specialized training that can provide insights into how humans react to particular ideas or events. Yet the social sciences have played little role in SETI and exobiology, even its broadened form represented by astrobiology. This undoubtedly reflects a variety of factors, including what C. P. Snow termed the "two cultures" phenomenon—the segregation of the natural and social sciences—as well as he increasing specialization already well developed in the early 1960s combined with the fact that there was no shortage of problems on Earth for social scientists to tackle. Thus, while the 1961 Green Bank conference on interstellar communication included astronomers, physicists, a biochemist, an engineer, and even a specialist on dolphin communication, no one represented the social sciences or humanities.

The social sciences are admittedly very broad, encompassing disciplines ranging from anthropology and archaeology to economics and political science, as well
as sociology and psychology. Sometimes history is even included in the social sciences (as by the U. S. National Research Council), but more often it is considered as part of the humanities, as by the National Endowment for the Humanities
in the United States. In any case there is a large amount of overlap between the
humanities and the social sciences. For our purposes here we focus on anthropology and sociology as exemplars of the potential role of the social sciences in the
problem of societal impacts of extraterrestrial life. They suffice to illustrate how
the problem leaves a wide scope for interdisciplinary research. They also highlight
the second principle stated in the introduction to this chapter, namely, that "society
is not monolithic, implying many impacts depending on the religious, cultural and
worldview aspects of each society." Certainly this is one of the main lessons of
anthropology, with its studies of the many cultures throughout terrestrial history.
Finally, increasingly our knowledge of the cosmos may affect culture, including
the social sciences (Dick and Lupisella 2009; Lupisella 2009).

12.3.1 Anthropology

Anthropologists can contribute to the problem of the societal impact of extraterrestrial life not only through their expertise in culture contact, cultural diffusion, and
the evolution of technological civilization (Denning 2009, 2011b), but also in a
more general way through their understanding of the impact of novel critical ideas
and events on cultures. In this volume both Lowrie (2013) and Weigel and Coe
(2013) make this point. Whether microbial or intelligent, the discovery of extraterrestrial life is certainly such an idea, with the event of *actual discovery* likely
impacting very differently from the impact of *possible discovery* as it now stands.

Already in the early 1960s two roles had been identified for anthropology in
the context of SETI: the study of human evolution models as analogies to extraterrestrial contact, and the study of the impact of such contact. In the first case two
authors, one an anthropologist the other a mathematician, suggested an "analogy
between prehistoric contact and exchange, and hypothesized extraterrestrial contact and exchange" (Ascher and Ascher 1963, 307). In the second case, a NASA-
commissioned study published in 1961 warned that substantial contact could be
seriously destablilizing: "Anthropological files contain many examples of societies, sure of their place in the universe, which have disintegrated when they had to
associate with previously unfamiliar societies espousing different ideas and different life ways; others that survived such an experience usually did so by paying the
price of changes in values and attitudes and behavior" (Committee on Science and
Astronautics 1961, 215–216). Both studies are early exemplars of the problems
and the promise of analogical thinking in the field.

The following decades saw sporadic SETI overtures to social science as well as
sporadic overtures in the opposite direction. At a landmark international meeting

on CETI (Communication with Extraterrestrial Intelligence), held in the Soviet Union in 1971, two anthropologists were included, as well as historian William H. McNeill of the University of Chicago (Sagan 1973). Their arguments with the natural scientists about the evolution of technological civilizations make for interesting reading, but almost by definition could not represent definitive conclusions. In the following decades at least token social science representation became quite common at gatherings where extraterrestrial intelligence was discussed, most notably at the series of workshops on SETI chaired by Philip Morrison in the mid-1970s. Part of that effort was a "workshop on cultural evolution" chaired by Nobel laureate Joshua Lederberg and including anthropologist Bernard Campbell. Among the conclusions in the subsequent NASA volume was that "our new knowledge has changed the attitude of many specialists about the generality of cultural evolution from one of skepticism to a belief that it is a natural consequence of evolution under many environmental circumstances, given enough time" (Morrison et al. 1977, 49).

At about the same time the American Anthropological Association held a symposium resulting in a book entitled *Cultures Beyond the Earth: The Role of Anthropology in Outer Space* (Maruyama and Harkins 1975). It contained some new and sophisticated ideas, at least in outline, as well as an afterword by anthropologist Sol Tax, who noted that "Only when we have comparisons with species that are cultural in nonhuman ways –some of them maybe far more advanced than we—will we approach full understanding of the possibilities and limitations of human cultures" (203). This is similar to the point we made above, that even the possibility of such nonhuman ways opens new conceptual spaces for discussion.

In the 1980s Ben Finney, an anthropologist at the University of Hawaii, almost single-handedly took up the challenges of some of these cultural issues, including working with the SETI community (Finney and Jones 1985; Finney 1990, 2000). The "Cultural Aspects of SETI" workshops led by John Billingham, the head of the NASA SETI program, around the time of the inauguration of the NASA SETI observations in 1992, represent a coordinated effort to discuss broader social science issues (Billingham et al. 1999). Since then a few individuals have tackled SETI from the social science aspect, including Harrison (1997), Michaud (2007), and Denning (2009, 2011a, b, c), while Vakoch has led the way in bringing the issues to the attention of anthropologists at the annual meeting of the American Anthropological Association (Vakoch 2009). A recent overview of "social evolution" by Denning (2009) is particularly nuanced in discussing the problems and promise of the social sciences for SETI, while Battaglia (2009) has contributed substantially to this literature with her volume *E.T. Culture: Anthropology in Outerspaces*. Given these pioneering efforts, it is likely that more anthropologists will join the discussion, an outcome highly desired.

12.3.2 Sociology

The idea of extraterrestrial life has already had an impact on the public, as witnessed by some of the most popular films of all time, as well as by science fiction literature, the UFO debate, and public interest in the question of life on Mars. Many people seem predisposed to believe in extraterrestrial life in some form, at least 60 % in the U. S. according to polls, a rather astonishing fact given that only circumstantial evidence exists (in contrast to the case for Darwinian evolution). What is the source of the public "will to believe" in extraterrestrial life, at least in the Western world, and how does it compare to other cultures? What are the social factors that play into individual scientists' belief in extraterrestrial life? What are the social factors that should enter into an interstellar message, and what are the possible universal factors?

These are the types of questions to which sociologists can contribute their expertise. Sociologist David Swift was in the forefront of this approach in his interviews with SETI pioneers (Swift 1990). William Sims Bainbridge has led in studying attitudes of the general public to extraterrestrials (Bainbridge 2011). And sociologists such as Donald Tarter have contributed to policy formulation for reply to extraterrestrials (Tarter 1996, 1997). Douglas Vakoch—a psychologist by training—has done more than anyone in applying social science principles to the problem of interstellar communication over the last two decades (Vakoch 1998, 1999).

More broadly a new discipline dubbed "astrosociology" has arisen in the last few years that addresses the societal impact of space exploration, including extraterrestrial life (Pass 2004, 2005, 2009, 2012). Astrosociology is defined as "the study of *astrosocial phenomena*, where astrosocial phenomena comprises a subset of all social, cultural, and behavioral phenomena... characterized by a relationship between human behavior and space phenomena." As James Pass, the sociologist who coined the term in 2004, puts it, "the astrosociological perspective brings the social sciences into the space age by fostering the creation and development of a field dedicated to the study of the impact of space exploration" (Pass 2012).

The impact of astrobiology is an explicit part of this new field. As the founders of the field put it "Even without an announcement of success forthcoming in the near future, and even without consideration of the implications if such an announcement became a reality, the very attempt to seek out life in an organized manner merits the attention of astrosociologists from a number of disciplines, including sociology, psychology, anthropology, and history. If this is the case, astrosociology must investigate this behavior along with the implications of long-term failure and success. The social and cultural implications of this work make it too important to ignore. In fact, it is imperative that astrosociologists participate alongside their space-community counterparts to attain comprehensive knowledge; both for its own sake and for practical application should some type of reaction prove necessary" (Pass 2012).

Other scholars have demonstrated the complex relation of space exploration to social, racial and political themes. One such study is De Witt Douglass Kilgore's recent book *Astrofuturism: Science, Race and Visions of Utopia in Space* (Kilgore 2003). In this book Kilgore examines the work of Wernher von Braun, Willy Ley, Robert Heinlein, Arthur C. Clarke, Gentry Lee, Gerard O'Neill and Ben Bova, among others in what he calls the tradition of American astrofuturism. Similar studies can be undertaken more explicitly for the theme of extraterrestrial life. Even more than anthropology, sociology remains ripe for pioneering explorations by expanding its conceptual space to extraterrestrials.

12.4 Impacts on Theology and Philosophy as Exemplars

In this section we switch our focus from approaches to the problem of studying the discovery of extraterrestrial life using history and the social sciences, to the substantive nature of its actual and potential impacts. As exemplars we address two disciplines: theology, one of the most discussed potential impacts, and philosophy, one of the least discussed. Together, they illustrate the profound effect our subject could have, not only on scientific endeavors, but on everyday life.

12.4.1 Theology

The question of the impact of extraterrestrial life on religion and theology has very deep roots, at least in the Western tradition. The problem was perceived already in the 15th century, in relation to the reconciliation of Christianity with the Aristotelian doctrine opposing a plurality of worlds. Most theologians by that time agreed that God could create other worlds. But if so, they wondered "whether Christ by dying on this earth could redeem the inhabitants of another world" (Dick 1982, 88). The standard answer was that he could, because Christ could not die again in another world. Very early in the Protestant tradition Martin Luther's supporter, Philip Melanchthon, not only objected to such a speculative idea but also used it as an argument against the Copernican theory: "It must not be imagined that there are many worlds, because it must not be imagined that Christ died and was resurrected more often, nor must it be thought that in any other world without the knowledge of the Son of God, that men would be restored to eternal life" (Dick 1982, 88–89). For Copernicans of any religious persuasion, the problem was a thorny one that extended beyond specific religious doctrine. Kepler stated the conundrum already in the early 17th century in more general terms that might equally apply to other religions of the world: "If there are globes in the heavens similar to our earth, do we vie with them over who occupies a better portion of the universe? For if their globes are nobler, we are not the noblest of rational creatures. Then how can all things be for man's sake. How can we be the masters of God's handiwork?" (Dick 1996, 515).

These provocative Keplerian questions were still alive at the end of the 19th century, when H. G. Wells quoted them as the prelude to his novel *War of the Worlds*. By that time Christianity had explored these implications quite substantially. Despite Scriptural objections raised during the 17th century, by the early 18th century the Anglican priest and Royal Society Fellow William Derham reflected accepted theological opinion when he incorporated extraterrestrial life into natural theology; it is in the sense of inhabited worlds reflecting the magnificence of God's universe that Derham wrote his book *Astro-Theology*. The matter did not rest there, however. Thomas Paine bluntly stated in his 1793 *Age of Reason* (Dick 1996, 515–516) that extraterrestrials and Christianity did not mix, and that "he who thinks that he believes in both has thought but little of either." In a history that would repay study by those interested in theological implications of an actual discovery of extraterrestrial intelligence, during the 19th century some writers rejected Christianity, others rejected plurality of worlds, and still others found ways to reconcile the two (Crowe 1986, 1997).

The 20th century thus inherited a considerable discussion of the theological implications of extraterrestrial life, mostly within the Christian tradition, inspired by the mere possibility of intelligence beyond the Earth. Although the relation between theology and plurality of worlds occasionally reached the level of sustained debate in the 18 and 19th centuries, by the mid-20th century this controversy echoed only faintly in the background as scientists began to contemplate the possibility of a search for extraterrestrial intelligence. In the 20th century Derham's "astrotheology" assumed new meaning in light of efforts to detect signals from extraterrestrial intelligence, efforts that, if successful, would surely affect traditional theology with its emphasis on the relation between God and humankind. Rather than focusing on confirming evidence of the glory of God in the best tradition of natural theology, astrotheology in the 20th century—or cosmotheology, as some have called it—came to describe the considerable modifications to theology and religion that might develop in the wake of the discovery of intelligence in the heavens.

While most religions would undoubtedly have preferred to remain silent on the subject, the issue was pushed into the public and theological consciousness by the approach of the Space Age. As Arthur C. Clarke, one of the prophets of the new Era, remarked in his popular book *The Exploration of Space*, some people "are afraid that the crossing of space, and above all contact with intelligent but non-human races, may destroy the foundations of their religious faith. They may be right, but in any event their attitude is one which does not bear logical examination—for a faith which cannot survive collision with the truth is not worth many regrets" (Clarke 1951, 191). Religion could not for long avoid such a common-sense challenge, whose force could only increase as rocketry neared reality.

In Christianity, the doctrine of Incarnation has been a central focus of discussion, and the consensus has been that a discovery of intelligence beyond the Earth would not prove fatal to the religion or its theology. In general, for Christians as well as for other religions, indigenous theologians see little problem, while those external to religion proclaim the fatal impact of extraterrestrials on Earth-bound theologies (Peters 2009, 2011, 2013).

The Catholic version of Christianity, like the Protestant, was remarkably open-minded on the subject (Vakoch 2000). Father Daniel C. Raible was typical of this open-mindedness when he wrote in the wake of Project Ozma "Yes, it would be possible for the Second Person of the Blessed Trinity to become a member of more than one human race. There is nothing at all repugnant in the idea of the same Divine Person taking on the nature of many human races. Conceivably, we may learn in heaven that there have been not one incarnation of God's son but many" (Raible 1960, 532–535). The Church also had an eye on history; quoting a cardinal that "one Galileo case is quite enough in the history of the Church," an editorial in one Catholic journal suggested that "today's theologians would welcome the implications that such a discovery might open—a vision of cosmic piety and the Noosphere even beyond that of a Teilhard de Chardin" (Anonymous 1964).

The most substantial theological discussion of the subject, and the closest the Roman Catholic Church came to an official position, was given by the priest Kenneth Delano in his book *Many Worlds, One God* (Delano 1977). Complete with the official "nihil obstat" and "imprimatur" sanctions, the author's position was that any person with a religious faith including "an adequate idea of the greatness of God's creative ability, of humanity's humble position in the universe, and of the limitless love and care God has for all His intelligent creatures," should not be afraid to examine the implications of intelligence in the universe (Delano 1977, xv). Delano characterized the fears of some in the religious community with regard to extraterrestrials as analogous to early Church skepticism that any humans could live in the terrestrial "antipodes" because none of Adam's descendants could have reached the Southern Hemisphere. Reacting to an early 20th century writer who claimed that "If he [man] is not the greatest, the grandest, the most important of created things, the one to whom all else is made to contribute, then the Bible writers have misrepresented entirely man's relation to God and the universe" Delano (1977, 9) pointed out that God was not obliged to reveal extraterrestrials in the Bible when it would have served no moral purpose.

Delano further emphasized that the Space Age requires a theology that is neither geocentric nor anthropomorphic, and it therefore follows that the Earth may not be the only planet that has seen an incarnation: "Any one or all three Divine Persons of the Holy Trinity may have chosen to become incarnated on one or more of the other inhabited worlds in the universe" (Delano 1977, 115). This he considered much more likely than a theory of the "cosmic Adam," in which the single redemptive act by Christ on Earth is applicable to the entire universe. On the other hand, humanity's "mission" could be to spread the Gospel among the inhabited planets, while refraining from any form of religious imperialism. The Church, while spreading the story of terrestrial redemption, might also encourage fallen races to seek salvation. Although Delano made it clear that Catholic opinion was not unanimous, he certainly reinforced the prevalent idea of flexibility toward a discovery of extraterrestrials in the Church doctrines.

The same flexibility was expressed in a study of religious implications of the problem for Jewish thought, where the primary concern was of course not the Incarnation, but the uniqueness of man and his relationship to God. Cautioning that extraterrestrial intelligence was far from proven, Rabbi Norman Lamm nevertheless pointed to precedents in medieval Jewish thought, and declared that in the spirit of open-mindedness toward new knowledge, it was prudent to explore "a Jewish exotheology, an authentic Jewish view of God and man in a universe in which man is not the only intelligent resident, and perhaps inferior to many other races" (Lamm 1978, 371). Medieval Jewish philosophy already rejected the uniqueness of man, Lamm pointed out, but non-singularity of man did not mean insignificance. Shapley and others, he argued, were "profoundly mistaken" in assuming that the number of intelligent species had any relation to the significance of man, and even more so in holding that a peripheral position in the galaxy implied metaphysical marginality and irrelevance. That "geography determines metaphysics" he called a "medieval bias" that should have disappeared with the collapse of geocentrism. Judaism, therefore, "could very well accept a scientific finding that man in not the only intelligent and bio-spiritual resident in God's world," as long as the insignificance of man was not an accompanying conclusion. Man could still be considered unique in "spiritual dignity," and the existence of innumerable intelligences does not lessen God's attention to man." A God who can exercise providence over one billion earthmen," Lamm concluded, "can do so for ten billion times that number of creatures throughout the universe. He is not troubled, one ought grant, by problems in communications, engineering, or the complexities of cosmic cybernetics. God is infinite, and He has an infinite amount of love and concern to extend to each and every one of his creatures" (Lamm 1978, 364).

Internal to various religions, therefore, the consensus was that terrestrial religions would adjust to extraterrestrials, an opinion echoed in several studies of religious attitudes (Ashkenazi 1992; Peters 2009, 2011). And, as one of the studies also pointed out, if the "Adamist religions" of Judaism, Christianity and Islam—those that share a view of the creation of man that links him directly to the godhead—can survive extraterrestrials, non-Adamist religions such as Buddhism, Hinduism or Taoism should have no trouble.

No systematic astrotheology was developed in the 20th century in the sense that new theological principles were created, or existing ones formally modified, to embrace other moral agents in the universe. While Freeman Dyson among others have argued that the age-old mystery of God will be little changed by new knowledge of the universe, others argue that the new universe not only could, but should, lead to a new "cosmotheology" (a term first used by Immanuel Kant), or a new "cosmophilosophy." Among the elements such a cosmotheology must take into account are (1) that humanity is in no way physically central to the universe, but located on a small planet circling a star on the outskirts of the Milky Way galaxy; (2) that humanity is probably not central biologically, even if our morphology may be unique; (3) that humanity is likely somewhere near the bottom, or at best midway, in the great chain of being—a likelihood that follows from the age of the universe and the youth of our species; (4) that we must be open to radically new conceptions of God grounded in cosmic evolution, including the idea of a "natural" rather than a "supernatural" God; and (5) that it must have a moral dimension, and respect for life that includes all species in the universe (Dick 2000c).

Each of these elements of cosmotheology provides vast scope for elaboration. Perhaps the most radical consequences stem from the fourth principle stating that we must be open to new conceptions of God, stemming from our advancing knowledge of cosmic evolution and the universe in general. As the God of the ancient Near East stemmed from ideas of supernaturalism, our concept of a modern God could stem from modern ideas divorced from supernaturalism. The billions of people attached to current theologies may consider this no theology at all, for a transcendent God above and beyond nature is the very definition of their theology. The supernatural God "meme," which we should remember is an historical idea the same as any other, has been very efficient in spreading over the last few thousand years, picking up new memes such as those accepted by Christianity and other religions. Nonetheless, the idea of a "natural" God in the sense of a superior intelligence is appealing to some. A natural God need not intervene in human history, nor be the cause for religious wars such as witnessed through human history. It remains an open question whether a natural God fulfills the apparent need that many have for "the Other". Such a "God" is different enough from tradition concepts that some may wish to call it a cosmophilosophy rather than a cosmotheology. In any case some will see it as an important part of religious naturalism. Over the next centuries or millennia, religions will likely adjust to these cosmotheological principles.

Although the mere possibility of extraterrestrial intelligence has thus generated sporadic attempts at a universal theology, systematic astrotheology, or cosmotheology, will probably be developed only when—and if—intelligence is discovered beyond the Earth. In the meantime, merely posing the problem demonstrates the anthropocentricity of our current conceptions of religion and theology, and suggests that they should be expanded beyond their parochial terrestrial bounds. Though theologians have gone some way toward addressing Clarke's challenge, even the theological legacy of the Space Age in a broader sense is as yet unfulfilled. And as C. S. Lewis suggested in his trilogy of space novels, if extraterrestrials are actually discovered, the problem will become much more urgent.

In the end, the effect on theology and religion may be quite different from any impact on the narrow religious doctrines that have been discussed during the 20th century. It may be that in learning of alien religions, of alien ways of relating to superior beings, that the scope of terrestrial religion will be greatly expanded in ways that we cannot foresee.

12.4.2 Philosophy

Related to this issue of impacts is the question of how philosophy itself would change if we confirm the existence of extraterrestrial intelligence. An historical approach to this question might ask how philosophers viewed the possibility of extraterrestrial life in the context of their philosophy, or conversely how a belief in extraterrestrial life has historically affected philosophy. As Crowe (1986) has shown, Immanuel Kant and many others believed in extraterrestrials and this belief was in the background of their respective philosophies.

A forward-looking approach would ask how standard philosophical problems would be affected by the discovery of extraterrestrial life. Philosophers have not been quick to address such questions, despite an early call in 1971 by Lewis White Beck to do so (Beck 1985), and sporadic efforts in that direction (Regis 1985). The problem of "objective knowledge," for example, is one of the oldest problems of philosophy, and forms a branch of that field known "epistemology," the nature, origin, scope, and limits of human knowledge. Hume, Kant and many other classical philosophers had much to say about the relation between the mind and external reality, as do modern philosophers. Nor is this an abstruse academic argument; the long-running "science wars" embody the question in the form of postmodernism and the social constructionism debate, one element of which claims that science, like everything else, is socially constructed, and thus there is no objective knowledge. While this seems to me very questionable in the terrestrial context, the epistemological question takes on new meaning in the context of extraterrestrial biologies and minds.

Contact with extraterrestrial intelligence would provide a major insight into the question of objective knowledge on a universal, not just a terrestrial, scale. The basic question is, "Do humans and putative extraterrestrials perceive the universe in the same way?" There are three cases in comparative terrestrial and extraterrestrial perception: (1) complete overlap, (2) partial overlap, and (3) zero overlap, graphically shown in Fig. 12.2. On one level, these sets may be taken to represent terrestrial and extraterrestrial knowledge, but more deeply they represent terrestrial and extraterrestrial ways of perceiving. Case 1, in which ETI perceives the same electromagnetic spectrum as we do, processes the information in the same way, and comes to the same conclusions, holds out hope for easy dialogue and objective agreement. Case 2, in which there may be differences to a greater or lesser degree in sensory organs and mental processes, implies some common basis for dialogue. In case 3, with no senses or mental processes in common, there may be no possibility of dialogue or objective knowledge. Vakoch has suggested a dialogue chain,

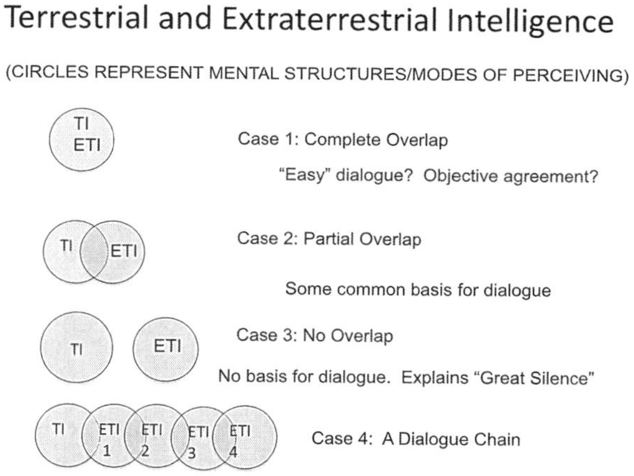

Terrestrial and Extraterrestrial Intelligence

(CIRCLES REPRESENT MENTAL STRUCTURES/MODES OF PERCEIVING)

Case 1: Complete Overlap

"Easy" dialogue? Objective agreement?

Case 2: Partial Overlap

Some common basis for dialogue

Case 3: No Overlap

No basis for dialogue. Explains "Great Silence"

Case 4: A Dialogue Chain

Fig. 12.2 Possible scenarios in the relationship between extraterrestrial intelligence (*ETI*) and terrestrial intelligence (*TI*). The *Venn diagrams* may be taken as mental structure or modes of perceiving and thinking. In *case 1* these overlap entirely, in which case dialogue may be relatively "easy." In *case 2* they may overlap only partially, yielding some common basis for dialogue. In *case 3* there is no overlap at all, in which case there is no overlap at all and thus no dialogue at all. In *case 4* a "dialogue chain" of partially overlapping mental structures may eventually enable dialogue

in which partially overlapping mental structures may eventually enable dialogue (Billingham 2000a, b; Dick 2000a).

The possibility exists that contact with extraterrestrial intelligence will result in the long-sought objective knowledge, by gleaning the common elements remaining after processing many sensory and mind systems independently evolved throughout the universe. Yet with few exceptions (Minsky 1985; Rescher 1985), no one has taken up the fundamental problem of objective knowledge in the extraterrestrial context. The problem is, however, central, for it bears on the possibility of communication, on the role of language, and on those aspects of the universe that have the possibility of verification. Knowledge must be distinguished from belief, which may have no basis in the objective world; one would not expect extraterrestrial religious belief, for example, to take the same form as on Earth, though the existence of God may be an objective question. If contact is successful, a major task over the next millennium will be to synthesize the knowledge of many worlds. The nature of this task will depend greatly on which of the three cases above turns out to be most common among galactic civilizations.

In summary, as in all astrobiological endeavors, we are only at the beginnings of a great investigation, perhaps a new field of study. At the turn of the 21st century, the societal impact of extraterrestrial life is the subject of increasing scholarship, even if a long way from coalescing with consensual conclusions. In keeping with the interdisciplinary nature of astrobiology, the humanities and social sciences have a prominent role to play. Surely, just as we plan for events large and small in the everyday world, it is better to plan ahead and lay out the scenarios of what might happen in the case of such a momentous event as the discovery of life beyond Earth.

12.5 Overview of Part III

In this section five authors deal with anthropological aspects of extraterrestrial life, one with media reactions, and two with theological reactions. In what has become, surprisingly but understandably, a central point of contention in this volume, anthropologist/archaeologist Kathryn Denning (2013) argues that the use of historical analogies of culture contacts on Earth may be "essentially useless or perhaps worse than useless," that predictions about contact on this basis are impossible and likely harmful, and that the humanities and social science scholars might better use their time to examine other issues at the intersection of astrobiology, SETI and society. She recognizes what is surely true—that the source analogue (contact among cultures on Earth) is often not well understood, that some authors have used this erroneous history uncritically in attempts to shed light on the target analogue (contact with extraterrestrials), and that as a first step the history of culture contacts on Earth must be better understood. But she then asks a more fundamental question: even with better history, what is the value of this activity at all?

Though she points to some possibly better cases of source analogues, such as the transmission of Greek and Arabic knowledge to the Latin West via the Arabs in the 13th century (discussed above), or the decipherment of Mayan glyphs as discussed in Finney and Bentley (1998), she suggests that all such activity is likely fruitless, at least if it is elaborated in ever more subtle detail, detail that is unlikely to be mirrored between the source and target analogue.

The other four anthropologists beg to differ, either implicitly or explicitly. While admitting the limitations of analogical argument, they believe it is still a useful, even a crucial, methodology for discussing the societal impact of discovering extraterrestrial life in the absence of other data. Ian Lowrie (2013) argues that by approaching the anthropological record of culture contact "with different epistemological premises, and shifting the focus from the material to the symbolic and cognitive dimensions of this contact, one can avoid many of the pitfalls of the analogical mode of argumentation, and provide a solid conceptual basis for the development of an adequate heuristic." In particular, he argues that historical and contemporary events will not mirror each other (a claim no serious scholar makes), but that modes of conceptualizing novel objects and phenomena would be the most profitable approach to the problem. To put it another way, he suggests we move away from the dynamics of contact to its conceptual and symbolic dimensions.

Klara Capova (2013) argues that our current perceptions of extraterrestrial contact, largely shaped not mostly by science but by science fiction and popular culture, will affect our reaction to the actual discovery. She characterizes it as a significant part of the modern worldview, what Karl Guthke 20 years ago called "the myth of modern times," using myth to mean that which culture holds to be part of its overarching worldview (Guthke 1990). In this she is in agreement with Denning's view that "contact has now been rehearsed so many times in popular culture that these representations and their dissemination in new media will be influential beyond almost any other factor" (Denning 2011a, 2013). Capova furthermore emphasizes how polls indicate belief in some form of "other life," whether microbial, humanoid or postbiological, hovers around 50 % of those populations polled. The idea about "other life," in the vast majority of that 50 %, is not informed by science but by the tropes of popular culture, ranging from belief in UFOs to themes in alien literature and film.

M. Margaret Weigel and Kathryn Coe (2013), also embracing analogical methodology, outline strategies honed by humans over millennia in response to threatening events, and then apply the results to the impact on developing countries of the detection of life beyond Earth. Drawing on a broad social science literature as well as on their own fieldwork, they find the strategies for coping with such events in developing countries (tsunamis, floods, volcanic eruptions; culture contacts; supposed contact with supernatural or celestial beings) include stories, rituals, song and dance that are passed down through many generations. The authors comment "The use of analogies to describe comparable events and the appropriate response made it possible to preserve an important and attractive lesson, transmitting it from one generation to the next, over a great many generations. This transmission and replication allowed ancestors to protect, consciously or unconsciously, multiple generations of their descendants." Their conclusions about the response to novel events are unusual in applying to the "traditional societies" found in developing countries, which they point out encompasses 85 % of the human population, in contrast to the developed countries usually discussed in this context. They go one step further in proposing a data-gathering program among these traditional societies, but also an education/preparedness program following the general principles of public health.

In contrast to Denning (2013), all these authors make the point, as have we in this chapter, that prediction of societal reaction to the discovery of extraterrestrial life, whether microbial or intelligent, is impossible, but that analogy employing Earth history can nevertheless serve as a useful guide when appropriate precautions are taken. Denning's title "Impossible Predictions of the Unprecedented: Analogy, History, and the Work of Prognostication," therefore seems to set up a straw man. The analogical process is not about prediction or prognostication, but about laying out a set of scenarios. Her call for precaution is well-taken, as well as her points about how stories about "superior" cultures contacting "inferior" cultures can often be wrong. But those stories can be, and to the extent possible are being, corrected by historians, anthropologists and social scientists, and in any case, culture contacts on Earth are likely not the best analogies, since contact would likely not be physical. It is arguably better to discuss possible scenarios with the best available information and approaches than to throw up our hands and say "nobody knows." As Denning points out, archaeology is largely an analogical enterprise; however, it is nevertheless a thriving and intellectually viable field. So too can be the study of the societal impact of discovering extraterrestrial life.

Analogy aside, other approaches exist to discussing societal impact, as exemplified in the two authors in this section who discuss religious reactions, Constance Bertka and Ted Peters. Concentrating on Christianity, which constitutes one-third of the world's religious communities, Bertka (2013) emphasizes the difficulties of generalizing about the Christian response to the discovery of life beyond Earth. She writes that "The variety in Christianity worldwide, both at the denominational level as well as at the level of individual experience, and the variety of options for relating science and religion, will combine to insure that integrating what SETI or astrobiology learns about the universe into Christian worldviews will at minimum be a long and convoluted process with more than one likely outcome." Moreover, she points out that academic conclusions on this subject among both astrobiologists and theologians are likely to differ from public responses. She points to a comprehensive study showing that over the last century the distribution of Christianity has dramatically shifted from the Global North (North America, Europe, Australia, Japan and New Zealand) to the Global South (Sub-Saharan Africa, Asia–Pacific, and Latin America). In the Global North, she points out, there is a well-developed typology for the relationship between science and religion, including conflict, independence, dialogue, and integration (Barbour 1997). This raises the question of whether there is any correlation between a person's idea of the relation between science and religion, and their response to extraterrestrial life, since a person who believes that science and religion are two non-overlapping areas could well have a different opinion than those who believe, for example, that science and religion are in conflict. In the case of the Global South, the more basic question of the responses of Christians to the idea of life beyond Earth remains uncharted territory, since that entire area is underrepresented in astrobiologial discourse.

Peters (2013) takes a broader view and deploys a different methodology, drawing on responses to his "ETI Religious Crisis Survey" that encompassed Orthodox Christians, Roman Catholics, mainline Protestants, Evangelical Protestants, Jews, Mormons, Buddhists and those who self-identify as non-religious. He finds that in all cases those outside religious traditions believe the discovery of extraterrestrial intelligence will precipitate a religious crisis, even the extinction of religion, while those who affirm religious belief think extraterrestrials can be incorporated into their world view. As Berka points, out, however, the questions of Peters' survey are sufficiently general so as to give no indication of "whether or not the respondents have considered the implications of discovery for the doctrines of their religious traditions in any depth, or if at the conclusion of that exercise they are confident that they can successfully 'integrate' the implications of ETI into their existing religious tradition." If they believe science and religion are independent endeavors addressing non-overlapping matters, no reflection on implications for specific doctrines is required. Moreover, as is well-known, a notoriously shortcoming of polls is that the wording of the questions matters greatly. Had the respondents been asked whether they believe in a planet-hopping Jesus, which is at the core of the matter for Christians in the form of the doctrines of Redemption and Incarnation, the results would likely have been significantly different.

Finally, in this section Morris Jones (2013) makes the valid point that the distribution of the news of the discovery of extraterrestrial life will affect public reactions and behavior. Focusing on media and communications behavior, he emphasizes the important role of the new social media such as Facebook and Twitter, as opposed to the mainstream media, not only in disseminating the news of discovery, but also in shaping the message and amplifying errors and distortions. Covering a wide variety of issues ranging from the changing media environment to crisis management, Jones concludes that the preparation of an appropriate media strategy is likely essential to the dissemination of accurate reporting in the event of the discovery of extraterrestrial life. This is surely a burning issue for those who hold out hope that the societal impact of such an Earth-shattering discovery may in some sense be controlled.

References

Achenbach, Joel. 1999. *Captured by Aliens: The Search for Life and Truth in a Very Large Universe.* New York: Simon and Schuster.

Almar, Ivan. 1995. "The Consequences of a Discovery: Different Scenarios." In *Progress in the Search for Extraterrestrial Life*, ed. G.Seth Shostak, 499–505. San Francisco: Astronomical Society of the Pacific.

Anonymous. 1964. "Messages from Space." *America* 111: 770.

Arnould, Jacques. 2008. "Does Extraterrestrial Intelligent Life Threaten Religion and Philosophy?" *Theology & Science* 6: 439–450.

Ascher, Robert, and Marcia Ascher. 1963. "Interstellar Communication and Human Evolution." In *Interstellar communication*, ed. A. G. W. Cameron, 306–308. New York: W. A. Benjamin.

Ashkenazi, Michael. 1992. "Not the Sons of Adam: Religious Responses to ETI." *Space Policy* 8: 341–350.

Bainbridge, William S. 2011. "Cultural Beliefs About Extraterrestrials: A Questionnaire Study." In *Civilizations Beyond Earth: Extraterrestrial Life and Society*, ed. Douglas A. Vakoch, and Albert A. Harrison, 118–140. New York: Berghahn Books.

Barbour, Ian G. 1997. *Religion and Science: Historical and Contemporary Issues*. San Francisco: Harper.

Bartha, Paul. 2010. *By Parallel Reasoning: The Construction and Evaluation of Analogical Arguments*. New York: Oxford Univ. Press.

Basalla, George. 2006. *Civilized Life in the Universe: Scientists on Intelligent Extraterrestrials*. Oxford: Oxford Univ. Press.

Battaglia, Debora. 2009. *E.T. Culture: Anthropology in Outerspaces*. Durham: Duke University Press.

Beck, Lewis W. 1985. "Extraterrestrial Intelligent Life." In *Extraterrestrials: Science and Alien Intelligence*, ed. Edward Regis, Jr., 3–18. Cambridge: Cambridge Univ. Press; Presidential address delivered before the Sixty-eighth Annual Eastern Meeting of the American Philosophical Association in New York City, December 28, 1971.

Berenzden, Richard, et al. 1976. *Man Discovers the Galaxies*. New York: Science History Publications.

Bertka, Constance M., ed. 2010. *Exploring the Origin, Extent, and Future of Life: Philosophical, Ethical and Theological Perspectives*. Cambridge: Cambridge Univ. Press.

Bertka, Constance M. 2013. "Christianity's Response to the Discovery of Extraterrestrial Intelligent Life: Insights from Science and Religion and the Sociology of Religion." In *Astrobiology, History, and Society: Life Beyond Earth and the Impact of Discovery*, ed. Douglas A. Vakoch. Heidelberg: Springer.

Biddanda, Bopaiah A., Stephen C. Nold, and Gregory J. Dick, et al. 2012. "Rock, Water, Microbes: Underwater Sinkholes in Lake Huron are Habitats for Ancient Microbial Life." *Nature Education Knowledge* 3: 13. http://www.nature.com/scitable/knowledge/library/rock-water-microbes-underwater-sinkholes-in-lake-25851285. Accessed 31 July 2012.

Billingham, John. 2000a. "Summary of Results of the Seminar on the Cultural Impact of Extraterrestrial Contact." In *Bioastronomy '99: A New Era in Bioastronomy*, ed. Guillermo A. Lemarchand, and Karen J. Meech, 667–675. San Francisco: Astronomical Society of the Pacific.

Billingham, John. 2000b. "Who Said What: A Summary and Eleven Conclusions." In *If SETI Succeeds: The Impact of High Information Contact*, ed. Allen Tough, 33–39. Bellevue: Foundation For the Future.

Billingham, John, Roger Heyns, David Milne, et al. 1999. *Societal Implications of the Detection of an Extraterrestrial Civilization*. Mountain View, CA: SETI Press.

Blumberg, Baruch S. 2003. "The NASA Astrobiology Institute: Early History and Organization." *Astrobiology* 3: 463–470.

Blumenberg, Hans. 1987. *The Genesis of Copernican World*. Trans. R. M. Wallace. Cambridge: MIT Press.

Bok, Bart. 1974. "Harlow Shapley and the Discovery of the Center of Our Galaxy." In *The Heritage of Copernicus: Theories Pleasing to the Mind*, ed. Jerzy Neyman, 26–62. Cambridge: MIT Press.

Bowler, Peter. 1989. *Evolution: The History of an Idea*. Berkeley: University of California Press.

Caneva, Kenneth. 2005. "'Discovery' as a Site for the Collective Construction of Scientific Knowledge." *Historical Studies in the Physical Sciences* 35: 175–291.

Cantril, Hadley. 1940. *The Invasion from Mars: A Study in the Psychology of Panic*. Princeton: Princeton Univ. Press. Reprint, 2005. Piscataway: Transaction Publishers.

Capova, Klara A. 2013. "The Detection of Extraterrestrial Life: Are We Ready?" In *Astrobiology, History, and Society: Life Beyond Earth and the Impact of Discovery*, ed. Douglas A. Vakoch. Heidelberg: Springer.

Capra, Fritjof. 1975. *The Astrobiological Landscape: Philosophical Foundations of the Study of Cosmic Life*. London: Wildwood House.

Clarke, Arthur C. 1951. *The Exploration of Space*. New York: Harper.

Committee on Science and Astronautics, U. S. House of Representatives. 1961. *Proposed Studies on the Implications of Peaceful Space Activities for Human Affairs*. Prepared for NASA by the Brookings Institute.

Crowe, Michael J. 1986. *The Extraterrestrial Life Debate, 1750–1900: The Idea of a Plurality of Worlds from Kant to Lowell*. Cambridge: Cambridge Univ. Press.

Crowe, Michael J. 1997. "A History of the Extraterrestrial Life Debate." *Zygon* 32: 147–162.

Danielson, Dennis. 2001. "The Great Copernican cliché." *American Journal of Physics* 69(10): 1029–1035.

Danielson, Dennis. 2013. "Early Modern ET, Reflexive Telescopics, and Their Relevance Today." In *Astrobiology, History, and Society: Life Beyond Earth and the Impact of Discovery*, ed. Douglas A. Vakoch. Heidelberg: Springer.

Davies, Paul. C. W. 1995. *Are We Alone?: Philosophical Implications of the Discovery of Extraterrestrial Life*. New York: Basic Books.

Delano, Kenneth. 1977. *Many Worlds, One God*. New York: Exposition Press.

Denning, Kathryn. 2009. "Social Evolution: State of the Field." In *Cosmos & Culture: Cultural Evolution in a Cosmic Context*, ed. Steven J. Dick, and Mark Lupisella, 63–124. Washington: NASA.

Denning, Kathryn. 2011a. "Is Life What We Make of It?" *Philosophical Transactions of the Royal Society A: Mathematical, Physical and Engineering Sciences* 369(1936): 669–678.

Denning, Kathryn. 2011b. "Being Technological." In *Searching for Extraterrestrial Intelligence: SETI Past, Present, and Future*, ed. H. Paul Shuch, 477–496. Heidelberg: Springer.

Denning, Kathryn. 2011c. "'L' on Earth." In *Civilizations Beyond Earth: Extraterrestrial Life and Society*, ed. Douglas A. Vakoch, and Albert A. Harrison, 74–83. New York: Berghahn Books.

Denning, Kathryn. 2013. "Impossible Predictions of the Unprecedented: Analogy, History, and the Work of Prognostication." In *Astrobiology, History, and Society: Life Beyond Earth and the Impact of Discovery*, ed. Douglas A. Vakoch. Heidelberg: Springer.

Des Marais, David J., Joseph A. Nuth, III, and Louis J. Allamandola, et al. 2008. "The NASA Astrobiology Roadmap." *Astrobiology* 8: 715–730.

Dick, Steven J. 1982. *Plurality of Worlds: The Origins of the Extraterrestrial Life Debate from Democritus to Kant*. Cambridge: Cambridge Univ. Press.

Dick, Steven J. 1995. "Consequences of Success in SETI: Lessons from the History of Science." In *Progress in the Search for Extraterrestrial Life*, ed. G. Seth Shostak, 521–532. San Francisco: Astronomical Society of the Pacific.

Dick, Steven J. 1996. *The Biological Universe: The Twentieth Century Extraterrestrial Life Debate and the Limits of Science*. Cambridge: Cambridge Univ. Press.

Dick, Steven J. 1998. *Life on Other Worlds*. Cambridge: Cambridge Univ. Press.

Dick, Steven J. 2000a. "Extraterrestrials and Objective Knowledge." In *If SETI Succeeds: The Impact of High Information Contact*, ed. Allen Tough, 47–48. Bellevue: Foundation For the Future.

Dick, Steven J. 2000b. "Cultural Aspects of Astrobiology: A Preliminary Reconnaissance at the Turn of the Millennium." In *Bioastronomy '99: A New Era in Bioastronomy*, ed. Guillermo A. Lemarchand, and Karen J. Meech, 649–659. San Francisco: Astronomical Society of the Pacific.

Dick, Steven J. (ed.). 2000c. *Many Worlds: The New Universe, Extraterrestrial Life and the Theological Implications*. Philadelphia: Templeton Press.

Dick, Steven J. 2013a. *Discovery and Classification in Astronomy*. Cambridge: Cambridge Univ. Press.

Dick, Steven J. 2013b. "The Twentieth Century History of the Extraterrestrial Life Debate: Major Themes and Lessons Learned." In *Astrobiology, History, and Society: Life Beyond Earth and the Impact of Discovery*, ed. Douglas A. Vakoch. Heidelberg: Springer.

Dick, Steven J., and Launius, Roger. 2007. *Societal Impact of Spaceflight*. Washington: NASA History Division.

Dick, Steven J., and Mark Lupisella, eds. 2009. *Cosmos & Culture: Cultural Evolution in a Cosmic Context*. Washington, DC: NASA. Also available online. http://history.nasa.gov/SP-4802.pdf. Accessed 29 July 2012.

Dominik, Martin, and John C. Zarnecki. 2011. "The Detection of Extra-terrestrial Life and the Consequences for Science and Society." *Philosophical Transactions of the Royal Society A: Mathematical, Physical and Engineering Sciences* 369(1936): 499–507.

Finney, Ben, and Eric M. Jones. 1985. *Interstellar Migration and the Human Experience*. Berkeley: University of California Press.

Finney, Ben. 1990. "The Impact of Contact." *Acta Astronautica* 21(2): 117–121.

Finney, Ben. 2000. "SETI, Consilience and the Unity of Knowledge." In *If SETI Succeeds: The Impact of High Information Contact*, ed. Allen Tough, 139–144. Bellevue: Foundation For the Future.

Finney, Ben, and Jerry Bentley. 1998. "A Tale of Two Analogues: Learning at a Distance from the Ancient Greeks and Maya and the Problem of Deciphering Extraterrestrial Radio Transmissions." *Acta Astronautica* 42(10–12): 691–696.

Gaddis, John Lewis. 2002. *The Landscape of History: How Historians Map the Past*. Oxford: Oxford Univ. Press.

Geertz, Clifford. 1973. *The Interpretation of Cultures*. New York: Basic Books.

Goodman, Matthew. 2008. *The Sun and the Moon: The Remarkable True Account of Hoaxers, Showmen, Dueling Journalists, and Lunar Man-Bats in Nineteenth-Century New York*. New York: Basic Books.

Grant, Edward. 1971. *Physical Science in the Middle Ages*. New York: Wiley.

Greene, John. 1981. *Science, Ideology and World View*. Berkeley: University of California Press.

Guthke, Karl S. 1983. *Der Mythos der Neuzeit. Das Thema der Mehrheit der Welten in der Literatur- und Geistesgechichte von der kopernikanischen Wende bis zur Science Fiction*. Bern: Franck; English translation 1990. *The Last Frontier: Imagining Other Worlds from the Copernican Revolution to Modern Science Fiction*. Ithaca: Cornell Univ. Press.

Harrison, Albert A. 1997. *After Contact: The Human Response to Extraterrestrial Life*. New York: Plenum.

Harrison, Albert A. 2007. *Starstruck: Cosmic Visions in Science, Religion, and Folklore*. New York: Berghahn Books.

Harrison, Albert A., John Billingham, and Steven J. Dick, et al. 2000. "The Role of Social Science in SETI." *IIf SETI Succeeds: The Impact of High Information Contact*, ed. Allen Tough, 71–85. Bellevue: Foundation For the Future.

Harrison, Albert A., and Steven J. Dick. 2000. "Contact: Long-term Implications for Humanity." In *If SETI Succeeds: The Impact of High Information Contact*, ed. Allen Tough, 7–31. Bellevue: Foundation For the Future.

Harrison, Albert A., and Kathleen Connell. 2001. *Workshop on the Societal Implications of Astrobiology*. Held at NASA Ames Research Center, November 16–17, 1999. Moffett Field, CA: NASA Ames Research Center. Available online. http://astrobiology.arc.nasa.gov/workshops/societal/. Accessed 27 July 2012.

Helmreich, Stefan. 2009. *Alien Ocean: Anthropological Voyages in Microbial Seas*. Berkeley: University of California Press.

Hesse, Mary B. 1966. *Models and Analogies in Science*. Notre Dame: University of Notre Dame Press.

Hofstadter, Douglas. 2001. "Epilogue: Analogy as the Core of Cognition." In *The Analogical Mind: Perspectives from Cognitive Science*, ed. Dedre Gentner, Keith J. Holyoak, and Boicho N. Kokinov, 499–538. Cambridge: The MIT Press.

Hull, David. 1973. *Darwin and His Critics: The Reception of Darwin's Theory of Evolution by the Scientific Community*. Cambridge: Harvard Univ. Press.

Human Genome Project. 2012. http://www.ornl.gov/sci/techresources/Human_Genome/elsi/elsi.s html. Accessed 29 July 2012.

Jones, Morris. 2013. "Mainstream Media and Social Media Reactions to the Discovery of Extraterrestrial Life." In *Astrobiology, History, and Society: Life Beyond Earth and the Impact of Discovery*, ed. Douglas A. Vakoch. Heidelberg: Springer.

Kilgore, Douglas D. 2003. *Astrofuturism: Science, Race and Visions of Utopia in Space*. Philadelphia: University of Pennsylvania Press.

Kuhn, Thomas S. 1957. *The Copernican Revolution*. Cambridge: Harvard Univ. Press.

Kuhn, Thomas S. 1962a. *The Structure of Scientific Revolutions*. Chicago: University of Chicago Press.

Kuhn, Thomas S. 1962b. "The Historical Structure of Scientific Discovery." *Science* 136: 760–764. Reprinted in Kuhn, Thomas S. 1977. *The Essential Tension: Selected Studies in Scientific Tradition and Change*, 165–177. Chicago: University of Chicago Press.

Lamm, Norman. 1978. "The Religious Implication of Extraterrestrial Life." In *Challenge: Torah Views on Science and Its Problems*, ed. Aryeh Cannell, and Cyril Domb, 354–398. Jerusalem: Feldheim Publishers.

Lemarchand, Guillermo A., and Karen J. Meech eds. 2000. *Bioastronomy '99: A New Era in Bioastronomy*. San Francisco: Astronomical Society of the Pacific.

Lindberg, David. 1978. "The Transmission of Greek and Arabic Learning to the West." In *Science in the Middle Ages*, ed. David Lindberg, 52–90. Chicago: University of Chicago Press.

Lindberg, David. 1992. *The Beginnings of Western Science*. Chicago: University of Chicago Press.

Lowrie, Ian. 2013. "Cultural Resources and Cognitive Frames: Keys to an Anthropological Approach to Prediction." In *Astrobiology, History, and Society: Life Beyond Earth and the Impact of Discovery*, ed. Douglas A. Vakoch. Heidelberg: Springer.

Lupisella, Mark. 2009. "Cosmocultural Evolution: The Coevolution of Culture and Cosmos and the Creation of Cosmic Value." In *Cosmos & Culture: Cultural Evolution in a Cosmic Context*, ed. Steven J. Dick, and Mark Lupisella, 321–356. Washington: NASA.

Maruyama, Magorah, and Arthur Harkins, eds. 1975. *Cultures Beyond the Earth: The Role of Anthropology in Outer Space*. New York: Vintage Books.

Mazlish, Bruce. 1965. *The Railroad and the Space Program: An Exploration in Historical Analogy*. Cambridge: MIT Press.

Meech, Karen J., Jacqueline V. Keane, Michael Mumma, et al., eds. 2009. *Bioastronomy 2007: Molecules, Microbes and Extraterrestrial Life*. San Francisco: Astronomical Society of the Pacific.

Michaud, Michael A. G. 2007. *Contact with Alien Civilizations: Our Hopes and Fears about Encountering Extraterrestrials*. New York: Copernicus.

Minsky, Marvin. 1985. "Why Intelligent Aliens Will Be Intelligible." In *Extraterrestrials: Science and Alien Intelligence*, ed. Edward Regis Jr, 117–128. Cambridge: Cambridge Univ. Press.

Morrison, Philip. 1973. "The Consequences of Contact." In *Communication with Extraterrestrial Intelligence (CETI)*, ed. Carl Sagan, 333–349. Cambridge: MIT Press.

Morrison, Philip. 1995. *Nothing is Too Wonderful to be True*. New York: American Institute of Physics.

Morrison, Philip, John Billingham, and John Wolfe. 1977. *The Search for Extraterrestrial Intelligence (SETI)*. Washington, DC: NASA.

Murphy, Cullen. 2007. *Are We Rome?: The Fall of an Empire and the Fate of America.*. Boston: Houghton Mifflin.

Pass, James. 2004. "Inaugural Essay: The Definition and Relevance of Astrosociology in the Twenty-First Century," and Part 2, "Relevance of Astrosociology As a New Subfield of Sociology." http://www.astrosociology.org/Library/Iessay/iessay_p2.pdf. Accessed 16 March 2012.

Pass, James. 2005. "The Sociology of SETI: An Astrosociological Perspective." http://www.astrosociology.org/Library/PDF/submissions/Sociology%20of%20SETI.pdf. Accessed 18 March 2012.

Pass, James. 2009. "Pioneers on the Astrosociological Frontier: Introduction to the First Symposium on Astrosociology." In *Space, Propulsion and Energy Sciences Forum— SPESIF-2009—SPESIF-2009*, ed. Glenn A. Robertson, 375–383. New York: American Institute of Physics. http://www.astrosociology.org/Library/PDF/Pass2009_Frontier_SPESI F2009.pdf. Accessed 12 March 2012.

Pass, James. 2012. "An Astrosociological Perspective on the Societal Impact of Spaceflight." In *Historical Studies in the Societal Impact of Spaceflight*, ed. Steven J. Dick. Washington: NASA.

Peters, Ted. 1995. "Exo-Theology: Speculations on Extraterrestrial Life." In *The Gods Have Landed: New Religions from Other Worlds*, ed. James R. Lewis, 187–206. Albany: State University of New York Press.

Peters, Ted. 2009. "Astrotheology and the ETI Myth." *Theology and Science* 7(1): 3–30.

Peters, Ted. 2011. "The Implications of the Discovery of Extra-terrestrial Life for Religion." *Philosophical Transactions of the Royal Society A: Mathematical, Physical and Engineering Sciences* 369(1936): 644–655.

Peters, Ted. 2013. "Would the Discovery of ETI Provoke a Religious Crisis?" In *Astrobiology, History, and Society: Life Beyond Earth and the Impact of Discovery*, ed. Douglas A. Vakoch. Heidelberg: Springer.

Pyle, Rod. 2012. *Destination Mars: New Explorations of the Red Planet*. New York: Prometheus Books.

Race, Margaret S. 2007. "Societal and Ethical Concerns." In *Planets and Life: The Emerging Science of Astrobiology*, ed. Woodruff T. Sullivan, III and John A. Baross, 483–497. Cambridge: Cambridge Univ. Press.

Race, Margaret S., Kathryn Denning, and Constance Bertka, et al. 2012. "Astrobiology and Society: Building an Interdisciplinary Research Community." *Astrobiology*, 12:958–965.

Raible, Daniel C. 1960. "Rational Life in Outer Space?" *America: National Catholic Weekly Review* 103: 532–535.

Randolph, Richard, Margaret Race, and Christopher McKay. 1997. "Reconsidering the Theological and Ethical Implications of Extraterrestrial Life." *Center for Theology and Natural Sciences Bulletin* 17(3): 1–8.

Rapport, Nigel, and Joanna Overing. 2000. *Social and Cultural Anthropology: The Key Concepts*. London: Routledge.

Regis, Edward, Jr., ed. 1985. *Extraterrestrials: Science and Alien Intelligence*. Cambridge: Cambridge University Press.

Rescher, Nicholas. 1985. "Extraterrestrial Science." In *Extraterrestrials: Science and Alien Intelligence*, ed. Edward Regis Jr, 83–116. Cambridge: Cambridge Univ. Press.

Ross, Joseph T. 2013. "Hegel, Analogy, and Extraterrestrial Life." In *Astrobiology, History, and Society: Life Beyond Earth and the Impact of Discovery*, ed. Douglas A. Vakoch. Heidelberg: Springer.

Sagan, Carl, ed. 1973. *Communication with Extraterrestrial Intelligence (CETI)*. Cambridge: MIT Press.

Shostak, G. Seth, ed. 1995. *Progress in the Search for Extraterrestrial Life*. San Francisco: Astronomical Society of the Pacific.

Shuch, H. Paul, ed. 2011. *Searching for Extraterrestrial Intelligence: SETI Past, Present, and Future*. Heidelberg: Springer.

Smith, Robert. 1982. *The Expanding Universe: Astronomy's "Great Debate."* Cambridge: Cambridge Univ. Press.

Stenger, Victor. 2011. *The Fallacy of Fine Tuning: Why the Universe is Not Designed for Us*. Amherst: Prometheus.

Stimson, Dorothy. 1972. *The Gradual Acceptance of the Copernican Universe*. Gloucester: Peter Smith.

Sullivan, Woodruff T., III. 2013. "Extraterrestrial Life as the Great Analogy, Two Centuries Ago and in Modern Astrobiology." In *Astrobiology, History, and Society: Life Beyond Earth and the Impact of Discovery*, ed. Douglas A. Vakoch. Heidelberg: Springer.

Sullivan, Woodruff T., III, and John A. Baross, eds. 2007. *Planets and Life: The Emerging Science of Astrobiology*. Cambridge: Cambridge Univ. Press.

Swift, David. 1990. *SETI Pioneers*. Tempe: University of Arizona Press.

Tarter, Donald. 1996. "Alternative Models for Detecting Very Advanced Extraterrestrial Civilizations." *Journal of the British Interplanetary Society* 49: 291–295.

Tarter, Donald. 1997. "Is Real-Time Communication Between Distant Civilizations in Space Possible?: A Call for Research." *Journal of the British Interplanetary Society* 50: 249–252.

Tatersall, Ian. 1995. *The Last Neanderthal: The Rise, Success, and Mysterious Extinction of Our Closet Human Relatives*. New York: MacMillan.

Tough, Allen (ed.). 2000. *When SETI Succeeds: The Impact of High-Information Contact*. Bellevue: Foundation For the Future.

Toynbee, Arnold. 1957. *A Study of History*. Abridgement by D. C. Sovervell, vol. 2. London: Oxford Univ. Press.

Vakoch, Douglas A. 1998. "Constructing Messages to Extraterrestrials: An Exosemiotic Perspective." *Acta Astronautica* 42: 697–704.

Vakoch, Douglas A. 1999. "The View from a Distant Star: Challenges of Interstellar Message-Making." *Mercury* 28: 26–32.

Vakoch, Douglas A. 2000. "Roman Catholic Views of Extraterrestrial Intelligence: Anticipating the Future by Examining the Past." In *If SETI Succeeds: The Impact of High Information Contact*, ed. Allen Tough, 165–174. Bellevue: Foundation For the Future.

Vakoch, Douglas A. 2009. "Anthropological Contributions to the Search for Extraterrestrial Intelligence." In *Bioastronomy 2007: Molecules, Microbes and Extraterrestrial Life*, ed. Karen J. Meech, Jacqueline V. Keane, Michael Mumma, et al., 421–427. San Francisco: Astronomical Society of the Pacific.

Vakoch, Douglas A., and Albert A. Harrison (eds.). 2011. *Civilizations Beyond Earth: Extraterrestrial Life and Society*. New York: Berghahn Books.

Vidal, Clement. 2007. "An Enduring Philosophical Agenda: Worldview Construction as a Philosophical Method." http://cogprints.org/6048. Accessed 29 July 2012.

Vidal, Clement. 2012. "Metaphilosophical Criteria for Worldview Comparison." *Metaphilosophy* 43(3): 306–347.

Vorzimmer, Peter. 1970. *Charles Darwin, the Years of Controversy: The Origins of Species and its Critics, 1859–1882*. Philadelphia: Temple Univ. Press.

Wason, Paul K. 2011. "Encountering Alternative Intelligences: Cognitive Archaeology and SETI." In *Civilizations Beyond Earth: Extraterrestrial Life and Society*, ed. Douglas A. Vakoch, and Albert A. Harrison, 42–59. New York: Berghahn Books.

Weigel, M. Margaret, and Kathryn Coe. 2013. "Impact of Extraterrestrial Life Discovery for Third World Societies: Anthropological and Public Health Considerations." In *Astrobiology, History, and Society: Life Beyond Earth and the Impact of Discovery*, ed. Douglas A. Vakoch. Heidelberg: Springer.

Westman, Robert S. 1975. *The Copernican Achievement*. Berkeley: University of California Press.

Westman, Robert S. 2011. *The Copernican Question: Prognostication, Skepticism, and Celestial Order*. Berkeley: University of California Press.

White, Frank. 1990. *The SETI Factor*. New York: Walker and Co.

Wilson, Edward O. 1998. *Consilience: The Unity of Knowledge*. New York: Knopf.

Chapter 13
Cultural Resources and Cognitive Frames: Keys to an Anthropological Approach to Prediction

Ian Lowrie

Abstract In this chapter, I suggest a methodological and theoretical framework for preliminary investigations designed to gauge the potential societal response to the discovery of either microbial or intelligent extraterrestrial life. The uncritical use of analogies to the ethnographic record of contact between societies and the discovery of extraterrestrial life has been, rightfully, the target of sharp criticism since the earliest days of the scientific search for this life. However, I argue that by approaching this record with different epistemological premises, and shifting the focus from the material to the symbolic and cognitive dimensions of this contact, one can avoid many of the pitfalls of the analogical mode of argumentation, and provide a solid conceptual basis for the development of an adequate heuristic. Specifically, I draw upon the germinal debate between Sahlins and Obeyesekere over the nature of human meaning-making in the face of radically other societies and their meanings to treat the discovery of an intelligent civilization. In parallel, I draw upon Sharp's discussion of the relationship between the changes in the symbolic order and the material organization of society to suggest that much of this analysis also applies to the discovery of extraterrestrial microbial life. In both cases, I do not argue for a one-to-one correspondence between the historical and the contemporary, but rather use these arguments as illustrations of what I see as particularly profitable modes of conceptualizing the universal human processes of making sense out of novel objects and phenomena. Finally, this chapter argues for a mixed-methods quantitative-qualitative investigation into the character and distribution of societal resources for understanding life and intelligence, rather than the extraterrestrial as such. The qualitative is advanced as a necessary adjunct to the quantitative, as the best method for gaining access to the repertoire of cultural frames upon which people more or less unconsciously draw in forming their understandings of the world. The focus on life and intelligence is justified both insofar as they are the categories which will be brought

I. Lowrie (✉)
Department of Anthropology, Rice University, Houston, TX, USA
e-mail: il4@rice.edu

D. A. Vakoch (ed.), *Astrobiology, History, and Society*, Advances in Astrobiology and Biogeophysics, DOI: 10.1007/978-3-642-35983-5_13,
© Springer-Verlag Berlin Heidelberg 2013

to bear on the extraterrestrial in terms of integrating it into people's worldviews, and insofar as these categories are substantially more implicated in both societal and personal stability than that of the extraterrestrial as such.

13.1 Introduction

Anthropologists have commented on, participated in, and, indeed, precipitated a wide range of first contacts between radically different societies. Given this long history of familiarity, and the seeming felicity of the analogy between these events and contact with extraterrestrial life, it makes a certain sense to begin this chapter with the knowledge we've gleaned therefrom. This is far from a novel suggestion. Indeed, contact has been put forward as one possible predictive model since the earliest days of the scientific search for extraterrestrial life. At the second Byurakan conference, for example, Philip Morrison briefly toyed with the model before concluding "that a message channel cannot open us to the sort of impact which we have often seen in history once contact is opened between two societies at very different levels of advance" (Sagan 1973, 337). More recently, Kathryn Denning (2013) has pointed out that reasoning from analogy with the human record of contact events is often based upon the false belief that contact necessarily entails a dynamic of conquest of the "inferior" by the "superior" civilization. More vigorously, she has also questioned the ethical, political, and indeed, epistemological implications of the use of reasoning from analogy with the ethnographic record as such. Both points are well taken. However, I think that by shifting the focus here from the *material* dynamics of contact to the symbolic and cognitive dimensions, anthropological theorization of societal responses to contact with the radically different and unexpected can be illuminating—keeping in mind, of course, that we are dealing with "the messy business of human agency and free will" (Denning 2009, 386). What follows, in my discussion of the Hawaiian and Australian contact events, are not exact or strict analogies in terms of their functioning in my argument, then; rather, they are examples from the panoply of human experience which inform rather than mechanically constrain my analysis.

To prefigure my argument somewhat, let us assume that there are two likely shapes which contact with extraterrestrial life by a human society might take. First, there might be the reception of an intentional, ordered signal, or the discovery of other evidence of technological development on the part of a highly developed extraterrestrial civilization—the discovery of another form of intelligence. Second, there might be either the direct encounter with, or incontrovertible secondary evidence of, biological life in which intelligence is absent or unknown—the discovery of another form of life. Neither is likely to lead to the sort of scenario of contact rejected as model by Morrison and Denning. However, I argue below that both will pose similar challenges to the societally-specific cognitive resources and cultural frames used by everyday people to understand their worlds and lives.

   It seems to me that the essential factor for conceptualizing the societal response
to either of these scenarios of contact is an understanding of understanding. How
are we to view people's relationship to the categories by which they make sense of
the world? How are these categories shaped by contact with the radically other? In
order to suggest answers to these questions, I rely upon examples drawn from the
ethnographic record in order to illuminate the broad conceptual framework which
I am advancing as explanatory of much of human social behavior. These examples are not to be taken as suggesting a homology between the intensely complicated and often intensely violent histories of encounter between the "West" and
the "rest" and the encounter between humanity and the extraterrestrial. Certainly,
I am not, for example, attempting to cast "the Hawaiians as us and Cook as the
alien" (Denning 2013). I wholeheartedly agree that that sort of analogical thinking implies the existence of a trans-historical, trans-particular structure of relationship between dominating and dominated halves of any given contact milieu. In
this mode of analysis, the particular valences of the relationships imputed to inhere
in the source or "known" case are brought to the case at hand as a priori factors
of analysis. While there are without doubt cases in which this is a perfectly constructive epistemological tool, I here eschew its use; the situations surveyed, both
historical and contemporary, are much too complex and, indeed, murky. Instead,
working loosely in the tradition of middle range theory (Merton 1968), I look to
these stories of contact primarily in order to inductively construct a workable heuristic for, rather than a grand theory of, predicting societal responses to an encounter with extreme alterity or generalized otherness. Of course, as each situation of
contact is unique, so too are the perspectives of their chroniclers; anthropologists
often disagree on both the historiographic particulars of and the theoretical explanations for the changes in societies wrought by contact. However, these arguments
themselves are often informative.

13.2   Intelligent Life

On his third voyage, having failed in his search for the fabled Northwest Passage,
Captain Cook headed south from the Arctic Ocean, eventually reaching the
Hawaiian Islands. He circled the largest of the islands for some time, before
choosing to moor his ships in Kealakekua Bay. His arrival at Kealakekua coin-
cided with the local *Makahiki* harvest festival, which heralded the annual return
of the god Lono; similarly, his clockwise circumnavigation of the island before
landing mirrored the *Makahiki* procession, and his ship and masts resembled
some of the traditional artifacts used in the *Makahiki*. In any event, when Cook
landed, he and his crew were accorded great status, with Cook being called Lono
and involved in a number of ceremonies. After passing a month in Hawaii, the
ship departed for home, but had to return shortly after, having broken a mast.
In contrast to the welcome Cook had initially received, this time the native
Hawaiians were surprised and hostile. Eventually, an escalating series of thefts and

altercations prompted Cook to attempt to take the Hawaiian King hostage; in the ensuing fracas, Cook was killed.

On these very broad outlines of the contact between Cook and the Hawaiians, the scholarly literature is in more or less agreement. For my discussion of the possible impact upon contemporary society of the discovery of intelligent life in the universe, however, I turn to a brief examination of a debate between two anthropological heavyweights over how to view the Hawaiians' response to the arrival of Captain Cook. Here, given my focus on building a framework for the understanding of understanding, I am not so interested in the historiographic quibbles which make up the bulk of Gananath Obeyesekere and Marshall Sahlins' dispute. In any event, they agree on the essential outlines of the brief historiography I related above. Instead, I aim to draw out the theoretical orientations from which their positions issue: the argument between Sahlins and Obeyesekere turns around whether or not Cook was *really* apotheosized by the locals when he arrived, and, ultimately, on the actual functioning of the cognitive and cultural mechanisms through which societies and their constituent individuals handle the introduction of novel phenomena into their worlds. My goal in examining this debate is not the creation of a mechanical model for explaining the contemporary response of society to the discovery of extraterrestrial intelligence. Rather, it is to build a workable model of how culture *as such* becomes an important factor in societal responses to disruptions issuing from beyond societies' horizons.

Marshall Sahlins' argument apropos of Cook's death is that, initially, the Hawaiians took him for a God. For Sahlins, the Hawaiians' primary mode of understanding the world was one of mythopoesis; they made sense of phenomena by integrating them into their time-tested set of categories, derived from their traditional myths. The priestly and aristocratic elite, charged in Hawaiian society with acting as the arbiters of spiritual and political meaning, actively worked to fit the strange-looking beings in a heretofore unprecedented conveyance into their existing worldview. Insofar as Cook's arrival was sufficiently novel to demand an extraordinary explanation, but felicitously in line with the myth of the return of Lono, the Hawaiians used the cognitive resources available to them to place him within the realm of the intelligible *as* Lono; Sahlins suggests that the elite's predication of Cook *qua* Lono was widely accepted among the laity as well. Sahlins' explanation of Cook's death hinges on his argument that "the diverse and delicate relationships between the two peoples had been ordered by the one salient interpretation of Cook as the Makahiki god which the Hawaiian authorities were able to reify, and with which the Great Navigator could comply" (1985, 128). When Cook disrupted that placement by returning out of sequence with the myth, "that reality began to dissolve" (1985, 128). The traditional *Makahiki* ended with the mythic departure of Lono, and the return to regnancy of the worldly Hawaiian elite. Sahlins suggests that Cook *qua* Lono's reappearance "was sinister because ... bringing the god ashore during the triumph of the King ... would reopen the whole issue of sovereignty" (1985, 128). In other words, the elites charged with interpretation could no longer, for political reasons, simply allow Cook's visit to play out; they had to take an active hand in accomplishing his departure *qua*

God, in order to restore the traditional narrative of the re-ascendance of temporal authority following the *Makahiki*; so they killed him.

This retelling of the encounter has had a number of sharp responses, but I'd like here to focus on the theoretical objections brought to bear by Obeyesekere. Perhaps the central argument he advances against Sahlins is that the reception of Cook as a God was in fact part of a *European* myth, common to Western popular and historiographic descriptions of any number of first encounters between "whites" and "natives," not a Hawaiian one. He argues that the Hawaiians, who have a tradition of divine kingship, honored Cook as a powerful leader by the appellation of Lono, but did not *actually* think that he was in fact a God. Obeyesekere faults Sahlins for falling prey to the "commonplace assumption of the savage mind that is given to prelogical or mystical thought and in turn is fundamentally opposed to the logical and rational ways of thinking of modern man." He does not "object to mythic thought per se but to the assumption of a lack of rational reflection implicit in the premise of ... mythic thought" (1997, 15). Obeyesekere instead advances a model of understanding in which rationality interacts positively with inherited cultural categories; there is a certain amount of mental distance to these categories even among so-called "traditional" societies. For him, "pragmatic rationality" is the hallmark of being human. He suggests that "human beings reflectively assess the implications of a problem in terms of practical criteria," rather than solely in terms of the myth-models advanced by and inherited from political and religious elites (1997, 19). While the Hawaiian elite, for political reasons, feted Cook as Lono, Obeyesekere suggests that the great bulk of both the elite and the laity nevertheless rationally understood him to be a mere man. Indeed, Obeyesekere suggests even the dominant elite interpretation of Cook as quasi-divine *qua* potent chief did not likely travel much further than their circles:

> Cook's arrival was a powerfully unsettling experience and people must have reacted to it in a variety of ways. It is, for example, difficult to believe that the women and lower classes shared the chiefly interpretations; but even if they did, owing to the power of the establishment and its Priests, they must have had other ideas about Cook and his crew (1997, 91).

Obeyesekere is arguing here that even in relatively simple societies there are fractures and cleavages in societies which produce a variety of competing interpretations: "the structures through which experiences is filtered are multiple; these structures are not mechanically followed but are manipulated in accordance with rational reflection" (1997, 175). Cook's murder, then, was not an acting out of the ceremonial death of Lono at the end of the *Makahiki*, but rather a common-sense response by an aristocratic elite to a foreign interloper attempting to kidnap one's king and fellow elite, and presciently, believed to be attempting to replace oneself as the regnant power in one's homeland.

The bulk of contemporary society is composed of individuals who appear to be much closer to Obeyesekere's rational actor that to Sahlins' myth-making subject. Nevertheless, in understanding the relationship of actors to their cultural frames, I'd like to suggest that we can learn from both figurations. Sahlins' view of contact as involving primarily the assimilation of novelty to already existing and only somewhat flexible cultural categories suggests an approach that understands the human response in terms of the felicity of the fit between the content or form of the extraterrestrial message with the current cognitive resources available for understanding intelligence and civilization. Obeyesekere's objection here, though, is well taken; complex societies have numerous fractures that make the identification of salient, wide-ranging cultural categories difficult. However, the rational actor hypothesis doesn't go far enough to explain what happens when people are confronted with things—such as the meanings contained in an extraterrestrial message—that by definition fall outside the boundaries of their rationality.

Moreover, the extra-rational, in this case, insofar as it appears in the form of an intelligent and presumably intelligible message, brings its *own* rationality to bear. Social scientists have been aware since at least Weber that our particular rationality is a result of our own historical trajectory, and is no more universal or absolute than our kinship structure (Luhmann 1998). I would suggest that without even similar neurobiological limitations imposed on cognition, extraterrestrial rationality is likely to be radically different than any present in the ethnographic record. The resulting intersection of more or less disparate ways of making sense out of the world is a hallmark of what Sahlins calls the "structure of the conjuncture" typical of scenes of contact. In the meeting of two rationalities, there is no clear-cut triumph of one over the other; instead, they meet, meld, repulse, borrow, and learn from one another, as well as undergo purely internal processes of self-evaluation and transformation. In short, my argument here is that the introduction of new modes of signification and sense-making implied by the reception and translation of a truly alien message has the potential to force substantive revisions in both academic, and ultimately, popular conceptions of rationality and intelligence as such.

Thus, my proposed bridge between these two models—Obeyesekere's and Sahlins'—is an investigation into the social geography of normative expectations about life, intelligence, and the extraterrestrial. This research would be a large-scale quantitative-qualitative inquiry, moving beyond merely attitude sampling. It must be coupled to a theoretical paradigm that acknowledges that people's responses to unexpected phenomena are substantially more prefigured by their cultural categories than we might otherwise think those of "moderns" to be. I return to this suggestion in the conclusion.

13.3 Life Itself

The second situation—the discovery of life unattached to intelligence—is still a matter of understanding the interaction of the extraterrestrial with extant cultural categories. Even though it may appear that the potential for destabilization is less in this scenario, given the lack of actual *interaction* between two radically different types of thinking being, I suggest that much of the same analysis applies. We are still dealing with a "'structure of the conjuncture': a set of historical relationships that at once reproduce the traditional cultural categories and give them new values out of the pragmatic context" (Sahlins 1985, 125); in this case, however, the relationship is simply with a thing rather than a person (cf. Bennett 2010).

An apposite example here is the impact upon Yir Yoront culture of the introduction of steel axes. Before contact with western missionaries, the aboriginal Australian Yir Yoront used stone axes for a myriad of purposes. These axes were the property of the elder men; the women and children could only borrow axes from men. Lauriston Sharp (1952) argues that these bilateral borrowing relations, insofar as they followed and reinforced hierarchical kin relationships, were a critical feature of sustained sociality among the Yir Yorant. The one-to-one relationships created by the borrowing of axes were the very substance, the essential building blocks, of their society. Additionally, the axe was the totem item of one of the Yir Yorant clans. As Sharp (1952, 19) explains,

> While individual members of such totemic classes or species might disappear or be destroyed, the class itself was obviously ever-present and indestructible. The totems, therefore, lent a permanence and ability to the Clans, to the groupings of human individuals who generation after generation were each associated with a set of items which distinguished one clan from another.

If the kin partnerships sustained by axe borrowing were the *stuff* of Yir Yorant society, the coherence of the clan system was one of the primary ways of *ordering* their society. Membership in clans gave the field of the social a clear, more or less ordered division into constitutive organs, with distinct ritual responsibilities. These divisions were viewed as an essential, timeless feature of the world, a reenactment of the ancestral past in the contemporary; similarly, the totem objects were imagined to be in a timeless relationship to their specific clans.

As part of their evangelistic and colonialist efforts, during the early 20th century missionaries began handing out a great number of steel axes to the Yir Yorant. Whether as gifts at mission festivals, or as payment for services rendered to missionaries by the Yir Yorant, these axes were distributed to young and old, man, woman, and child alike:

> As a result a woman would refer to the axe as "mine," a possessive form she was never able to use of the stone axe. In same fashion, young men or even boys also obtained steel directly from the mission, with the result that older men no longer had a complete monopoly of all the axes in the bush unity. All this led to a revolutionary confusion of sex, age, kinship roles (Sharp 1952, 21)

While, Sharp argues, the "practical effect on the native standard of living was negligible," the disruption of the cultural category of "stone axes" had a devastating social and cultural effect (1952, 20). The axes disrupted the binary relations between kin, as well as the coherence of the totem system, throwing the internal order of Yir Yorant society into disarray. Sharp argues that "the result was the erection of a mental and moral void which foreshadowed the collapse and destruction of all Yir Yoront culture, if not, indeed, the extinction of biological group itself" (1952, 21).

Obviously, we have a category in place to explain the discovery of extraterrestrial creatures or microbes—that of life. However, the Yir Yorant also had a category (of "axes" in general) in place to understand the new steel axes. Yet, the seemingly incidental predicate actually entailed substantive, if subtle, changes in the subject. That is to say, the danger to the Yir Yorant social structure posed by the *steel* axe inhered in the fact that these new objects *appeared* to be able to be categorized as axes, when in fact, what an axe "meant" for the Yir Yoront was dependent upon a complex of social processes and values which could only superficially be mapped onto the new axes and their mode of introduction. The new axes, ultimately, destabilized the Yir Yorant system insofar as they tried to *assimilate* them to their categories, *failed*, and ultimately *distanced* themselves from their now-vitiated categories. This is particularly salient given the unstable and potentially explosive position of the category of "life" within contemporary society (see, e.g., Helmreich 2009). As anthropologists and social theorists have been showing us for some time now, life itself and its definition are intimately tied to the self-understandings of both people (Rabinow 2005) and collectives such as states (Rose 2006). Indeed, Rose and Novas (2005) have argued that citizenship, arguably the most salient category of belonging in much of contemporary society, is increasingly tied to our understandings of life. My modest suggestion here is that should the attempt to integrate extraterrestrial life, whatever its actual biochemistry,within already extant cultural frames similarly fail, and the category itself become sufficiently thrown into question, the effects could be far-reaching.

13.4 Conclusion

To reiterate, the basic model I am suggesting here is primarily methodological. It does not provide any concrete answers to the question of how society will react to the discovery of extraterrestrial life. Instead, it suggests a starting point for asking this question: qualitative and quantitative research into the distribution and character of existing cultural frames and cognitive resources with which people might attempt to make sense of the discovery of extraterrestrial life and intelligence.

This research cannot be solely quantitative for a variety of reasons. First, I would argue that much of what folks actually have at hand in the way of cultural categories through which to interpret the world isn't actually available to self-reflection. This is not simply a reiteration of the tragic view of humanity as always already trapped in a web of mystification and false consciousness; rather, it is the more humble claim that a certain amount of opacity is characteristic of the way that we think about our own thought. It's not that this opacity isn't amenable to clearing up through intensive self-reflection, such as what we might find in meditative practice or the psychoanalytic experience: it is that it is unlikely to dissipate suddenly during the administration of a short, quantitative survey instrument. Indeed, as Weigel and Coe (2013) point out, certain populations are unlikely to provide reliable survey data, such as those characterized by a "high proportion of peasants, tribal members, or people with low literacy who distrust outside scholars or high status social groups even within their own society." While quantitative survey methods have their place within any large-scale research enterprise, an astute qualitative approach, using interviews as well as content and discourse analysis is necessary: both to get at the social facts not immediately apparent to self-reflection and as part of the iterative process of producing better survey instruments. As Pierre Bourdieu, a French sociologist, cautions us, "one has explained nothing and understood nothing by establishing the existence of a correlation between an 'independent' variable and a 'dependent' variable. Until one has determined what is designated in ... each particular relationship, by each term in the relationship, ... the statistical relationship ... remains a pure datum, devoid of meaning" (1984, 16). Qualitative research is the key to determining what is designated by these terms whose relationship we are attempting to quantify, and thus the meaning of the quantification as such.

My experience with qualitative social analysis has led me to argue that the quantitative instruments used in this research should not be solely designed to solicit people's self-predictions about their response to the discovery of extraterrestrial life. Indeed, I argue that they should not primarily be directed towards eliciting respondent's opinions about the extraterrestrial at all [although Peters (2013) demonstrates the value of such research programs for answering finely honed questions about specific populations' responses]. Instead, it seems to me evident that the questions must be designed to produce data on the distribution and character of the salient categories brought to bear *on* the extraterrestrial—that is to say, on the ways which our contemporaries engage with and understand life and intelligence *broadly*. These are the categories through which the extraterrestrial, depending upon its mode of *praesentia*, will be made sense of; critically, they are also the categories which might be disrupted by their ill fit with the realities of the extraterrestrial.

I see two additional problems with focusing on the category of the "extraterrestrial" in this inquiry. First, I would suggest that the category itself, and the cultural frames used to make sense of it in contemporary society, do not occupy the same critical place in most people's and collectivities' self-conceptions as I have argued do those for understanding life and intelligence more broadly. Second, the "extraterrestrial" is a highly capacious, fluid, and polyvalent category, not amenable to being "broken" in the same way that I have argued these latter are (see, e.g., Battaglia 2006). By its very nature, given the contemporary lack of an actual empirical referent which would impose even the loosest strictures on people's imagination of its signification, the category itself is *marked* by a protentive holding open of its final meaning. That is to say, on the one hand, people use and think of the category of extraterrestrial beings in a variety of ways, and can draw upon a wide range of cultural materials, both scientific and popular, as cognitive resources. On the other, however, the category itself is *self-consciously* open to confirmation or disconfirmation of its myriad, potential significations by the discovery of an actual extraterrestrial referent: as Denning astutely points out in her contribution to this volume, the actual character of the life or intelligence discovered may not actually be the most salient feature for understanding people's initial responses to it. However, in the middle- and long-term, as sense is being made of the extraterrestrial by the expert communities devoted to understanding it, and circulated more widely through educational and media programs, it seems to me that the meaning of the category "extraterrestrial" will become more and more constrained by its actual material referent. Precisely because of the previously polyvalent meaning and structurally marginal placement of this category within the cultural edifice of modernity, however, this ossification is unlikely to pose any particularly threatening cognitive difficulties, from a societal point of view.

This discussion has drawn upon anthropological thought about the nature of cultural frames and cognitive resources to suggest a model for guiding our prognostications about the societal response to the discovery of extraterrestrial intelligence. I have argued that a large-scale investigation into the distribution of understandings of life and intelligence is critical to the formulation of specific possible outcomes. However, I believe the examples above will allow me to make a modest predictive suggestion: if the life or intelligence discovered is either quite similar to or very different from the cultural imaginations of life and intelligence *tout court*, the disruption at a cultural level is likely to be minimal.[1] However, I am not as sanguine as Lamb, who suggests that "the knowledge that we are not unique is unlikely to have any destabilizing influence" (2001, 194). I agree that "years of popular science and SF have prepared the public mind for contact" (2001, 194) with both *expected* and *radically unexpected* extraterrestrial life; but I believe that my discussion has suggested that the middle range of the uncanny—that which both assimilates to and confounds expectations—appears to hold a particular challenge to myth and rationality alike, by infiltrating and subverting the available cultural categories for understanding life and intelligence as such. This is where anthropology has a particular role to play within the development of these predictive models, in offering theoretical and interpretive explanations of understanding based upon long experience with the symbolic and cultural aspects of societies' encounters with the radically foreign or other.

References

Battaglia, Debbora. 2006. *E.T. Culture: Anthropology in Outerspaces*. Durham, NC: Duke University Press.
Bennett, Jane. 2010. *Vibrant Matter*. Durham, NC: Duke University Press.
Bourdieu, Pierre. 1984. *Distinction*. Cambridge: Harvard University Press
Denning, Kathryn. 2009. "Ten Thousand Revolutions: Conjectures about Civilizations" *Acta Astronautica* 68: 381–388.

---

[1] In the former case, of course, there could be substantial disruption at the *social* level if the life appears to confirm pop-culture horror stories about extraterrestrial life, but this would not pose a cognitive challenge.

Denning, Kathryn. 2013. "Impossible Predictions of the Unprecedented: Analogy, History, and the Work of Prognostication." In *Astrobiology, History, and Society: Life Beyond Earth and the Impact of Discovery*, ed. Douglas A. Vakoch. Heidelberg: Springer.

Helmreich, Stefan. 2009. *Alien Ocean: Anthropological Voyages in Microbial Seas*. Berkeley: University of California Press.

Lamb, David. 2001. *The Search for Extraterrestrial Intelligence: A Philosophical Inquiry*. London: Routledge.

Luhmann, Niklas. 1998. *Observations on Modernity*. Stanford, CA: Stanford University Press.

Merton, Robert. 1968. *Social Theory and Social Structure*. New York: The Free Press.

Obeyesekere, Gananath. 1997. *The Apotheosis of Captain Cook: European Mythmaking in the Pacific*. Princeton, NJ: Princeton University Press.

Peters, Ted. 2013. "Would the Discovery of ETI Provoke a Religious Crisis?" In *Astrobiology, History, and Society: Life Beyond Earth and the Impact of Discovery*, ed. Douglas A. Vakoch. Heidelberg: Springer.

Rabinow, Paul. 2005. "Artificiality and Enlightenment: From Sociobiology to Biosociality." In *Anthropologies of Modernity: Foucault, Governmentality, and Life Politics*, ed. Jonathan Inda, 181–193. New York: Wiley-Blackwell.

Rose, Nikolas. 2006. *The Politics of Life Itself, Biomedicine, Power, and Subjectivity in the Twenty-First Century*. Princeton, NJ: Princeton University Press.

Rose, Nikolas, and Carlos Novas. 2005. "Biological Citizenship." In *Global Assemblages*, eds. Aihwa Ong and Stephen Collier, 439–463. Oxford: Blackwell.

Sagan, Carl, ed. 1973. *Communication with Extraterrestrial Intelligence*. Cambridge: MIT Press.

Sahlins, Marshall. 1985. *Islands of History*. Chicago: University of Chicago Press.

Sharp, Lauriston. 1952. "Steel Axes for Stone Age Australians." *Human Organization* 11 (2): 17–22.

Weigel, M. Margaret, and Kathryn Coe. 2013. "Impact of Extraterrestrial Life Discovery for Third World Societies: Anthropological and Public Health Considerations." In *Astrobiology, History, and Society: Life Beyond Earth and the Impact of Discovery*, ed. Douglas A. Vakoch. Heidelberg: Springer.

Chapter 14
The Detection of Extraterrestrial Life: Are We Ready?

Klara Anna Capova

Abstract This chapter offers a sociocultural perspective on the scientific search for life beyond Earth. It sheds light on the ways in which alien life is imagined and theorized in order to assess the possible societal response to the detection of *other life*. This chapter is based on the findings of research conducted over two years in the UK, which conceptualizes the extraterrestrial life hypothesis as a significant part of the general worldview, constantly shaped by the work and discoveries of science. Based on these data, the chapter offers insights into the current Western concepts of other life as understood, perceived, and interpreted by the scientific community and popular culture. The post-detection scenarios currently discussed deal mostly with a profound cultural shock following discovery of a superior extraterrestrial civilization. In contrast, the most recent scientific quest for *other life* now operates with a distinctly different concept of extraterrestrial life that ushers in other possible reactions to a detection or a contact. To establish current concepts of *other life* then seems to be crucial for predicting the societal response to a first contact. The chapter presents an overview of multiple conceptions of *other life* in science and science fiction to outline the potential variety of responses. The aim of this chapter is to suggest that the societal readiness and overall acceptance of the *other life* hypothesis needs to be taken into account and that the actual response to the discovery of *other life* will be determined by the actual form or type of life detected. This chapter will present examples from science fiction and other ethnographic material collected during fieldwork to demonstrate how popular culture has adapted the *other life* idea and how the presupposed *other life* is perceived.

K. A. Capova (✉)
Department of Anthropology, Durham University, Durham, UK
e-mail: ka.capova@gmail.com

D. A. Vakoch (ed.), *Astrobiology, History, and Society*, Advances in Astrobiology and Biogeophysics, DOI: 10.1007/978-3-642-35983-5_14,
© Springer-Verlag Berlin Heidelberg 2013

14.1 Is There Anybody Out There?

The past five decades have seen the rapid development of the scientific search for life beyond Earth, providing new insights to help determine whether we are alone in the cosmos. Inevitably many questions have been raised about the impact of the discovery of extraterrestrial life (ETL) on society. In recent years, there has been an increasing amount of literature published on the role and contribution of the social sciences to anticipating the societal response to the detection of ETL (Dick 2006). The role of anthropology in deciphering messages from extraterrestrials and dealing with the social consequences of detection has also been recognized (Denning 2011).

This chapter will focus on the cultural universe formed around extraterrestrial affairs as reflected in the imagination of Earthlings, in the world of fantasy where the close encounters with *other life* take place. Throughout this chapter the generic term *other life* will refer to ETL concepts that are commonly employed in science and science fiction, both embedded in popular culture. The diverse, sometimes contradictory, elements of the imagined Otherness listed below will introduce a variety of *other life* forms narrated in our stories.

This current contribution uses ethnographic examples of imagined contact with the other to examine post-detection scenarios. I address the importance of the human imagination in popular culture to present ETL as reflected within contemporary Western society. In the pages that follow, I will argue that current imagery and perceptions of ETL may play the key role following detection. My question is "Have the narratives about the encounter with aliens made the possibility of *other life* appear to be something to which one can get accustomed?"

14.2 Science and Science Fiction: The War of the Worlds

In 1898, H.G. Wells—whose work has been described as a turning point of the science fiction tradition (Suvin in Waites 1982)—published his famous book *The War of the Worlds*. One year later, Nicola Tesla, staying up late in his laboratory in Colorado Springs detected a suspicious sound; Tesla believed the signal originated from Mars. Shortly after that, in 1913, Edmund Ferrier summarized contemporary discoveries and discussed the evidence from Percival Lowell's observations of Mars surface: "Does all this mean that there are no inhabitants in the planet Mars? No. Mars is certainly inhabited. The collapse of the fairy world constructed by bold imaginations on the base of the canals of Schiaparelli disposes only of the wonderful engineers of whom Mr. H. G. Wells has given us, in his *War of the Worlds*, such a fantastic and captivating description" (Ferrier 1913, 108).

Martians became even more popular in 1939. The famous radio play *War of the Worlds* produced by Orson Welles caused a nationwide panic amongst at least one million of its listeners in the US (Cantril 1966). The on-air dramatization of alien attack demonstrated the "power of a narrative" (Berger 1997, 138) as well as the compelling power of mass media. Moreover, it provided a blueprint for a detection scenario that has shaped mainstream contemporary expectations of the societal response to the discovery of ETL: contact with an alien race will be a distressing event with global impact. The alien superiority is clear from the first lines of H.G. Wells' book: "And we men, the creatures who inhabit this earth, must be to them at least as alien and lowly as are the monkeys and lemurs to us" (Wells 1898).

Since the invention of cinematography, more than three hundred movies presenting a non-human life form have been produced, including more adaptations of *The War of the Worlds*. The film of that title from 1953 directed by Byron Haskins was released in the US and ten European countries between 1953 and 1955 (IMDb 2012a). Steven Spielberg's *War of the Worlds* was distributed to 67 countries worldwide in 2005, showing on more than four thousand screens in the US and UK alone (IMDb 2012b). Reaching a global audience, the fictional invasion of a superior alien race became a popular part of the modern West's narrative history.

And so has the UFO phenomenon. Although regarded as irrational, UFO sightings and abduction stories developed into folk mythology and are reflected in cinematography and mass media. The nine seasons of the TV series *The X Files* from 1993, narrating the story of an FBI agent who investigates paranormal activities and seeks "the truth out there" were followed by a film *The X Files* (1998) where "Mulder and Scully must fight the government in a conspiracy and find the truth about an alien colonization of Earth" (IMDb 2012c). Notably in the filmography of Steven Spielberg we find references to alien abductions. In his *Close Encounters of the Third Kind* (USA 1977), the first humanoids to leave the alien spaceship after successful contact were the abducted pilots.

On the scientific front line of search for *other life*, there was much activity. A turning point in the history of the scientific search for ETL occurred in 1960, when Frank Drake performed his famous radio experiment in the Search for Extraterrestrial Intelligence (SETI), Project Ozma. This was shortly followed by other search (SETI) and messaging (METI) activities including the Arecibo Broadcast and Voyager's Golden Record, followed later by the Cosmic Call and Teen Age Messages. Since then messaging to extraterrestrials has grown into the public sphere: A Message From Earth was created "democratically via the internet, made up of pictures and words from Bebo users" (A Message From Earth 2012) and broadcast in 2008 from the Evpatoria radar facility.

The year 2008 opened a new era in METI as well as in marketing: a Doritos commercial was broadcast toward the Great Bear constellation. Labeled as "First space ad targets hungry aliens" (Barras 2008) or "How to make a bad first impression" (McGovern 2008), the Doritos advertisement became the first commercial ever transmitted to the universe.

14.3 Our Place in the Universe

"Space… the Final Frontier. These are the voyages of the starship Enterprise. Its 5-year mission: to explore strange new worlds, to seek out new life and new civilizations, to boldly go where no one has gone before." *Star Trek* (opening narration) (IMDB 2012d).

There is no doubt that the scientific, fictional, and mythological concepts of *other life*—that is, ETL, science-fiction characters, and UFO sightings and abduction experiences, respectively— are focused around one cosmological question: Are we alone in the universe? Not only do they seem to be different facets of one thing, but all three of them are regarded as essentially cultural.

The anthropological dimension of UFO phenomenon had been highlighted by Grunloh (1977), who described the UFO sighting as a contemporary cult related to the religious visions of the past. Battaglia (2006) introduced the notion of the cultural universe and recognized extraterrestrial culture as a field of anthropological enquiry in the volume *E.T. Culture*: Anthropology *in Outerspaces*. Particularly the understanding of knowledge production and diffusion enables us to boldly explore the strange, new socio-cultural worlds formed around scientific activities and alien life, the "galaxies of discourse" (Battaglia 2006, 2).

The anthropology of science enables us to access the vast space where socio-scientific interactions take place. Works of science are cultural practices that exist in a social context (Martin 1998). Natives of Western culture are accustomed to scientific culture as they are scientifically literate (Harding 1991), and have access to scientific knowledge and science fiction, both shared via mass media. Here we arrive at the key premise of this chapter. The extraterrestrial life hypothesis is conceptualized as a significant part of the contemporary worldview, constantly shaped by the work and discoveries of science.[1] The concept of *other life* is not an element that destabilizes the belief system of a Western culture natives, but rather is in varying degrees embedded into their worldview.

14.4 Changing Perspectives: The Vision of New Worlds

More than a century has passed since the pioneering time of science fiction, as well as the scientific quest for other life forms. This century yielded developments, inventions, and scientific progress that brought humans to the Moon and transformed the Western world politically, economically, institutionally, and culturally.

Since the outset of the Space Age a whole set of new disciplines and institutions emerged: the office of outer space affairs, planetary protection, outer space treaties, and the prospect of space tourism. In 1987 the UN Report of the World Commission on Environment and Development (Brundtland Report) was addressed to "our common future":

> In the middle of the 20th century, we saw our planet from space for the first time. Historians may eventually find that this vision had a greater impact on thought than did the Copernican revolution of the 16th century, which upset the human self-image by revealing that the Earth is not the centre of the universe. From space, we see a small

---

[1] This chapter is based on data gathered from multiple sources during ethnographic research project *In Search of the Inhabited Universe*, conducted through the Department of Anthropology, Durham University.

and fragile ball dominated not by human activity and edifice but by a pattern of clouds, oceans, greenery, and soils. Humanity's inability to fit its activities into that pattern is changing planetary systems, fundamentally. Many such changes are accompanied by life-threatening hazards. This new reality, from which there is no escape, must be recognised—and managed (UN 1987).

The opening lines of the Brundtland report leave us with no doubt: the global vision facilitated by the view-from-outside of "us" is the result of space exploration. Similarly, the Apollo 8 astronaut Frank Borman reported after seeing Earth from outside: "When you're finally up at the moon looking back on earth, all those differences and nationalistic traits are pretty well going to blend, and you're going to get a concept that maybe this really is one world and why the hell can't we learn to live together like decent people" (Space Quotations 2012). The problems we face by virtue of our sharing one planet are highlighted through the experience of the "one-world" and "whole-earth" (Cosgrove 1994).

The identity question plays a key role in any anthropological writing. In this case, the inhabitants of "one-world"—Earthlings—have been inscribed with a new identity that unties boundaries of culture and introduces loose boundaries of being human. When describing human beings from planet Earth, as seen *sub specie aeternitatis* ("under the aspect of eternity"), we can borrow a description provided on the Voyager Record. This "scientific narrative about human beings" (Capova 2008, 77) displays a generic human identity that rests primarily in biological factors and is further based upon cultural universals, elements common to all members of our species, for instance language and reproduction. The Earth citizen of the 21st century, removed from the boundaries and traits of his/her native culture, is ready to set off into the new inhabited universe and introduce the non-anthropocentric identity, the "interstellar humanity" (Dick 2000).

No longer at the edge of science and society, and far from being marginal or heretical, the idea of *other life* seems to be focused around important topics of human identity, related to the origin of life and addressing questions of contemporary cosmology. Particularly after NASA launched the Origins program, another important question came to place—our extraterrestrial origins: Are we all children of the universe? The works of science have changed and are changing our understanding the world—in other words, our worldview—and also our understanding of ourselves.

14.5   ET Life in Numbers: Yes or No?

Now we turn to the quantitative evidence on the popular understanding of ETL. Following the emergence of the ETL idea in public spaces, several surveys have been undertaken to map public opinion. In this chapter, we examine views from the UK and US, the latter being a center of SETI activities.

In a 2008 survey in the UK, 43 % of respondents stated that they "have never seen a UFO but believe they exist" and 9 % of respondents reported that they have seen a UFO (YouGov 2008). While 36 % of people said they don't believe UFOs exist, almost the same percentage of people said that both the British and

US governments have information on UFOs and extraterrestrials that they are concealing.

Two years later, in the survey commissioned by the Royal Society, nearly half of the respondents said that they believed extraterrestrial life exists, while 28 % said they do not, and the same percentage of respondents didn't know (YouGov 2010a). Approximately one quarter of those surveyed expressed the belief that scientists should not be actively searching for and attempting to make contact with extraterrestrial life. Another YouGov survey from 2010 (YouGov 2010b) explored the key areas of social life related to scientific discoveries including global warming, public health, and religion. That survey found that public opinion was evenly split on whether evidence of life elsewhere in the universe would be found.

The Gallup Poll from 2005 exploring paranormal beliefs in the US, Canada, and UK showed that on average 21 % of respondents believed that "extraterrestrial beings have visited Earth at some time in the past" (Lyons 2005). According to the Ipsos international survey conducted in 2010 in 22 countries on behalf of Thompson Reuters (Ipsos 2010), 20 % of respondents agreed that "alien beings have come to earth and walk amongst us in our communities disguised as us." With the highest percentages in this survey being in India (45 %) and China (42 %), the question of cultural differences in ETL perceptions arises. Because ETL detection is likely to have global consequences, post-contact activities need to be managed on a multicultural and multinational basis, taking into account sociometric factors such as gender, religion, age, and educational level of respondents.

14.6 E.T. or ETI?: On the History of Ideas and Confusion

Arguably, the belief in *other life* oscillates around 50 % when the definition of *other life* surveys is not entirely explicit and includes "extraterrestrial life," "aliens," and "alien beings." In the rhetoric of science, however, we can identify two fundamental concepts of *other life*. The first possibility arises from the tradition of SETI searches: an intelligent, detectable, and inherently peaceful (scientific) civilization that initiates contact. The second, more recent concept of microbiological life is the subject matter of astrobiology. Key concepts of *other life* as presented and worked with in the scientific search for ETL include the following:

- advanced lifeforms in an advanced stage of technological development

  - advanced civilization (SETI)
  - postbiological civilization (SETI)
  - remnants of an extinct civilization (SETI)

- life at an early stage of development

  - traces of microbiological life (astrobiology)
  - evidence of past microbiological life (astrobiology)
  - habitable environments (SETI, astrobiology)

- other

  – unimaginable and unpredictable (within the rational frame)

As I observed during my fieldwork, these are the two key ETL concepts that are currently recognized by the scientific community. At the same time, however, during my inquiry into activities related to search of life beyond Earth, I realized that my research design would benefit from rethinking the popular concepts of *other life* presented in the media and the public sphere in conjunction with the scientific concepts.

In popular understanding, is there a clear distinction between UFOs, aliens, and ETL? When searching Google.com one gets nearly three million search results for the phrase "extraterrestrial life," but the number increases to fifty-one million for the phrase "UFO" and nearly forty million when entering the phrase "aliens."[2] This greater volume of searches for "UFO" and "aliens" than for "extraterrestrial life" is consistent with other variants of extraterrestrial species in popular culture:

Aliens	Little grey men (grey aliens)
Visitors	UFOs
ETI	There must be something out there

Science fiction produces a wider variety of imagined life forms. Wikipedia's (2012) entry "Fictional Extraterrestrials" is now arranged in alphabetical list of fictional aliens that classifies hundreds of species, divided into the following lists: humanoid, mammalian, non-sentient, reptilian, parasitic, aquatic, exotic, arthropod, robotic, and plant species, and intergalactic communities.

Having reviewed the variety of imagined others, we now turn to imagined contact scenarios. In the language of science fiction studies, each contact scenario follows a set of conventions in popular culture genres (Berger 1997, 127). The movie *Contact* (USA 1997), based on the novel by Carl Sagan, provides a rather scientific example of another common contact scenario, in which the receipt of a message from an advanced civilization is followed by a strong societal response. This fictional contact takes place in a political context, receives widespread coverage in mass media, and religious considerations arise when selecting the right candidate to make contact with the extraterrestrial civilization.

Contact in the *War of the Worlds* adaptations is portrayed as a struggle for survival, accompanied by panic and fear, a scenario similarly seen in the movies *Independence Day* (1996) and *Cloverfield* (2008). A contact scenario with a totally unknown and potentially harmful ETL is seen in the film *Sphere* (USA 1998) when the team of contactees is being briefed about the action plan:

[08:03] "We think there is an alien life form on the spacecraft and that is why you are here. You are the human contactees that were recommended by the Goodman Report. We have a biochemist to assess the physiology of the Unknown Life Form. We have a mathematician because that will probably be our common language. And we have an astrophysicist to locate its place in the cosmos."

---

[2] Valid as of April 28, 2012. In the case of the phrase "aliens" the high volume of searches may be elevated following the release of the Alien vs Predator film and personal computer game.

...

"Listen up [reading, quoting a report]: 'Contactee meeting an unknown life form, or U.L.F., must be prepared for severe psychological impact. Stress reaction when confronted by unknown life form has not been sufficiently studied and cannot be entirely predicted in advance. But the most likely consequence of contact is absolute terror. That's from Goodman's report" [08:48].

The latest works within the genre of science fiction bring new perspectives on contact. Notably in *Avatar* (USA 2009), contact is upgraded to a new level by showing humans as a more advanced civilization infiltrating a peaceful tribe on an alien planet. In the social commentary *District 9* (USA 2009), an alien population is moved out to a ghetto, allowing the viewer to see first contact from the perspective of apartheid.

On a less serious note, in the *Star Trek* parody *Galaxy Quest* (USA 1999) the representatives of an alien race visiting Earth are initially ignored because, in their human-like forms, they look too like the other fans of the defunct *Galaxy Quest* television series. Another lifeform is, sadly, eaten by a dog in *Hitchhikers Guide to the Galaxy* (2005).

The most influential works presented above cover only a small portion of the extensive body of science fiction, which do not begin to explore cinematography produced in other parts of the world. Nevertheless, this brief review illustrates ways that the extraterrestrial life hypothesis has been examined, not only as a subject of scientific inquiry, but also as part of a virtual world and our current narratives. The works of science fiction present not only conventional views based mostly on the binary opposites of hostile and peaceful alien race but also offer alternative perspectives on the contact situation.

14.7 Are We Ready for Contact?

On a few occasions during my fieldwork, people told me that they wouldn't be surprised to learn that the stories about alien visits to Earth and conspiracies are based on actual events. Similarly we have seen in public polls that a considerable number of respondents believe that aliens were or are present on Earth. The central argument of this contribution lays in showing the idea of *other life* in a different perspective and from a sociocultural perspective it is highly relevant to include popular conceptions about other life and science fiction stories in the ETL debate. While in the rhetoric of science the concepts of ETL are clearly defined, public opinion includes multiple and less articulate concepts of *other life* featuring various degrees of otherness. If the ETL debate is to be moved forward, a better understanding needs to be developed of the cultural landscapes from which the reaction of the public to the detection of extraterrestrial life arises.

We can speculate on the possible wider implications of our narratives about the encounter with aliens. But to be clear, we must be cautious about making any generalizations independent of the specific contact situation. The immediate societal response to the detection of extraterrestrial life will be cultural as well as individual, but above all contextual, and in any case influenced by the type of life discovered. Since it is impossible to anticipate the nature of unknown life forms, we cannot reliably predict how contact will unfold also because a variety of cultural, social, and historical factors will shape both short-term and long-term responses.

However, we should consider the sociocultural evolution and ask if the generation of *Star Trek* fans, familiar with the fictional idea of an inhabited universe, would be shocked to find out there is bacterial life outside of Earth. Would the day we discover microbiological life beyond Earth be the day the Earth stood still, particularly after NASA's 1996 announcement of evidence for primitive bacterial life in a meteorite from Mars (NASA 1996) and following the study week on astrobiology held by the Pontifical Academy of Sciences in the Vatican (Pontifical Academy of Sciences 2009)?

As is typical for this topic, we get more questions than answers. Despite the Great Silence, after fifty years of actively search for other life and science fiction works, westernized Earthlings seem to be more receptive to the idea *other life* than at any point in history. Alien life is vividly imagined and publicly discussed and as such embedded in popular culture and social conversations. The mass media play a key role as a dispersing tool, broadcasting not only fantastic stories about the imagined other but also scientific information and subsequently influencing public opinion globally. The others are described in our stories; they take part in our TV shows, films, science fiction, folk mythology; they are embedded in popular culture; and they reflect our hopes, fears, and anxieties.

Appendix I. List of Films

Avatar. Directed by James Cameron. USA: Twentieth Century Fox Film Corporation, 2009.
Contact. Directed by Robert Zemeckis. USA: Warner Bros. Pictures, 1997.
Cloverfield. Directed by Matt Reeves. USA: Paramount Pictures, 2008.
Close Encounters of the Third Kind. Directed by Steven Spielberg. USA: Columbia Pictures Corporation, 1977.
District 9. Directed by Neill Blomkamp. USA: TriStar Pictures, 2009.
Galaxy Quest. Directed by Dean Parisot. USA: DreamWorks SKG, 1999.
Hitchhikers Guide to the Galaxy. Directed by Garth Jennings. USA, UK: Touchstone Pictures, 2005.
Independence Day. Directed by Roland Emmerich. USA: Fox Home Entertainment, 1996.
Sphere. Directed by Barry Levinson. USA: Warner Bros. Pictures, 1998.
The X Files. Directed by Rob Bowman. USA: Twentieth Century Fox Film Corporation, 1998.
War of the Worlds. Directed by Steven Spielberg. USA: Paramount Pictures, 2005
War of the Worlds. Directed by Byron Haskin. USA, Paramount Pictures, 1953.
Star Trek: The Next Generation. Created by Gene Roddenberry. USA: Paramount Television, 1987–1994.

References

Barras, Colin. 2008. "First Space Ad Targets Hungry Aliens." *New Scientist*, June 12. http://www.newscientist.com/article/dn14130-first-space-ad-targets-hungry-aliens.html.

Battaglia, Debbora, ed. 2006. *E.T. Culture: Anthropology in Outerspaces*. Durham, NC: Duke University Press.

Berger, Arthur Asa. 1997. *Narratives in Popular Culture, Media, and Everyday Life*. London: Sage Publications.

Capova, Klara Anna. 2008. "Voyager Message." Master's (Mgr.) Thesis, Charles University, Prague, Czech Republic.

Cantril, Hadley. 1966. *The Invasion From Mars: A Study in the Psychology of Panic*. Princeton, NJ: Princeton University Press. Reprint, 2009. Piscataway, NJ: Transaction Publishers.

Cosgrove, Denis. 1994. "Contested Global Visions: One-World, Whole-Earth, and the Apollo Space Photographs." *Annals of the Association of American Geographers* 84 (2): 270–294.

Denning, Kathryn. 2011. "Is Life What We Make of It?" *Philosophical Transactions of the Royal Society A: Mathematical, Physical and Engineering Sciences*, 369 (1936): 669–678.

Dick, Steven J. 2006. "Anthropology and the Search for Extraterrestrial Intelligence: An Historical View." *Anthropology Today* 22: 3–7.

Dick, Steven J. 2000. "Interstellar Humanity." *Futures* 32 (6): 555–556.

Ferrier, Edmund. 1913. "What is Life on Mars Like?" *The North American Review*, 197 (686): 105–111.

Grunloh, Ronald L. 1977. "Flying Saucers." *RAIN, Royal Anthropological Inst Great Britain and Ireland* 23 (December): 1–4.

Harding, Sandra G. 1991. *Whose Science? Whose Knowledge?: Thinking from Women's Lives*. Ithaca, N.Y: Cornell University Press.

IMDb. 2012a. "Release Dates for *The War of the Worlds* (1953)." Accessed July 20. http://www.imdb.com/title/tt0046534/releaseinfo.

IMDb. 2012b. "Release Dates for *War of the Worlds* (1985)." Accessed July 20. http://www.imdb.com/title/tt0407304/releaseinfo.

IMDb. 2012c. *"The X Files."* Accessed July 20. http://www.imdb.com/title/tt0120902/.

IMDb. 2012d. *"Memorable Quotes for 'Star Trek'"* 1966. Accessed July 24. http://www.imdb.com/title/tt0060028/quotes.

Ipsos. 2010. "One in Five (20 %) Global Citizens Believe that Alien Beings Have Come Down to Earth and Walk Amongst Us in Our Communities Disguised as Humans." http://www.ipsos-na.com/news-polls/pressrelease.aspx?id=4742.

Lyons, Lynda. 2005. "Paranormal Beliefs Come (Super) Naturally to Some." Gallup. Accessed May 29. http://www.gallup.com/poll/19558/paranormal-beliefs-come-supernaturally-some.aspx.

Martin, Emily. 1998. "Anthropology and the Cultural Study of Science." *Science Technology Human Values* 23 (1): 24–44.

A Message from Earth. 2012. "Space Communication Timeline." Accessed July 20. http://projects.lessrain.net/public/downloads/amfe/A_Message_From_Earth_Timeline.pdf.

McGovern, Jeremy. 2008. "How to Make a Bad First Impression." *Astronomy*, June 13 http://cs.astronomy.com/asy/b/astronomy/archive/2008/06/13/how-not-to-win-friends.aspx.

Pontifical Academy of Sciences, Astrobiology, Study Week 6–10 November, 2009 http://www.casinapioiv.va/content/accademia/en/events/2009/astrobiology.html.

Space Quotations.com. 2012. "Looking Back at the Earth Quotes: Frank Borman, Apollo 8, *Newsweek*, 23 December 1968." Accessed July 20. http://www.spacequotations.com/earth.html.

UN, Report of the World Commission on Environment and Development. Development and International Economic Co-Operation: Environment. 1987. G. A. Document, United Nations.

Waites, Bernard, Tony Bennett, and Graham Martin, eds. 1982. *Popular Culture, Past and Present: A Reader*. London: Taylor and Francis.

Wells, H.G., 1898. *The War of the Worlds.*. "Book One: The Coming of the Martians, Chapter One: The Eve of the War." Project Gutenberg: HTML e-book, 2004. Accessed July 20, 2012. http://www.gutenberg.org/files/36/36-h/36-h.htm.

Wikipedia. 2012. "List of Fictional Extraterrestrials." Accessed February 26. http://
  en.wikipedia.org/wiki/List_of_fictional_extraterrestrials.
YouGov. 2008. *YouGov/The Sun Survey Results*. Accessed July 20, 2012. http://
  d25d2506sfb94s.cloudfront.net/today_uk_import/YG-Archives-lif-sun-UFO-080728.pdf.
YouGov. 2010a. *YouGov Survey Results*. Accessed July 20, 2012. http://cdn.yougov.com/
  today_uk_import/YG_Archives_Life_ColmanGetty_RoyalSoc_Science_041010.pdf.
YouGov. 2010b. *YouGov Survey Results*. Accessed January 18, 2013. http://d25d2506sfb94s.cloudfront.
  net/today_uk_import/YG_Archives_Life_LikelyToHappen_130810.pdf

Chapter 15
Impact of Extraterrestrial Life Discovery for Third World Societies: Anthropological and Public Health Considerations

M. Margaret Weigel and Kathryn Coe

Abstract In this chapter we focus on the strategies developed by humans, living around the world, to prevent the injuries and death that can occur when confronted by random, but not totally unanticipated disasters. While the occurrence of events such as tsunamis, floods, volcanic eruptions, drought, and hostility from a distant tribe may be forgotten, the ancestral memory of these events, and strategies for coping with them, are retained in such things as stories and rituals. These seemingly unimportant cultural strategies made it possible for individuals to respond to such events with immediate and appropriate actions, thus providing those people with significant survival advantages. We begin by outlining some of the strategies honed by humans over centuries and millennia that proved to be successful in responding to potentially threatening events and that informed future generations about these events and the strategies needed to address them. These strategies, which continue to be practiced in traditional groups, include the use of such things as stories, parables, song and dance. We then apply this thinking to develop a research design for studying the response of individuals living in developing countries to information about possible contact with extraterrestrial complex or intelligent life. We conclude this chapter by outlining a justification for such a study.

15.1 Introduction

Around the world, and throughout prehistory and history, humans have experienced unusual and random events that often presented significant danger. These events have included such things as natural disasters—tsunamis, floods, volcanic

M. M. Weigel (✉)
Department of Public Health Sciences, College of Health Sciences, The University of Texas, El Paso, TX, USA
e-mail: mmweigel@utep.edu

K. Coe
Department of Public Health, Fairbanks School of Public Health, Indiana University, Indianapolis, IN, USA

D. A. Vakoch (ed.), *Astrobiology, History, and Society*, Advances in Astrobiology and Biogeophysics, DOI: 10.1007/978-3-642-35983-5_15, © Springer-Verlag Berlin Heidelberg 2013

eruptions, and drought; contact with other human groups, who all too often presented serious threats; and even reported contact with supernatural or celestial beings. As many of these events posed serious threats to survival and livelihood, an immediate and appropriate response would have offered significant survival advantages. In this chapter, we discuss some of the strategies that humans honed over centuries and millennia that proved to be successful in responding to potentially threatening events and that informed future generations about these events and the actions needed to address them. These strategies, which continue to be practiced in many traditional groups around the world, include the use of such things as stories, parables, song and dance. The use of analogies to describe comparable events and the appropriate response made it possible to preserve an important and attractive lesson, transmitting it from one generation to the next, over a great many generations. This transmission and replication allowed ancestors to protect, consciously or unconsciously, multiple generations of their descendants. We begin this chapter by discussing the characteristics of traditional societies and then we briefly outline the traditional cultural strategies utilized for disaster preparedness. We then apply this thinking to outline a research design for studying the response of individuals living in developing countries to information about extraterrestrial complex or intelligent life. We conclude this chapter by outlining a justification for such a study.

15.2 Background

More than 85% of the world's seven billion inhabitants live in developing countries and 99% of global population growth is occurring in these countries (Population Reference Bureau [PRB] 2011). Within these developing countries there has been significant cultural variation. Today, in all these countries, there are, some residents who remain highly traditional and others who are highly westernized in their behavior. We are probably safe in assuming that the response of traditional and geographically-isolated groups will be distinct from that of urban and westernized populations. We also may be safe in assuming that there was significant cultural variation within traditional groups. This, however, may not always be a valid assumption. As As Eaton, Cordain and Lindeberg (2002, 122) have pointed out when discussing foragers, "the differences between them were minor compared with their essential similarities." Our particular aim in writing this chapter was to identify and outline how these traditional groups prepared for and responded to potential disaster of the groups that live in more remote areas and that tend to follow age-old traditions. These groups are usually characterized by shared kinship and ancestry, unique creation myths, narratives, social norms, social organization and political structure, moral systems and behavioral codes, informal and formal systems of education, and even strategies for warning future generations of possible catastrophic events.

Today, as these groups tend to live in areas that are remote geographically, both science/technology literacy and access to modern communication technologies are low to non-existent and group members still continue to follow many traditions that are assumed to provide important strategies for dealing with the contingencies of life. Furthermore, most of these isolated groups rely on foraging combined with subsistence farming or small-scale industry (e.g., carpentry, pottery and basket making), skills they acquired from their ancestors, to provide for their families. While their religious practices often involve complex pantheons of supernatural beings, their immediate and long-term priorities not only draw upon strategies used by their ancestors, but are focused on where their family's next meal is going to come from, how to protect their family, crops, and domesticated animals from predators or thieves; how to prevent illness and deal with unexpected events, such as acts of nature that damage crops, and illnesses that affect their children. These people, in sum, are what we call traditional people.

The use of the terms "traditional society" implies that in the midst of the seeming chaos of cultural diversity in the world, a recognizable dichotomy exists between traditional and non-traditional societies, whether those societies are referred to as tribal or peasant. Although this dichotomy is actually a continuum, we define traditional societies as those in which cultural behaviors tend to have been copied from ancestors for many generations. Traditional behaviors, to quote Osaghae (2010, 204), are "the legacy of the past." These copied behaviors include not only the rituals that are recognized as being stereotyped and repeated from one generation to the next, but also those ordinary practices related to subsistence, social interaction, and preparedness for possibly dangerous events.

15.3 Human Cultural Response to Novel Events

All human societies develop cultural explanations for novel events. For much of history, based on inferences with hunter-gatherer populations soon after contact, humans have drawn upon the teachings of ancestors and the religion they inherited from those ancestors to warn of, respond to, and explain such events. Such cultural strategies include religious teachings, the use of news carriers to spread news, and the use of stories to keep ancient knowledge alive, warning of the possibility of such events, describing how to respond to such events, and explaining why such events occur.

15.3.1 Religion and the Creation of the Earth, Heavens and People

While not all social groups have had a creator god, or gods, all known social groups have had religion (Steadman and Palmer 2010). Central to these religions are supernatural claims, including claims that supernatural beings are actively involved in the lives of the living. Supernatural claim are those that cannot be identified through empirical study or use of the senses. There currently is no way to prove or disprove that such claims are true; they are taken on faith. One of the most ancient cross-cultural claims is that the Earth and even the heavens were created by ancestors and that the ancestors continue to be interested in and interact with their living descendants (Steadman and Palmer 2010).

To provide one example, in southern Africa, it is claimed that "[w]hen someone dies, the person is believed to transform into an ancestral spirit…Death does not make a person cease to belong to his or her social unit, family, clan, tribe, village or nation" (Bojuwoye 2005, 62). The claim was often made that there are several types of ancestors: the distant ancestor, who as Mosha (2000) explains, is the Great Ancestor and Supreme Creator, who created the Earth and heavens and all that included in them, and distant ancestors who may only be known by name, and ancestors who were known but have recently died. The role of the minor ancestors was to regularly watch over their descendants and guide their social behavior. It was assumed that they interceded in nearly all aspects of life, "including helping with marital and interpersonal relationship conflicts, bringing about good health, averting illness, assisting in obtaining good fortune and averting natural disasters and accidents" (Bojuwoye 2005, 62–63). In such societies, "if the rules and values set down by ancestors are ignored or broken, as Kopytoff (1971) explains, "the ancestors will be angry and calamity—illness, failure of crops, floods, and drought—will fall." Keeping the ancestors happy, as Daniel (2010, 25) explains, was a major responsibility and task of living kin.

Religion, in this context, offers an explanation for novel events, specifically disasters, one that places blame on humans and their inappropriate actions, and offers a practical strategy, i.e., behave appropriately, if you wish to prevent disasters that affect individuals and/or their social groups. Traditional religions as the guardian of traditions helped keep these explanations alive generation after generation. Religions hold, as Daniel (2010, 24) writes, "society in its fixed pattern." In other words, the continuity of many traditions, including those focused on disaster preparedness, depends upon the stability of religious beliefs and practices.

Before we conclude this brief discussion of religion, it is important to point out that accounts of supernatural beings interacting with humans are not unique to traditional groups in developing countries. Such accounts are common to most religions (Harvey 2002). Christianity, as explained in accounts found in the Bible, its written guide, has an abundance of supernatural beings, such as angels and the Son of God, who are not of this world and who possess supernatural powers and who interact with or have interacted with humans. The same claim can be made in regard to other major religious traditions including Islam, Judaism, Hinduism, and Mormonism (Matthews 2011).

15.3.2 News Carriers

Social groups around the world have regularly had individuals whose role was to relay news (Coe, Aiken, and Palmer 2006). For example, among many American

Indian tribes, news was carried by criers, or "walking newspapers." These news carriers are described in a great many North American ethnographies, including those of the Blackfoot (Hungry Wolf 1977; Wissler 1918), Delaware (Newcomb 1956), Lakota (Amiotte 1992), Objibwa (Tanner 1830), Pawnee (Grinnell 1889), Seminole (Sattler 1987), Tlingit (Emmons 1991) and Yokut (Gayton 1940). News carriers were men who were known to be trustworthy and who could provide plainspoken information of importance in decision-making (Firth 1956; Rodnick 1937). They spoke in a distinct style. Among the Pawnee, their "wording was in a precise telegraphic style so that it could be clearly understood by everyone" (Weltfish 1965, 25). Among the Cooper Inuit, when these news carriers relayed information, they would begin by describing the crisis and then working forward to causes of the crisis. Among the Assiniboine and Western Apache, the news was often relayed only to the chief, who, it was felt, could interpret it dispassionately and accurately, before relaying it to others in his clan (Rodnick 1937; Basso 1971).

To summarize this short discussion, a function of news carriers was to provide accurate and plainspoken information that could be used to help the recipients make important decisions regarding potentially threatening events. To be able to do so, the news carriers had to be trained to provide information accurately and dispassionately. As Firth (1956, 123) described in his study of the Tikopians, journalistic narrative involved, "ideally the reporting of verified [or verifiable] events while stories… can involve the reporting of unverified [or unverifiable] events." There was, he claimed, "very great interest in the truth" of news accounts, but that stories involved the suspension of judgment on the veracity of accuracy of the account" (Firth 1956, 123). As stories may lack veracity, the question we must address is why they might provide a useful strategy for disaster preparedness?

15.3.3 Traditional Stories

While some scholars argue that everything we read or write qualifies as a story (Dawes 2000, 147), most scholars are more exclusive in their definitions. Stories are defined as a form of narrative or text that includes form or structure, context, attractiveness (e.g., it attracts and hold our attention) and memorability. Stories are made more attractive and memorable when they use what the Dogon refer to as poetic or pleasant speech (Calame-Griaule 1986), which is distinguished from ordinary speech by the use of things such as euphony or pleasant sounds (Elgin 1998), or mnemonic devices: assonance, alliteration, repetition, rhyme, metaphor, similar, harmony between the sounds of adjoining words, and so forth (Biesele 1993; Calame-Griaule 1986; Dawes 2000). The Dogon refer to these as the "spellbinding" elements of stories as they have the capacity to please the listener, hold their attention, and stick in their memories (Calame-Griaule 1986, 559).

Accounts of supernatural beings and acts are included in many of the traditional stories, not only those that are part of religious practices. The actions of the supernatural beings also are depicted in rituals and the visual arts (Matthews

2011; Harvey 2002). As part of religion, most, if not all, cultures have developed explanations for the origin or creation of the heavens, the land, and all who inhabit the land and these accounts share a number of common characteristics (See discussions in Eliade 1963; Johnson 2009; Leeming 2010; Long 1963). First, when these stories are associated with religion they are considered to be sacred stories set sometime in the very distant past. In addition, they usually involve a plot in which a supernatural being or ancestor creates the world and what surrounds it. These characteristics hold true across cultures, even western countries, where major world's religions discuss creation as accomplished through actions performed by supernatural beings. Further, these supernatural or fantastic beings or forces regularly interact with humans and communicate with them. These forces, objects or beings can include animals and animal-like creatures that typically are combinations of parts from actual animals, vegetable matter, the elements (e.g., Sun, Moon, winds), and/or physical features of the environment (Coe, Aiken, and Palmer 2006). Often the supernatural beings are depicted as animals and the immediate effect of giving animals human characteristics seems to be that it is successful in attracting and holding the attention of the audience (Calamae-Griaule 1986), thus making the stories more memorable and easier to remember.

Oftentimes the supernatural or fantastic beings in stories are human-like beings that can be corporeal or incorporeal and possess special supernatural abilities. Examples include ghosts and deceased ancestors, gods, and good or evil spirits. These supernatural beings are said to have the ability to interact with the living. Many of the human-like supernatural beings described in stories appear to be ancestors or associated with them (Steadman, Palmer, and Tilley 1996). Pedro Sarmiento de Gamboa (1532–1592) related a traditional myth the Inca told about the return of a celestial being, Viracocha, who, long ago, had created the universe, Sun, Moon, stars, time, and human beings. Viracocha, the myths related, would be physically distinct, being white skinned, but he should be welcomed as he would come in time of trouble to solve problems. Not all supernatural beings are described as helpful or even benign. A myth, believed to have been passed down from pre-Colombian times, describes less benign supernatural beings, "Pishtacos," as tall, shadowy figures with blond hair and blue eyes who stalk Andean peasants along dark roads, kill them and drain them of their fat. Another myth, widespread through Latin America and the Caribbean, concerns "Chupacabras," creatures with a reptile-like appearance and red glowing eyes who kill livestock and drain them of their blood.

Stories, however, can do more. They can describe behaviors considered to be important and point out the consequences of misbehaving. They also can contain important knowledge of how to respond to events. Basso (1990, 103) writes that the stories of the Western Apache "promote compliance with standards for acceptable social behavior and the moral values that support them. Among the Dogon, "every narrative is a pretext for a lesson in social ethics" (Calame-Griaule 1986, 570).

It is one thing to say that stories influence behavior and outline strategies for respond to novel events and another to demonstrate this influence (see discussion in Coe, Palmer, Aiken, and Cassady 2006). This point may be particularly important when we are talking about stories, that exaggerate and embellish, as methods for preventing harm. The attractive qualities of stories are what keeps people interested in them and repeating them generation after generation. The potential usefulness of traditional stories for responding to novel events is well illustrated by the Bafmen of Cameroon who believe that lakes harbor ancestors and spirits and, sometimes, death. The ethnographer Kevin Krajick describes one Bafman story about an exploding lake. This story influences them to build their houses on high ground, above the danger posed by the lake (Krajick 2003).

On August 21, 1986, approximately 1,800 villagers living on the lower slopes of Lake Nyos in African Cameroon died of carbon dioxide asphyxiation when the lake literally exploded into the surrounding air. Apparently, carbon dioxide is fed into the lake from underwater springs where it is kept by the pressure of the water from forming bubbles just the same as a cap keeps soda from fizzing in a bottle. Because the lake is located along the equator, the seasonal changes in temperature are minor and, as the water is not mixed, the deepest water might remain stagnated for centuries (Krajick 2003). However, on the night of August 21, something disrupted the surface of the water (perhaps a boulder rolled down the slope into the lake) and blew the "cap" off. A cloud of deadly gas then filled the valley formed by the lake and the people, cattle, birds, and insects living on the lower slopes all died quickly. However, people and animals living on high ground survived the event (Krajick 2003). The high slopes around Lake Nyos were occupied by the Bafmen whose ancestors have lived there for hundreds of years and had passed this story down since earlier times. In contrast, those who died on the lower slopes were from other clans or tribes who began moving into the area only 60 years ago. These did not have the exploding lake story in their traditional arsenal of tales.

Another example of the power of traditional stories that may help individuals respond appropriately to novel events comes from the devastating tsunami of December 26, 2004. Survivors report that stories coming from their ancestors, as told to them by their parents and grandparents, caused them to respond to the first Earth tremors by fleeing to higher ground. On the March 20, 2005 telecast of CBS's "60 minutes" news program, the Moken people, who were saved by getting to high land, reported that they have a legend about the wave that eats people. The Sentinelese people also explained they ran to the hills because this legend, passed down from their forefathers, told them that they should do so when the Earth shakes; the shaking means the sea will rise up onto the land (Mukerjee 2005). Consequently, many people who had lived for generations in the area hit by the tsunami were saved because of stories that told of such events by their ancestors.

These and similar myths are still being told in remote areas of the developing world. Such myths set the stage for how individuals in these groups behavior, how these groups would interpret the nature of first contact with unknown beings, and how they might draw upon stories to provide strategies for responding. For example, natives from the New Guinea highland did not consider the white European outsiders who unexpectedly appeared among them as humans like themselves but instead as "sky people" or spirits (Winzeler 2008).

The question we now turn to is how knowledge of these strategies might help prepare individuals in third world countries for a novel event, namely contact with extraterrestrial life and how we might collect more stories and other strategies upon which we could build an intervention that is culturally tailored, appropriate, and effective in preventing panic.

15.4 Preparing for the Possibility of Extraterrestrial Life

Astronomers are now able to detect planets orbiting stars other than the Sun where life may exist, and living generations could see the signatures of extra-terrestrial life being detected. Should it turn out that we are not alone in the Universe, it will fundamentally affect how humanity understands itself—and we need to be prepared for the consequences (Dominik and Zarnecki 2011, 499).

The increasingly rapid pace of new exoplanets and other astrobiological discoveries makes it likely that evidence of extraterrestrial life, and possibly intelligent life, will be confirmed within the next several decades. The most likely scenario is that astrobiologists will detect some sort of electromagnetic, chemical, or other signatures indicating the existence of life forms on planets or their moons at far enough away distances so that physical contact is prevented. Direct contact with intelligent life in the future, although a more distant possibility, is not an improbability.

The realization that evidence of intelligent life and extraterrestrial contact may lie not too far in the future has led scholars to issue a call for research that can help us to understand how different social groups may respond to such contact and, thus, develop evidence-based programs that can help prepare humanity for such contact. Although we have no way of predicting how contact might be made, or the aim underlying such contact, we can focus on our own possible response. Consequently, cross-cultural research to explore the diversity of human beliefs and assumptions about extraterrestrial life is needed (Harrison and Vakoch 2011). This is worthwhile from a theoretical perspective because it can help us to better understand ourselves and the possible range of responses that characterize our species. Cross-cultural studies of the attitudes, beliefs, and possible responses of diverse human societies also has high potential practical value because these findings can be applied to help anticipate, educate and inform, and prepare many different populations for the time when such contact occurs.

A number of research surveys conducted over the past several decades have attempted to identify human assumptions, perceptions, knowledge, beliefs about and possible responses to extraterrestrial life discovery and/or contact (Bainbridge 2011; Pettinico 2011; Peters 2011; Alexander 2003; Sabadell and Salamero 1996). The findings from these surveys have provided us with important insights into these questions. However, almost all of the studies identified in the literature were conducted among relatively well-educated persons in North American and European countries who ostensibly had ready access to telephones, the internet, and international popular science publications (e.g., *Discover, National Geographic, Popular Science*). To the best of our knowledge, only one study has been published which surveyed individuals from a non-western society but, even in this case, the respondents were educated Taiwanese college students with access to modern science and technology information and communication devices (Vakoch and Lee 2000). Thus, very little information is available on the potential assumptions, perceptions, knowledge, beliefs about and possible responses to extraterrestrial life discovery and/or contact among persons living in rural or other remote areas of the developing world.

The principal aim of this section of this chapter is to propose a research approach that incorporates both the applied social sciences and public health to revisit the question of how the discovery of extraterrestrial life would affect humans and their societies in order to plan for such an event in the short and long term. Towards that end, our major focus will be on rural tribal and peasant societies who live in developing countries and who have limited or no access to modern technology or science information, not only because such knowledge and access will bias responses, but because it often presents a serious challenge to traditions, resulting in their sudden loss. Several assumptions underlie this discussion.

First, the type of first contact (i.e., direct, indirect) and the information source (trusted insider, outsiders) will influence reactions. Secondly, we also assume, based upon a large amount of cross-cultural data, that most adult humans have a predisposition to react to strangers, particularly those who look and act differently, in a xenophobic or intolerant fashion. This is true whether these "strangers" are viewed first hand in person, through pictures or drawings, or recounted by news carriers and story-tellers. A third assumption is that given this propensity, first contact could evoke responses similar to that produced by the 1938 Orson Welles radio broadcast about an alien invasion from Mars. Widespread fear, anxiety, stress and widespread panic could result potentially affecting the mental and emotional health and social well-being of populations. If these responses occur, they bring with them serious issues that must be addressed by the public health system.

However, a fourth assumption, which may affect the third assumption, is that individuals living in more traditional rural societies in developing countries with little or no access to complex information will respond differently to first contact when compared to same-country urbanites or persons from more developed countries. This is because, to provide only one reason, seeing rituals and listening to myths and stories about contact with supernatural beings is a regular part of life. For this reason, it is important to focus research on these distinct traditional groups as they may present very unique profiles.

Human societies are exceeding diverse and we would be committing a serious error if we did not consider that diversity. We propose to do so by first focusing on developing research for identifying how individuals residing in more traditional rural societies in developing countries in Latin America, Asia, and Africa might respond to such contact, based on knowledge of their unique cultural understandings and possible cultural responses to such an event. Thus, the programs that prepare these communities for this event would most likely be very different from those developed for more educated urbanites living in the same country as well as westernized populations elsewhere.

15.4.1 Potential Impact of Extraterrestrial Life Discovery for Third World Countries

The authors' experience working with rural Ecuadorian groups suggests that while many are well aware of stars, as they can see vast quantities of them at night, and often have stories about individual stars and groups of stars and the beings who inhabit them, they are either unaware or only vaguely aware of the existence of other planets and, unless their myths claim that the origin of a certain supernatural being in their pantheon of supernatural beings was from the heavens or a particular star, they may not have ever considered the possibility of extraterrestrial life. In many cases, they may not have even thought about life existing in other countries outside the Andean or upper Amazon region. Many, in fact, believe their tribe or country to be the figurative center of the universe.

15.4.2 Collection of Data for the Development of Educational Programs

Building on the descriptions provided earlier in this chapter, when developing educational programs informing people that extraterrestrial contact has occurred or is likely to occur in the near future, one might well predict that in westernized groups, with communication systems that can rapidly spread and distort information and an often intense distrust of government, a panicked reaction could occur. This response could resemble that which occurred on October 30, 1948, when Orson Welles presented a series of simulated news bulletins reporting that on an apparent in-progress invasion by Martians. This same response predictably could occur in urbanites who have technology access living in third world countries. It is possible, however, that in traditional, rural, and hard-to reach social groups, who have long heard stories about supernatural beings, the response would probably be quite different, creating only a minimum wave over the long term.

The approach for planning educational programs for highly traditional people, who have religions that maintain ancient myths about creation and visiting celestial beings, will most likely differ significantly from those developed for westernized groups. Educational programs should utilize strategies that have been shown to be effective in facilitating responses and build upon extant assumptions regarding the supernatural beings that are already said to be affecting their lives since these would be much less likely to cause panic, anxiety, stress, and anguish. However, we cannot automatically assume that this will be the response across all groups or that all people utilize similar strategies for preparing future generations. Without information on these social groups, it is difficult to predict and strong educational programs cannot be developed. We turn now to a discussion of methods for increasing our information about these populations.

15.4.3 Ethnographic (Qualitative) Methods

The important question of the societal impact of extraterrestrial life can and should be systematically studied in different societies and a systematic collection made of strategies that appear to contain messages about dealing with future random and possibly dangerous events. We propose that such studies should be initially undertaken using ethnographic methods rather than population surveys. Surveys are not a particularly useful assessment methodology for studying individuals living in developing societies composed largely of peasants, tribal members, and people with low literacy who distrust outside scholars. The same is also true for even those from the same society but who speak, act, or dress differently, are better educated or possess other attributes that distinguish them. Surveys using Likert scales are difficult for them to understand and answer. It is not uncommon for investigators in such situations to hear comments like, "your question doesn't make any sense" and "why are you asking me that?"

It is important to begin any study by first working with local leaders and community members to build trust. Without trust, truths will never be revealed. Participatory action research is an applied collaborative approach developed by social scientists and used in many fields including anthropology, public health and community development. Using this collaborative approach, investigators and their community partners work together as a team to combine their knowledge, ideas, and actions in order to develop models and approaches to building communication, mutual trust, and capacity (NIH 2012).

Qualitative ethnographic methodologies can help us to avoid preconceived notions and more accurately identify not only the stories being told and the meanings behind dances, parables, and songs, but also which questions we should ask, with whom we need to speak, and how to interpret the answers that are given. Such methods are typically employed to collect in-depth information regarding an individual's reported knowledge, attitudes, perceptions, and opinions on specific topics and usually involve participatory approaches. This type of data provides a rich type of contextual detail that enables an understanding of meanings, processes, and reasons. Studies that investigate the possible impact of extraterrestrial life discovery on societies should be conducted in the same way that one would attack any anthropological problem involving investigation of ideology, influences on behavior, or social questions.

Examples of ethnographic methods that would be particularly appropriate for studies seeking to understand how developing country societies may respond to the discovery of extraterrestrial life include key informant interviews, focus groups, community discussions, and photo-voice. Study findings using these qualitative methods are strengthened when triangulation is employed, i.e., employing more than one type of data collection. These act as a check on the validity of findings from the various methods. By using data collected multiple qualitative methods we can, if desired, eventually design quantitative questionnaires built on the qualitative data and thus be able to triangulate the data to develop an even greater understanding of the topic of interest.

15.4.3.1 Direct Observation

Direct observation is a method in which a researcher or teams of observers record what they see or hear at a site or community using a detailed observation form. Observations can focus on physical surroundings, activities, processes, and/or social interactions, including how interactions unfold when novel events occur. A major advantage of the method is that an event, institution, facility, or process can be studied in its natural setting, thereby providing a richer understanding of the subject. We could, for example, observe a story-telling session, documenting who tells the story and how they tell it (seeing it as performance), where the story is told (e.g., told at night, around a fire), to whom it is told, and the response of those listening. This can provide us with background on how useful a storytelling format might be and what that format involves.

15.4.3.2 Participant Observation

Participant observation, widely used in anthropology, involves the researcher living with and participating in activities with residents of a community. This allows the researcher to collect data that may be more accurate as the study population is less likely to modify their responses or behavior as they often do in response to data collected by persons less well known or accepted by the community.

15.4.3.3 In-Depth Interviews

This method involves conducting intensive structured, unstructured, or semi-structured interviews with a small number of respondents for the purpose of exploring their reported knowledge, beliefs, attitudes, views and perspectives on a particular topic. Interviewees are usually individuals who possess special knowledge and skills or access to needed information (Hogle and Sweat 1996). In-depth interviews provide a private confidential atmosphere which allows interviewees provide sensitive, detailed information on the target topic about their personal experiences, views, and behavior in a setting where peers do not directly influence their responses. This method is easily adaptable to different circumstances, permitting researchers to collect data in geographically dispersed populations in both rural and urban settings.

15.4.3.4 Key Informant Interviews

Key informant interviews are a form of in-depth interviews. They involve one-on-one in-depth interviews of 15–35 individuals selected for their first-hand or expert knowledge on a specific topic in the context of the community. Depending on the topic(s) of interest, key informants may include the elderly, storytellers and herbalists, community leaders, traditional healers, teachers, local government representatives, and other persons with expertise. Key informant interviews rely upon a prepared list of questions that address issues to be discussed and have a relatively loose structure resembling a conversation where there is a free flow of ideas and topical information. This method is especially appropriate when there is a need to explore knowledge, attitudes, beliefs, and perceptions on a specific topic or provide recommendations for educational and other programs. It also is useful for framing issues and providing preliminary information for designing future quantitative studies.

15.4.3.5 Focus Groups

This group interview method involves discussions on a specific topic or topics among a homogenous group of 8–12 participants who are selected based on certain common characteristics, e.g., shamans or other religious leaders, elder status, mothers. A moderator/facilitator introduces the topic and raises issues identified in a discussion guide and uses probing techniques to solicit the group's views, ideas, feelings, and preferences on the topic while a tape recorder or another researcher documents the discussion. This flexible format permits exploration of unanticipated issues and fosters interaction among participants. The group-setting provides social checks and balances, thus minimizing false or extreme views. Participatory techniques such as listing, mapping, and ranking are often used to ensure all group members contribute to the conversation.

15.4.3.6 Community Discussion (Charlas)

Community discussions are public meetings conducted with a diverse array of community members including men, women, adolescents, the elderly, children, and other subgroups. These differ from focus groups as their main goal is to get answers and opinions on a topic of importance to the whole community. For this reason, meetings are open to all community members and anyone might participate actively. Community discussions can be successfully conducted with large groups of up to one hundred individuals. Although the moderator/facilitator asks questions using a detailed interview guide to lead the discussion, the primary interaction in this type of method is among participants whose responses, verbal and behavioral, are recorded by a second investigator using notes, audio- or videotapes.

15.4.3.7 Photovoice

Photovoice is an innovative community participation research process that involves providing community members with disposable cameras and asking them to document a specific concept or issue. This method is particularly valuable as it allows investigators and other outsiders to view the question of interest through the eyes of the community participants themselves. Photovoice could be used to ask community members to document the role of religious traditions involving supernatural in their lives, where supernatural beings live, and the stories that they have been told about supernatural beings. If it is difficult to use cameras, it could be of value to ask community participants to document these or draw their conceptions of extraterrestrial life.

15.4.4 Caveats

Although the above described approaches are more appropriate than other methods for studies seeking to understand how third world societies, especially tribal and peasant groups, may respond to the discovery of extraterrestrial life, there are caveats. First, what people say is not always what they believe and how they actually behave may be quite different from how they say they will react. We also cannot assume that all individuals living in a particular social group or geographic area hold the same beliefs or would respond in the same way. In addition, our interpretation of what they say or how they act may be biased by our own world view and experiences. Furthermore, given that even remote societies are increasing being exposed to western societies, even "good" information may have a limited shelf-life. However, despite the effects of westernization and other sources of culture change, the information we collect will provide us with a great deal of useful information that can be used to anticipate and plan for the eventual discovery of extraterrestrial life and first contact. Just as important, they can help us better understand human nature as many of these societies resemble more closely than our own, those of our distant ancestors and can tell us about the way our ancient ancestors acted and reacted and how they prepared generations of individuals for contact with outsiders who look and act differently.

15.4.5 The Next Step: Application of Survey Results to Educational/Preparedness Programs

We propose that once the qualitative study data have been gathered and interpreted, they can be used to help construct preparedness programs following the general principles and practices employed in public health. The timing, nature, and scope of the event of the discovery of extraterrestrial life is unpredictable similar to what human populations often encounter in the event of many natural and man-made public health emergencies, crises, disasters and other events. However, it seems reasonable to suggest that many of the challenges associated with such an event can be mitigated using public health preparedness principles. The keys to any type of effective public health preparedness are anticipating the event/crisis/threat, educating and informing the public (WHO 2007) and involving all sectors in the process from the household to community to national and international sectors. Further, preparing for and mounting an effective response depends on having the capacity to reach and assist all community members including the most vulnerable.

The planning processes and other tools necessary for emergency preparedness, mitigation and response are similar regardless of the nature of the hazard or event It would be reasonable to piggyback extraterrestrial life discovery preparedness onto existing emergency/disaster preparedness programs (e.g., volcanoes, floods, forest fires, famine, drought) or new/emerging threats (e.g., influenza pandemics, bioterrorism) or vice versa, in the case of communities without extant programs.

15.5 Conclusion

Over the course of prehistory and history, human societies have accumulated information that has been passed down for generation after generation, i.e., culture, to make sense of the world and help guide their responses to events caused by natural or supernatural forces. Much of this cultural information transmitted from the ancestors employs the use of analogies. These allow individuals and their societies to identify and respond to novel events, as the best inference is from the known to the unknown, and have developed cultural strategies to explain such events and respond appropriately to them. We propose that the qualitative research approaches used by anthropologists and other social scientists can be employed to gather useful information on these cultural strategies to better understand the possible responses of different third world societies to the discovery of complex or intelligent extraterrestrial life. This information is invaluable since not only can it be used to anticipate and plan for the eventual discovery of extraterrestrial life and first contact but it can also advance our understanding of the strategies that humans have evolved to deal with and plan for the unknown.

References

Alexander, Victoria. 2003. "Extraterrestrial Life and Religion: The Alexander UFO Religious Crisis Survey." In *UFO Religions*, ed. James R. Lewis, 359–370. Amherst, NY: Prometheus Books.
Amiotte, Arthur. 1992. "The Call to Remember." *Parabola* 17 (3): 29–35.
Bainbridge, William Sims. 2011. "Cultural Beliefs about Extraterrestrials: A Questionnaire Study." In *Civilizations Beyond Earth: Extraterrestrial Life and Society*, ed. Douglas A. Vakoch, and Albert A. Harrison, 118–140. New York: Berghahn Books.

Basso, Keith. 1971. *Apachean Culture History and Ethnology*. Tucson: University of Arizona Press.

Basso, Keith. 1990. *Western Spache Language and Culture: Essays in Linguistic Anthropology*. Tucson: University of Arizona Press.

Bieselle, Megan. 1993. *Women Like Meat: The Folklore and Foraging Ideology of the Kalahari Ju/'hoan*. Bloomington: Indiana Univ. Press.

Bojuwoye, Olaniyi. 2005. "Traditional Healing Practices in Southern Africa: Ancestral Spirits, Ritual Ceremonies, and Holistic Healing." In *Integrating Traditional Healing Practices into Counseling and Psychotherapy*, ed. Ray Moodley, and William West, 61–72. Thousand Oaks: Sage Publications.

Calame-Griaule, Genevieve. 1986. *Words in the Dogon World*. Philadelphia: Institute for the Study of Human Issues.

Coe, Kathryn, Nancy Aiken, and Craig T. Palmer. 2006. "Once upon a Tme: Ancestors and the Evolutionary Significance of Stories." *Anthropological Forum* 16 (1): 21–40.

Daniel, Kasomo. 2010. "The Position of African Traditional Religion in Conflict Prevention." *International Journal of Sociology and Anthropology* 2 (2): 23–28.

Dawes, Milton. 2000. "Science, Religion and God." *Review of General Semantics* 57 (2): 147–153.

Dominiki, Martin, and John Zarnecki. 2011. "The Detection of Extra-terrestrial Life and the Consequences for Science and Society." *Philosophical Transactions of the Royal Society A* 369 (1936): 499–507. doi:10.1098/rsta.2010.0236.

Eaton, S. Boyd, Loren Cordain, and Staffan Lindeberg. 2002. "Evolutionary Health Promotion." *Preventive Medicine* 34: 119–123.

Eghosa Osaghae, E. 2010. "Applying Traditional Methods to Modern Conflict." In *Traditional Cures for Modern Conflicts: African Conflict Medicine*, ed. I. William Zartman, 183–200. Boulder, CO: Lynne Rienner Publishers, Inc.

Elgin, Suzette. 1998. *The Grandmother Principles*. New York: Abbeville Press.

Eliade, Mircea. 1963. *Patterns in Comparative Religion*. New York: The New American Library-Meridian Books.

Emmons, George. 1991. *The Tlingit Indians*. Seattle: University of Washington Press.

Firth, Raymond. 1956. "Rumor in a Primitive Society." *The Journal of Abnormal and Social Psychology* 53: 122–132.

Gayton, Anna. 1940. *Yokuts and Western Mono Myths*. Berkeley: University of California Press.

Grinnell, George B. 1889. *Pawnee Hero Stories and Folk Tales, with Notes on the Origin, customs and Character of the Pawnee People*. New York: Forest and Stream.

Harrison, Albert A., and Douglas A. Vakoch. 2011. "Introduction. The Search for Extraterrestrial Intelligence as an Interdisciplinary Effort." In *Civilizations Beyond Earth. Extraterrestrial Life and Society*, ed. Douglas A. Vakoch, and Albert A. Harrison, 1–29. New York: Berghahn Books.

Harvey, Graham. 2002. *Readings in Indigenous Religions*. Continuum International Publishing Group.

Hogle, Jan, and Michael Sweat. 1996. *FHI/AIDSCAP Evaluation Tools: Qualitative Methods for Evaluation Research in HIV/AIDS Prevention Programming*. Chapel Hill, NC: Family Health International.

Hungry Wolf, Adolf. 1977. *The Blood People: A Division of the Blackfoot Confederacy*. New York: Harper and Row.

Johnson, Susan A. 2009. *Religion, Myth, and Magic: The Anthropology of Religion*. Prince Frederick, MD: Recorded Books, LLC.

Kopytoff, Igor. 1971. "Ancestors as Elders in Africa." *Africa: Journal of the International African Institute* 41 (2): 129–142. http://lucy.ukc.ac.uk/era/ancestors/.

Krajick, Kevin. 2003. "Defusing Africa's Killer Lakes." *Smithsonian* 34 (6): 46–55.

Leeming, David A. 2010. *Creation Myths of the World*, (2nd edition). Santa Barbara: ABC-CLIO.

Long, Charles H. 1963. *Alpha: The Myths of Creation*. New York: George Braziller.

Matthews, Warren. 2011. *World Religions*, (7th edition). Belmant, CA: Wadsworth Cengage Learning

Mosha, R. Sambuli. 2000. *Heartbeat of Indigenous Africa*. New York: Garland Publishing, Inc.

Mukerjee M. "The Scarred Earth. News Scan. Geophysics." *Scientific American*, March 2005. Accessed at: http://www.indiana.edu/~volcano/notes/mukerjee.pdf.

National Institute of Health (NIH). 2012. *Community-Based Participatory Research*. http://obssr. od.nih.gov/scientific_areas/methodology/community_based_participatory_research/ index.aspx.

Newcomb, William. 1956. *The Culture and Acculturation of the Delaware Indians*. Ann Arbor: University of Michigan Press.

Peters, Ted. 2011. "The Implications of Extraterrestrial Life for Religion." *Philosophical Transactions of the Royal Society A* 369 (1936): 644–655. doi:10.1098/rsta.2010.0234.

Pettinico, George. 2011. "American Attitudes About Life Beyond Earth: Beliefs, Concerns, and the Role of Religion and Education in Shaping Public Perceptions." In *Civilizations Beyond Earth. Extraterrestrial Life and Society*, ed. Douglas A. Vakoch, and Albert A. Harrison, 102–117. New York: Berghahn Books.

Population Reference Bureau (PRB). 2011. *The World at 7 Billion*. http://www.prb.org/ Publications/Datasheets/2011/world-population-data-sheet/world-map.aspx#/map/population.

Rodnick, David. 1937. "Political Structure and Status Among the Assinboine Indians." *American Anthropologist* 39: 408–416.

Sabadell, Miguel A. and Fernando, Salamero. 1996. "How Do People Feel about Contact with ETIs?" *Proc. of SPIE* 2704, 172–183. http://dx.doi.org/10.1117/12.243433.

Sattler, Richard. 1987. *Seminoli Italwa: Socio-political Change Among the Oklahoma Seminoles Between Removal and Allotment, 1936–1905*. Ann Arbor, MI: University Microfilms International.

Steadman, Lyle, and Craig T. Palmer. 2010. *The Supernatural and Natural Selection: Religion and Evolutionary Success*. New York: Paradigm Publishers.

Steadman Lyle B., Craig T. Palmer, and Christopher F. Tilley. 1996. "The Universality of Ancestor Worship." *Ethnology* 35 (1): 63–76.

Tanner, John. 1830. *A Narrative of Captivity and Adventures of John Tanner*. New York: G. and C. and H. Carvill.

Vakoch, Douglas A., and Yuh-Shiow Lee. 2000. "Reactions to Receipt of a Message from Extraterrestrial Intelligence: a Cross-Cultural Empirical Survey." *Acta Astronautica* 46 (10–12): 737–744.

Winzeler, Robert L. 2008. *Anthropology and Religion*. Lanham, MD: AltaMira Press.

Weltfish, Gene. 1965. *The Lost Universe*. New York: Basic Books.

World Health Organization. 2007. *Risk Reduction and Emergency Preparedness. WHO Six-year Strategy*. http://www.who.int/hac/techguidance/preparedness/emergency_preparedness_eng.pdf.

Wissler, Clark. 1918. *The Sun Dance of the Blackfoot Indians*. New York: The American Museum of Natural History Anthropological Papers.

Chapter 16
Impossible Predictions of the Unprecedented: Analogy, History, and the Work of Prognostication

Kathryn Denning

Abstract At the beginning of exobiology and SETI as research programs circa 1960, it was reasonable and responsible for scientists and others to consider the potential effects of a detection of other life, or contact with it, upon humanity. It is no coincidence that this was a time of reckoning with the power of science and technology. The Cold War was settling in, space programs were beginning, and the technologies of war and those of discovery were then, as now, intertwined, in a way that made Carl Sagan, Philip Morrison, Joshua Lederberg, and others, concerned for humanity's future, and the future of life. Those concerns are as well-founded as ever. However, 50 years on, after half a century of predictions and untested hypotheses, we still only know that a detection of extraterrestrial life could come tomorrow, in the next century, or never. Many potential scenarios have been identified and explored, planetary protection protocols have been implemented for astrobiology, policy concerning SETI detections has been created and debated, and some valuable empirical work has been done concerning potential cultural reactions. We might now reasonably ask: what are our real goals here? And do they match what we are actually accomplishing? Are these exercises still beneficial, or are they reaching the point of diminishing returns? Might there be undesirable effects of prognostications about detection and contact? Elsewhere, I have discussed at some length what I think can sensibly be done to prepare for a detection. This leaves me with a further argument to make here: first, that the use of historical analogies of intercultural contact on Earth to predict or explore the potential consequences of contact with ETI may now be essentially useless or perhaps worse than useless; second, that the long-standing practice of prediction about contact now also invites scrutiny in terms of its utility; and third, that turning our attention to pressing topics at the intersection of astrobiology, SETI, and society, could be worthwhile for scholars of humanity.

K. Denning (✉)
Department of Anthropology and Science and Technology Studies Program, York University, Toronto, ON, Canada
e-mail: kdenning@yorku.ca

D. A. Vakoch (ed.), *Astrobiology, History, and Society*, Advances in Astrobiology and Biogeophysics, DOI: 10.1007/978-3-642-35983-5_16,
© Springer-Verlag Berlin Heidelberg 2013

16.1 Introduction

Discussions of the potential risks and benefits of inter-species contact with ETI frequently invoke the history of inter-cultural contacts on Earth. Earth's cultural history is thus a potent contributor to debates concerning the "public policy" aspects of SETI (e.g. active SETI, and post-detection protocols), and this invites closer examination. Accordingly, this chapter will first briefly look at the use of analogies. Then, I will examine positions identifying problems with the use of historical analogies within SETI discourse, and will then point towards a deeper problem of historical understanding which is embedded within the case studies themselves. Finally, I will turn to the matter of prognostication itself.

Elsewhere, I have suggested that really, in advance of a detection, all that scientists can do is keep abreast of the rapidly evolving media environment, and do their best to provide accurate and reliable information to the world, and that what scholars in other fields can do is help with the policy questions on issues that affect everyone (Denning 2011). I also contend that really, any matter of policy can only be sensibly based on an acknowledgement that we do not know the risks of contact, rather than efforts to calculate the unimaginable (Denning 2010b). Further, I consider that in the majority of imaginable detection or contact scenarios, people will not be dealing with the facts or other life forms, but rather, with our distorted and refracted and fractured cultural representations of those facts and entities. I think it can be further reasonably stated that contact has now been rehearsed so many times in popular culture that these representations and their dissemination in new media will be influential beyond almost any other factor (Denning 2011).

So, then, what to make of the use of historical analogies in our anticipations of contact? This requires, first, a close look at analogy.

16.2 The Use of Analogies

Analogies are fraught. Humans frequently employ analogical thinking: along with narrative, metonymy, anthropomorphism, and other cognitive gymnastics, it's an element in our battery of strategies for understanding a complex world with information that is simultaneously too abundant and too incomplete to easily comprehend.

But analogy carries its own baggage. When we use one thing as a basis for understanding another, this can illuminate but also cast shadows and confound. It all depends on how they are used. The analogue (source) and the target take different forms and are differently related, depending on the science.

For example, analogues are integral to astrobiology, in exploring the potential for life in the universe. Of course a range of strategies must be used to understand the origins of life on Earth; proxies are often necessary. But for other aspects of astrobiology, quite often, the targets are there, and in our sights: they are just very hard to get to. So, astrobiologists study Earth deserts to better understand how to explore Mars, not as a permanent substitution for studying Mars itself. Or they study a wide range of environments or life forms to understand what the range of possibilities might be for "weird life" and its homes, and to generate hypotheses for testing, and parameters that guide future missions—not as a permanent substitution for actually looking for those data elsewhere. In astrobiology, one might say, the use of analogues makes efficient use of scientists' time and energy while spacecraft are built, helps to guide the deployment of those spacecraft, and has the added benefit of deepening our understanding of life on Earth. No doubt there are some downsides to the practice—for example, the provisional redefinition of some Earth places as "alien" in nature may have some negative effects downstream—but in the meantime, it is a robust scientific strategy that makes good use of what we have easily available to guide study of our targets. And, even better, both the source and the target end up being better understood. The interplay is constructive.

However, analogies work differently in other fields. For example, in my discipline of archaeology, we have a quite tortured relationship with analogies (Wylie 2002). Basically, archaeology is so difficult as to be daft—it really is fundamentally impossible[1]—and this makes us use absolutely any method we can. Our target, a full understanding of ancient human behavior, is permanently inaccessible. We actually can't get there from here *at all*, without a time machine, because most human behavior simply does not preserve. The problem is not that the data are hard to get to: it is that *they are not there*. Furthermore, archaeological sites present a conundrum: sets of data which result from human actions but also from various taphonomic processes (rearrangement due to forces ranging from biological to geological to meteorological) which degrade those patterns over time, removing some data completely and altering others. So, we have to work in recognition of our very acute limits, work hard to map the distortions and absences, keep pushing new methods to recover and amplify unimaginably faint traces to be analyzed, and basically do the best we can by throwing absolutely every method available at the data we do have. One method is ethnographic analogy—that is, using the behavior of contemporary (or recent) human beings to help explain the patterns observed in an archaeological site.

This analogical approach is inevitable, and can certainly be helpful, but nonetheless, the problems with it are many: they can include the introduction of significant distortion into interpretations of the sites, but also, importantly, a "revenge effect" on the source. For example, when a contemporary traditional hunting-foraging culture in Africa is used to explain an archaeological site in Europe from 30,000 years ago, this constitutes a conflation of time and space, and means that modern people are defined as "Stone Age"... and, by extension, as less culturally evolved than their agriculturalist or industrialist neighbors, and overdue to be superseded. This, of course, mirrors the rationale of nation-states who have forcibly resettled and "modernized" these groups, in a calculated obliteration of their cultures, languages, and uniquenesses (Davis 2009). This matters—arguably, much more than understanding a 30,000 year old site in Paleolithic Europe matters.

---

[1] As David Clarke famously said, archaeology is "the discipline with the theory and practice for the recovery of unobservable hominid behavior patterns from indirect traces in bad samples" (1973, 17). This both produces and explains the eccentricity of many archaeologists.

My point here is simply this: that the use of analogies can be a messy business, and we should be alert to their limitations and detrimental effects. Some key questions to ask would seem to be these:

- What exactly is the analogue good for?
- How well do we know the analogue (source)?
- Will this process also improve our knowledge of that source?
- All things considered, is it likely that the analogue will be similar enough to the target to be a useful model?
- Will our impressions of the source analogue over-determine our understanding of the phenomenon we want to study?
- Given that this is a two-way relationship of reflection between the known and unknown… are there detriments to the source?

This all becomes more complicated when the phenomenon we want to study is comprehensively absent—when we're simply projecting into a void. This is, so far, the case for intelligent extraterrestrial life. Does reckoning from analogy then become recruited into an increasingly elaborate web of speculation, unmoored from its original purpose? Perhaps. It brings to mind the now-elaborate and extensive discourse concerning the Fermi Paradox—it is, quite literally, a substantial body of analysis about nothing, which is now evolving into metaanalysis of nothing. I would not suggest that these intellectual projects are without value, but one can legitimately ask what exactly that value is, and what the discussion is now really about.

This is, perhaps, a trap that is easy to fall into when we cannot yet tell whether our target is hard to get to (like Mars), or simply not there (like ancient human behavior). What can be said from the vantage point of archaeology is that there is wisdom in knowing the empty spaces—in knowing the difference between what we might someday recover and what we never will. We will never see an ancient parent's smile or hear their child's laugh, except in a dream. We must simply live with that silence. Once in a long while, we discover small red handprints on a cave ceiling that tell us that long ago, a 5 year old was once lifted up on someone's shoulders to reach up high, or we find a carefully made toy which bears the marks of baby teeth. Those traces of love and laughter have to be enough. And the wisdom archaeologists gain through experience is a matter of feel: it lies in knowing where to push into the faint traces because maybe there's a little more we can know, and when to use analogues from the present to fill in holes in the past, and when to just leave those spaces open, resist filling in the blanks, and say: *we do not know, and we never will*. And it lies in having the courage to say that beyond this point, speculation is nothing but fiction, and to recognize that stories about the past can take on lives of their own and do things in this world that are not merely benign.

So, then, when it comes to using historical analogues to think about the potential impacts of detection or contact, perhaps the wisdom for us now lies in knowing when to use the past to predict the future, and when to simply say we do not know what might happen, and instead turn our attention to what we *do* know and *can* observe. But first, let us unpack some more.

16.3 What SETI Thinkers Have Written About Using Earth's Intercultural Contacts as Analogies for SETI Contact

History is a tricky business. It is difficult enough to understand past events in all their fullness, and it is even more difficult to understand the relationship of the past to our own present and future (Denning 2010a). And yet, Earth's history provides us with essential raw material that we constantly use for thinking about problems that we face now, or may face someday. The arena of SETI is no exception: historical contacts between Earth's societies are frequently mentioned in speculations about what contact with ETI might be like. This is of considerable interest to a North American anthropologist/archaeologist like me, because much of what we do involves deconstructing, interrogating, and reconstructing narratives of contact: the story of the Americas over the past 600 years can only be understood through this lens.

Most recently, Stephen Hawking has been much-quoted as suggesting that aliens could well be hostile, and perhaps we should keep our heads down: "If aliens visit us, the outcome would be much as when Columbus landed in America, which didn't turn out well for the Native Americans," he said in his show *Into The Universe with Stephen Hawking* (Grier 2010).

This is only one recent example of an invocation of Earth history, specifically a contact episode, in service of an argument about hypothetical future contact between human beings and extraterrestrial intelligence. Much of the time, such references to episodes of contact on Earth are casual and rhetorical—and yet, they are powerful in shaping arguments about how and whether contact should be sought, e.g., via Active SETI. Accordingly, a closer look is worthwhile. As noted, some SETI thinkers have considered, with care and skepticism, the use of Earth analogues for hypothetical contact with ETI.

The 1999 volume *Social Implications of the Detection of an Extraterrestrial Civilization*, by Billingham et al. (1999) is an essential synthesis covering SETI and history, probable human behavior post-detection, and SETI and policy, amongst other topics. Their position was that direct contact is unlikely, analogies are invariably problematic, and that "analogies drawn from history should be considered as prompts for thinking about the future, not reliable guides to its course," since "Nothing in human history is fully analogous to the type of encounter to which SETI research may lead us" (Heilbron et al. 1999, 34). Their exploration of historical analogues included those involving just transmission of ideas (rediscovery of Greek science, and the impact of Darwinian and Copernican theory), and also what might be learned from direct "confrontations between cultures" on Earth. In the latter category, the authors observed that popular ideas about natives being overwhelmed by the more "technologically advanced" ways of their colonizers are forgetting about the violence of colonization and the impacts of disease. They also note that cultures tend to misunderstand each other profoundly, illustrating with the example of cargo cults in Melanesia. But finally, they optimistically note that not all contacts are disastrous, through two examples. First, contact between China and Japan and the West was initially "wrenching," but these nations are now economically very powerful internationally. Second, contact between Europeans/Australians and the indigenous peoples of highland New Guinea in the 1930s, though tremendously disruptive to the traditional cultures, was "much more benign than meetings of earlier centuries," due to better medical management and colonial administration (Heilbron et al. 1999, 56–59). This is one of the more detailed and optimistic assessments to date of Earth intercultural encounters as analogues for SETI contacts.

Rather less optimistically, anthropologist Ben Finney (1990) suggested, on the basis of "cultural misunderstanding between closely related human groups," that radio contact would result in a hugely difficult task of "intercivilizational comprehension." Following up in 1998, Finney and Bentley argued that the case of ancient Greek science entering medieval Western Europe is perhaps too easy because the problems of translation were comparatively minor. A more appropriate analogue can be found, they suggested, in the difficult case study of modern scholars struggling to decipher and comprehend ancient Mayan inscriptions. They conclude that "Examination of the Maya case suggests that if we are to employ terrestrial examples to help us think about extraterrestrial knowledge transmission, we should explore the range of human experience and not just focus upon those examples which support our hopes" (1998, 691).

In another assessment of Earth intercultural contacts as analogues for SETI contact, Al Harrison summarized work to date on planning for contact—going back to 1961, shortly after Project Ozma, and continuing through the activities of the IAA SETI Permanent Committee—and noted that "Much planning draws on historical events involving the meeting of two radically different cultures, such as the arrival of Spaniards in the New World, the British and Dutch in Africa, the British in Australia, and the Americans in Japan… In these comparisons it is assumed that the extraterrestrial society is vastly advanced in terms of science and technology; hence, "they" are cast in the role of the wealthy and powerful Europeans. Humans are cast in the technologically inferior roles" (2007, 183). However, Harrison noted, we must ask whether these precedents are useful to us, and concluded that "Historical precedents give us our only experiential basis right now, but we must be aware that extrapolations yield only rough approximations of what might happen following real contact, so we have to take these approximations with a grain of salt" (Harrison 2007, 183).

Michael Michaud went further in his work, *Contact with Alien Civilizations*. Like the scholars above, he accepted the inevitable influence of analogies in our thinking, but cautioned that such analogies are problematic. But in addition, sketching out the positions of "contact optimists" and "contact pessimists," Michaud noted that each side tends to choose historical Earth analogues that support their own position (2007, 213). Michaud himself also invokes Earth analogues, for example to consider the nature of empires (2007, 321).

All these positions are sensible: one-way informational contact over considerable time/distance seems more probable than direct contact with ETI; social impact may thus be in terms of knowledge transmission; it may be very difficult to understand the content of any transmissions; historical terrestrial analogues can surely only give us an approximation of a real Earth—ETI contact scenario; and the choice of historical analogues used in exposition about SETI can support either a profoundly optimistic view or a deeply pessimistic view of contact with ETI.

My own assessment is that certainly the direct contact historical analogies are likely to be useless, and indeed, most popular historical accounts are not good (and this is not accidental). Casting the Hawaiians as us and Cook as the alien, or the Aztecs as us and Cortez as the alien, or the New World as us and Columbus as the alien, is neither particularly useful, nor innocent. The triumph of the Old World over the New was categorically the result of demographic collapse due to extraordinary epidemic mortality, not due to civilizational superiority of the Old World. This is completely elided in the casual remarks of Hawking and others.

The analogies of information transfer, as from ancient Greece into Renaissance Europe, or the decipherment of Mayan glyphs in the twentieth century, would certainly seem more appropriate than the 'direct contact' analogies. But although they are better, this does not necessarily mean that they are useful beyond the general illustration of 'it would be more like this than like that'. The contours are useful but I doubt the details are. Our society has been changed by those very precursors, is pretty well-prepared for the concept of ETI and radical contact, and has a completely different information environment. We have surveys of what people think and we have models of information movement in the present. What does the analogy really contribute?

However, there is another layer here that is worth excavating. It is certainly worthwhile to ask how useful Earth intercultural contacts are as analogues for contact with ETI: what relationship do these historical precedents really have to the question before us (Harrison 2007, 183, and Denning 2010a)? But it is also worthwhile, now, to go further and also ask: how good has our information been so far about intercultural contacts on Earth? That is: how good is our understanding, really, of what has happened when one culture has encountered another here on our own world? It seems established that there are built-in problems with the intellectual exercise of drawing an analogy between our known and the unknown, but are there also built-in problems within our analogue cases? *How well do we even know our 'known'?*

16.4 A Deeper Problem

The main problem, of course, with popular versions of world history is that they are deeply politicized, significantly slanted, and tend to reduce complex truths to simple stories (see Denning 2010a and 2011 for further discussions). If a person only reads a couple of books on the history of a particular region, or watches a few TV series on world history, then chances are that their understanding will be very broad and very biased. But our problem here is even worse: the scholarship upon which popular accounts are based has *itself* has been significantly biased.

A useful example is the encounter of the Old World and the New. The general European understanding of the peoples of the Americas started off as appallingly subjective and did not improve quickly. Early accounts of islands inhabited by dog-headed cannibals prevailed a long while, and there was a systematic disregard for the possibility that the inhabitants of the Americas were possessed of intellects, languages, customs, religions, cultures, and governments equivalent to those of Europeans. This is a remarkable kind of double-think, given that the earliest colonial encounters involved much negotiation, not just hostility, and that the riches and sophistication of the New World civilizations matched or surpassed those of medieval Europe (Zamora 1999; Wright 1992; 2008). Much of post-contact European thought derives from texts foundational to the disciplines of history and political philosophy like those of Adam Smith and G. W. Hegel, who were working from such a poor knowledge of the societies of the New World that they dismissed them as having had any significance at all (Chomsky 1993, 4). Again, this is a remarkable kind of double-think given that the Inca, for example, had one of the most successful empires ever seen on Earth, and more wealth than any European king at the time—wealth which ultimately fuelled the Industrial Revolution (Wright 1992, 82; 2008, 35).

Indeed, this kind of bias was still clearly visible in the late twentieth century even among anthropologists and archaeologists. For example, a comparison of atlases of archaeology from the 1970s and 80s shows that at least half of the coverage was devoted to Europe and the Near East, with the Americas receiving only 13.5% of the coverage, sub-Saharan Africa only 10%, and India only 4.5% (Scarre 1990, 14). And there has been another problem too: historical and social disciplines long tended to focus inquiry on discrete groups of people, rather than the interactions between them. As Eric Wolf famously argued in 1982, "Historians, economists, and political scientists take separate nations as their basic framework of inquiry. Sociology continues to divide the world into separate societies. Even anthropology, once greatly concerned with how culture traits diffused around the world, divides its subject matter into distinctive cases: each society with its characteristic culture, conceived as an integrated and bounded system, set off against other equally bounded systems" (1982, 4).

This is now changing: much scholarship is now concerned with interconnections, past and present. But still, it is only comparatively recently that anthropologists, archaeologists, ethnohistorians, and other scholars have pieced together comprehensive accounts of New World societies before contact, around the time of contact, and in the years afterwards (Kehoe 2002; Mann 2005; Wright 1992). The histories of those societies have yet to be fully integrated into the bigger picture. As Kehoe and others have pointed out, "These histories do not suddenly stop short in 1492 or 1607; they are strands in the fabric of United States (and Canada and Latin America) history. Indigenous nations literally shaped the land and resources taken over by European invaders and their descendants. Their labor and production constituted significant components of the United States economy, generally left out or underestimated" (Kehoe 2002, 252). This is not to deny that the nations of the New World were decimated by their encounter with the people and diseases of the Old World, but to say that the story does not end there; they were not obliterated, and their cultures and political agency are significant forces today.

This work of actually understanding contact and its sequelae continues. Scholars have recently been assembling new accounts of contacts in Earth's history, from the increasingly abundant data at our disposal—from oral history, archaeology, pre-contact writing, post-contact narratives both from the Native American civilizations and from the European invaders (conquistadors, missionaries, and eventually administrators/colonists etc). All of these forms of data are challenging to obtain and interpret, and we continue to fill in our understanding. For example, John Ralston Saul's iconoclastic book, *A Fair Country: Telling Truths About Canada* (2008) demonstrated that Canada is in fact a Métis nation, a truly hybrid society—not, as the dominant view would have it, an essentially European society built on the extinguished remnants of a previous Native order. Similarly, Ronald Wright's recent work, *What Is America?*, argued that the true nature of the United States of America has been staggeringly misunderstood, and remarked upon the "vast and enduring" legacies of the civilizations of the New World (2008, 20). The fact that these books are being written now—that these assessments of the histories of the New World are *new*—says something about how far we have recently come in the understanding of historical episodes of contact on Earth, and about how limited our understanding has been until now. More such new analyses are being published, including the classic contact case of the Aztec and Cortez (covered at some length in Denning 2010c).

What does this mean for SETI discourse and the way we consider hypothetical contact with ETI? It does not tell us which examples of contact are most appropriate to use as analogues, and it does not tell us *what* to think about contact. But it may tell us something about *how* to think about contact. It tells us that the stories we've been raised on, stories about technology and power and politics and how civilizations meet, and what happens next, require our critical attention instead of our unthinking allegiance. It tells us that we need to rethink our assumptions about how contact works, and about which variables make the most difference to outcomes. And it suggests that when we are addressing a new question—like, what can we learn from Earth's historical record about the potential consequences of contact with ETI?—then it is wise to go back to the record itself instead of to tertiary-level historical syntheses, which may not really answer the questions which are relevant to a SETI contact scenario.

That is, of course… if it's really a priority to do that work at all. This brings me to my last question.

16.5 What Are We Doing When We Prognosticate About the Potential Impacts of a Detection or Contact? And Might We Do Something Else?

There is value in exploring these subjects and figuring out what scientists ought to do to discharge their responsibilities, but the notion that the sequelae can be accurately predicted and by extension controlled is, from my perspective, highly contestable.

And if societal impact is a concern of any kind, what are we actually going to do about it? As I noted earlier, it seems that all we can do is encourage researchers and research agencies to take responsible and proactive information dissemination measures, and engage in policy discussions. So, then, what work is this prognostication really doing?

At least three things are worthy of remark. The first is the general human tradition of prognostication and relationship to the future. The second is the reinscription of history. The third is the specific implications of prognostication at the edge of techno-scientific change.

On the first: the practice of attempting to predict the future is ancient, and we might be justified in wondering whether our methods today are really much better than the ancient Mediterranean and Mesopotamian conventions of extispicy (divination using the entrails of unfortunate animals), the Shang Dynasty's use of oracle bones, the intoxicated priestesses of ancient Greece, or Mayan rites of bloodletting, etc.

Might our efforts to predict the impact of a detection be the modern, secular, academic equivalent of ancient rites of forecasting? Scholars are expected to demonstrate their wisdom and worth in much the same way as astrologers of old. Predictions are seen as inherently valuable. Furthermore, it's worth considering that due to its Judeo-Christian substrate, Western civilization exists in apocalyptic time, always waiting for a major change to descend, always prophesying something Earth-shattering just over the horizon. It has been 2000 years of something big "coming soon."

Second, every time we tell a story about what will happen in the event of a detection, or contact, we retell the story of contact here on Earth. And, as noted above, that matters.

On the third: what was the original spirit of this practice, and how might we carry it forward? As noted, it was important when the search for life in the universe began, for scientists and scholars of humanity to consider the potential impacts of detection or contact (whether biological or informational and cultural). It was important to begin the conversations of planetary protection, to engage in a larger conversation with the world about how this might affect them, and to explore previous human experiences of contact.

But also, scientists cared because they realized that they were doing or proposing edgy work with potential social consequences, and they wanted to officially discharge their responsibilities and demonstrate that their work was worthy of public support, that it was good and right to think about space as a place with life and liveliness of its own, instead of just our dominion, and to say that although this work might be disruptive, it would be *constructively* so. It was the Cold War, and the scientists involved were wrestling with the realities of the nuclear age and their own contributions to it, attempting damage control, pleading for rationality, and also struggling to get people thinking about a long-term future and their place in the cosmos, for very urgent reasons.

Thinking about that future was itself an act of hope. Perhaps it still is. But I want to suggest something else here: that the best way to take that legacy forward is not to keep asking the same questions and elaborating on answers, the contours of which have long been established, and the details of which cannot be filled in until and unless a detection is confirmed. Perhaps this work is nearly done— because of all the effort that has been put in, such as that represented by this present volume.

To continue predicting would be in keeping with the tradition… the liturgy, the ritual. But I believe it is drifting away from the spirit of the original question. I think it can provide false assurances that we have answers which we do not, and that we know the consequences of our technologies and our science, when we can not. It may be comfortable, but I think we need to ask what exactly that comfort is worth.

Further, I would like to suggest that the original spirit of this line of inquiry can guide us to look instead at immediate, near-horizon concerns instead of hypothetical detection scenarios that may or may not transpire.

16.6 Conclusion

Once again, I will take a lesson from archaeology and suggest that it may be applicable here. One thing all contemporary archaeologists know is how precious our sites are, and how threatened these fragmentary data about the past are—threatened not just by the passage of time, but by human actions ranging from road development to looting to intentional destruction. Accordingly, archaeologists consider it part of our jobs to work for the protection of sites, both those already discovered, and those not yet known. This is part of our way of looking after the spaces in our knowledge—holding open the possibility of discoveries that might fill them, the possibility of learning things we really want to know about others distant from us in time.

Similarly, in this new era of much wider access to outer space, which will be shared with a variety of stakeholders, both public and private, those who value the search for life and those who don't, perhaps scholars of society and the extraterrestrial could make their most meaningful contribution in this way: instead of using past social conditions to make guesses about what would happen if a detection occurred, we might use our knowledge of present social conditions to help ensure that the science can continue to be done.

References

Billingham, John, Roger Heyns, and David Milne, et al. 1999. *Social Implications of the Detection of an Extraterrestrial Civilization*. Mountain View, CA: SETI Press.

Chomsky, Noam. 1993. *Year 501: The Conquest Continues*. Montreal: Black Rose Books.

Grier, Peter. 2010. "Stephen Hawking Aliens Warning: Should We Hide?" *Christian Science Monitor*, April 26. www.csmonitor.com/Science/2010/0426/Stephen-Hawking-aliens-warning-Should-we-hide.

Clarke, David. 1973. "Archaeology: The Loss of Innocence." *Antiquity* 47 (185): 6–18.

Davis, Wade. 2009. *The Wayfinders: Why Ancient Wisdom Matters in a Modern World*. Toronto: House of Anansi Press.

Denning, Kathryn. 2011. "Is Life What We Make of It?" *Phil. Trans. R. Soc. A* 369 (1936): 669–678. DOI:10.1098/rsta.2010.0230.

Denning, Kathryn. 2010a. "Social Evolution: State of the Field." In *Cosmos and Culture: Cultural Evolution in a Cosmic Context*, ed. Steven J. Dick and Mark L. Lupisella, 63–124. Washington, DC: NASA. NASA SP-2009-4802. Available online:http://history.nasa. gov/SP-4802.pdf.

Denning, Kathryn. 2010b. "Unpacking the Great Transmission Debate." *Acta Astronautica* 67: 1399–1405. doi:10.1016/j.actaastro.2010.02.024. Reprinted as "Unpacking the Great Transmission Debate." 2011. *Communication with Extraterrestrial Intelligence*, ed. Douglas A. Vakoch, 237–252. Albany, NY: SUNY Press.

Denning, Kathryn. 2010c. "The History of Contact on Earth: Analogies, Myths, Misconceptions." IAC-10-A4.2.2. On conference DVD and in IAC online archives. 61st International Astronautical Congress. Prague, Czech Republic, September.

Finney, Ben. 1990. "The Impact of Contact." *Acta Astronautica* 21 (2): 117–121.

Finney, Ben, and Jerry Bentley. 1998. "A Tale of Two Analogues: Learning at a Distance from the Ancient Greeks and Maya and the Problem of Deciphering Extraterrestrial Radio Transmissions." *Acta Astronautica* 42 (10–12): 691–696.

Harrison, Albert. 2007. *Starstruck: Cosmic Visions in Science, Religion, and Folklore*. New York: Berghahn Books.

Heilbron, John L., Jill Conway, D. Kent Cullers, Steven J. Dick, Ben Finney, Karl S. Guthke, and Kenneth Kenniston. 1999. "History and SETI." In *Social Implications of the Detection of an Extraterrestrial Civilization*, ed. John Billingham et al. Mountain View: SETI Press.

Kehoe, Alice Beck. 2002. *America before the European Invasions*. London: Longman.

Mann, Charles. 2005. *1491: New Revelations of the Americas before Columbus*. New York: Vintage Books.

Michaud, Michael. 2007. *Contact with Alien Civilizations: Our Hopes and Fears about Encountering Extraterrestrials*. New York: Springer Science.

Saul, John Ralston. 2008. *A Fair Country: Telling Truths about Canada*. Toronto: Penguin.

Scarre, Christopher. 1990. "The Western World View in Archaeological Atlases." In *The Politics of the Past*, ed. Peter Gathercole and David Lowenthal, 11–17. London: Unwin Hyman.

Wolf, Eric. 1982. *Europe and the People Without History*. Berkeley: University of California Press.

Wright, Ronald. 1992. *Stolen Continents: The "New World" Through Indian Eyes*. London: Penguin Books.

Wright, Ronald. 2008. *What Is America?* Toronto: Vintage Canada.

Wylie, Alison. 2002. *Thinking from Things: Essays in the Philosophy of Archaeology*. Berkeley: University of California Press.

Zamora, Margarita. 1999. "'If Cahonaboa Learns to Speak …': Amerindian Voice in the Discourse of Discovery." *Colonial Latin American Review* 8 (2): 191–205. DOI:10.1080/10609169984629.

Chapter 17
Mainstream Media and Social Media Reactions to the Discovery of Extraterrestrial Life

Morris Jones

Abstract The rise of online social media (such as Facebook and Twitter) has over-turned traditional top–down and stovepiped channels for mass communications. As social media have risen, traditional media sources have been steadily crippled by economic problems, resulting in a loss of capabilities and credibility. Information can propagate rapidly without the inclusion of traditional editorial checks and controls. Mass communications strategies for any type of major announcement must account for this new media landscape. Scientists announcing the discovery of extraterrestrial life will trigger a multifaceted and unpredictable percolation of the story through the public sphere. They will also potentially struggle with mis-information, rumours and hoaxes. The interplay of official announcements with the discussions of an extraterrestrial discovery on social media has parallels with traditional theories of mass communications. A wide spectrum of different mes-sages is likely to be received by different segments of the community, based on their usage patterns of various media and online communications. The presentation and interpretation of a discovery will be hotly debated and contested within online media environments. In extreme cases, this could lead to "editorial wars" on col-laborative media projects as well as cyber-attacks on certain online services and individuals. It is unlikely that a clear and coherent message can be propagated to a near-universal level. This has the potential to contribute to inappropriate reactions in some sectors of the community. Preventing unnecessary panic will be a prior-ity. In turn, the monitoring of online and social media will provide a useful tool for assessing public reactions to a discovery of extraterrestrial life. This will help to calibrate public communications strategies following in the wake of an initial announcement.

M. Jones (✉)
3/81 Queens Rd, Hurstville, NSW, Australia
e-mail: morrisjones@hotmail.com

D. A. Vakoch (ed.), *Astrobiology, History, and Society*, Advances in Astrobiology and Biogeophysics, DOI: 10.1007/978-3-642-35983-5_17, © Springer-Verlag Berlin Heidelberg 2013

17.1 Introduction

The discovery of life beyond Earth seems to be a strong possibility for the twenty-first century. Projects in astrobiology and the Search for Extraterrestrial Intelligence are progressing steadily. If we are not alone in the universe, it is probable that extraterrestrial life will be found within a few decades. News of a discovery will be conveyed to most of the world through the media and other channels of mass communications. It is thus essential to understand how these systems would respond to such a discovery. The study of mass communications in a rapidly evolving "new media" environment is critical. Changes in the way people communicate will not only affect the transmission of the news of a discovery. They will also influence public reactions and behavior. To a large extent, the propagation of the news itself will be a major segment of the overall public reaction to such an announcement. Through the use of social media, the general public will partially control the dissemination of the news, and shape the content of the message. This chapter thus acknowledges an increasingly strong overlap between communications and reactions to the discovery of extraterrestrial life. However, this chapter is not about general behavioral reactions to a discovery. The focus of this chapter is primarily on how media and communications behavior will contribute to the overall post-discovery scenario.

Exactly how the human race would respond to such a discovery is the subject of much speculation. Researchers from various disciplines have advanced theories and models for the consequences of such an announcement. Some have explored the possible effects on specific social groups such as religious leaders (Peters 2013) and certain cultures within third-world societies (Weigel and Coe 2013). There has also been attention on the practices used to generate such predictions, which can draw on various disciplines such as history (Dick 2013) and anthropology (Capova 2013; Denning 2013; Lowrie 2013). The lack of actual post-discovery scenarios poses challenges to such research, and necessitates approaches such as the use of analogy with previous events. Considerable debate exists within the astrobiology community on the likelihood of various outcomes, or the strength of the methodologies behind them.

This chapter focuses on a more easily documented environment. The mainstream media is clearly observable, both by its output and its internal operations. It operates according to its own clearly articulated principles. Consistency in its overall activity is generated by the industrial nature of most media activity. Social media is more chaotic, but generic trends can be discerned from the clearly visible usage patterns of hundreds of millions of people. The documentation of mainstream and social media behavior is thus akin to cartography of a visible landscape.

The availability of this direct observation is fortunate, as scholarly media theories are frequently not as advanced as sociological or anthropological insight into other areas of human activity. Some theories and paradigms are inconsistent with observable trends. Others are simply untestable or impractical for planning media-related activity. Media theories cited in this chapter are considered to be among the more respected in the field.

The objectives of this chapter are also different from much post-detection research, which complies with the scientific behavioral norm of "disinterestedness" outlined by the sociologist Robert Merton (Marks 1996). This states that scientists should separate their personal wishes and interests from the acquisition of scientific knowledge, which should be obtained in a detached, objective fashion. By contrast, this chapter is unapologetically prescriptive, and seeks to actively influence post-discovery behavior. This chapter thus focuses less on theory and more on practice.

17.2 Preconceptions for This Chapter

The theories and models outlined in this chapter are based on a set of preconceptions concerning the media and the nature of a potential discovery of extraterrestrial life. It is assumed that the operations of the mainstream media and of social media will remain relatively similar to their status at the time of writing. It is also assumed that both forms of media are subjected to the same sociological, legal and political conditions that currently exist in most developed nations.

This chapter also assumes that the discovery of extraterrestrial life will occur in one of two models. The first model would be the discovery of microbial life, or possibly fossils, somewhere in our solar system. The second model would involve the reception of an artificial transmission in the electromagnetic spectrum from an intelligent extraterrestrial civilization. There are alternative ways that humanity could discover the existence of extraterrestrial life, but current thinking in astrobiology circles suggests that the aforementioned two models are the most likely.

Changes in the media, or the socio-political factors that affect them, could influence the relevance of the response models outlined in this chapter. New theories and discoveries in areas such as psychology, sociology and media theory could also influence their credibility.

Thus, it is likely that the concepts outlined in this chapter could seem dated in another two decades. This could be due to changes in technology, society, sociological research or the actual confirmed discovery of extraterrestrial life.

17.3 The Fall of Traditional Media

In most sectors of the developed world, traditional media sources such as newspapers, radio and television are struggling to remain effective as channels for news delivery. Recent years have seen declines in advertising revenue for many of these organisations, which has always been their principal source of livelihood. The rise of new channels of communication (such as the Internet) has also caused audience and readership levels to fall, further weakening the appeal of these channels to advertisers, and decreasing revenue obtained through subscriptions. Conventional print newspapers, in particular, have manifested a steady decline in circulation for several decades.

This has prompted the retrenchment of many journalists, editors and researchers. The quality of journalism has suffered. Newsrooms are increasingly staffed by young, relatively inexperienced journalists who often lack specialised knowledge of specific areas of reportage. Science journalism has suffered enormously. There are few full-time specialist reporters around the world who are experts in astrobiology and related disciplines.

In an effort to broaden their readership levels, and thus increase revenue, news outlets are increasingly adopting tabloid news values in their selection of stories. There is an emphasis on celebrity news and scandals. Serious topics are sometimes avoided because they are thought to have limited appeal to a mass audience, or are too complex for a threadbare team of journalists to address suitably.

Declining levels of science education in schools are producing scientifically-illiterate journalists and media workers. These problems with education also produce media audiences without a firm grounding in basic science. Reportage on these subjects suffers from a both a lack of transmission, and a lack of reception.

The overall lack of resources within media organisations is strengthening the dependence of journalists on external providers of stories, regardless of the subject. Public relations firms, lobbyists and advocacy groups regularly "spoonfeed" the media. Journalists are happy to receive stories that are simply "fed" to them, without the need for the journalists themselves to go out and hunt for news. The stories are often not investigated any further or checked for accuracy. In some cases, media outlets will simply reproduce media releases verbatim, without even changing or editing their contents.

Journalism has its own formal method of neutralising the dilemmas of blindly quoting "spoonfeed" sources. In a process dubbed the "Web of Facticity" by the sociologist Gaye Tuchman (1978, 82), statements are simply repeated, with attribution to the source. The journalist does not endorse any statements as truth, but simply reports them in a "he said, she said" framework. The accuracy in the reporting lies in the fact that the statement was made by the attributed source, and that, at least, is probably true.

Media content is also increasingly driven by feeds from syndicated news organizations. These organisations station reporters in different sites around the world, and produce "wholesale" news that can be republished by news outlets. Some of these organisations are themselves suffering from budget cutbacks and poor staff performance. An error made by such an organisation can be propagated to news channels throughout the world.

Mass media are increasingly electronic in nature. The rise of online and digital media is merely the latest chapter in a story that began with the introduction of radio in the early twentieth century. We have also witnessed the rise of 24 hour news channels on radio and television. A breaking news story is reported almost immediately. This fast transmission of news has its advantages in emergencies or crises. However, it also influences the operation of news organisations and the behavior of journalists. Getting a story out quickly is sometimes more important than getting the story right. This severely reduces the time that journalists can use to collect news, check facts, or perform in-depth research. Simple events can be easily covered in this compressed format, but a story that requires more consideration or thought can be garbled.

The shortcomings of the mainstream media pose a serious challenge for reportage on any major announcement relating to astrobiology or the Search for Extraterrestrial Intelligence (SETI). Already, we have witnessed inadequate or spurious reporting on some research projects and announcements.

17.4 Skepticism

An announcement of an extraterrestrial discovery from an accredited scientific institution would be expected to be treated as factual, and faithfully reported by the media. However, a growing tally of "false alarms" or questionable claims by scientists has probably bred skepticism in some quality media circles. Noteworthy events include the high-profile claims of potential Martian microfossils in the meteorite ALH 84001, which made global headlines in 1996 (Savage, Hartsfeld and Salisbury 1996). Scientists have also claimed that bacteria retrieved from the stratosphere are potentially extraterrestrial in origin (Wainwright, Wickramasinghe, Narlikar and Rajaratnam 2002), and that "red rain" possibly contains extraterrestrial bacteria (Louis and Kumar 2006). In 2012, a member of the Russian Academy of Sciences claimed to have discovered evidence of life on Venus in photographs taken by a Soviet robot lander in 1982 (RIA Novosti 2012).

None of the claims listed above have wide-ranging support within the scientific community, yet all have been presented by qualified scientists to the media. Some have gained minimal levels of media coverage, followed by strong denials issued by other researchers in follow-up stories.

A skeptical attitude is a healthy trait for the media. It will help to filter spurious claims, and will also give an organization behind a genuine announcement more time to deal with the percolation of the story through the media.

17.5 Words and Pictures

An announcement of the discovery of extraterrestrial life must be handled in a different fashion to the release of other scientific stories. It is likely, and perfectly reasonable, that a conventional media conference will be staged. The conference will present scientists and officials connected to the discovery. Most scientific organizations involved in astrobiology and SETI research have the skills and personnel required to carry out such an exercise. However, the organization behind the discovery will need to do more than making an announcement. The media are increasingly visual in nature and will expect pictures to accompany their words. While there will be strong interest in hearing the statements of scientists connected to the discovery, this will not be enough to fill a mainstream news report.

If the discovery concerns microbial life, it is more than likely that there will be images of the lifeforms. We can expect microscopic images of these creatures, as well as images of the world of their discovery, such as Mars. It would also be interesting to supply images of the hardware used to make the discovery, such as a robotic spacecraft, sample-return capsule or the interior of a sample analysis laboratory. Stock footage and images of these subjects will constitute an essential part of a media strategy.

The interception of an artificial transmission from an extraterrestrial civilization would be more challenging for a multimedia presentation, as well as posing other challenges to the media. An artificial signal is more difficult to represent in visual terms that are accessible to mainstream audiences and journalists. It would be possible to supply images of radio telescopes, scientists sitting at consoles, reception equipment, and a "spike" or line on a graphical representation of the reception. However, these will not completely satisfy the expectations of editors or viewers.

Like the scientists behind the discovery, audiences and journalists will be considering an obvious question: What do the extraterrestrials look like? A SETI-related discovery is unlikely to supply any answer to that question, at least not immediately. Astrobiologists have produced some well-informed speculation on extraterrestrial physiology, but it remains only speculation.

It would be useful for SETI-related organizations to prepare a set of graphical representations of intelligent extraterrestrials in advance of a potential discovery. This could include models, illustrations and computer animation. The material should be of a high quality, both in terms of scientific input and media production. This "stock" footage could be distributed at short notice, in the event of a discovery.

This strategy could potentially alarm some scientists, who would be concerned about making unverified claims about the nature of the extraterrestrials. A rigid scientific approach would suggest that nothing was known for sure, and speculative pictures were unfit for publication by a reputable scientific institution. It is noteworthy that SETI-related texts and journals rarely attempt to visually depict extraterrestrials, despite the discussions of their biology.

This writer concedes that speculative pictures do not meet the normal standards of mainstream science. However, releasing such pictures with a note on their speculative nature would be better than some alternatives. If the media are not supplied with illustrations of the extraterrestrials, they will search for substitutes. Images of aliens from Hollywood movies or the black-eyed aliens that feature prominently in popular culture will abound. This would certainly cause more dismay to the scientific community than their own speculations.

One of many examples from the mainstream media is a 2012 article in the British newspaper *The Economist* (The Economist 2012). This is one of the world's quality news sources. The article, discussing the debut of the "SETI Live" project by the SETI Institute, ignored any graphics relating to the project itself. Instead, readers were treated to a stock image of little green men with big black eyes. Other news sources of lesser integrity could be expected to use similar images.

The publication of "alien" images from popular culture also carries other risks. Audiences could either consciously or unconsciously import the behavioral characteristics of fictional extraterrestrials into their perspectives on the announcement. This has the potential to contribute to negative opinions and actions by certain people.

17.6 Traffic Jams and Cybersecurity

Although most people will use the mainstream media as their primary source of information on an extraterrestrial discovery, there will certainly be a large amount of attention directed at the organization that makes the announcement.

Apart from implementing an effective media and communications strategy, there will be a need to address physical and security-related issues relating to information technology. Web servers related to the organization are likely to experience a traffic jam, which could cause them to collapse. Possible solutions include the creation of "mirror" sites that duplicate an organization's own Web site on other servers, helping to spread traffic elsewhere. An organization could also increase its own local server capacity and bandwidth. A special stripped-down Web site, filled with minimal graphics and effects, but retaining the most important messages, could also be substituted. This would download more quickly. Telephones, faxes and other channels of communication are likely to be besieged.

Cyberattacks are now a major global security threat. The scope and magnitude of this problem is generally underappreciated outside of intelligence and computing circles. The US Department of Homeland Security explains that online systems are "under perpetual attack by a variety of sources, from novice hackers to sophisticated groups that seek to gain or deny access to, disrupt, degrade or destroy the systems and the data contained therein" (Department of Homeland Security 2010).

Normal countermeasures such as antivirus software and firewalls will be insufficient to repel a dedicated cyberattack. Such an attack is almost guaranteed in the wake of a discovery announcement, due to the controversial nature of the issue. Servers, Web sites, email accounts and other systems could be hijacked to publish spurious and malicious messages. Malicious software could be installed on these servers, ensuring that they will spread to other users. Shostak (2002) notes that a hacker attack on a SETI-related organisation in 1999 caused a spurious story of an alleged discovery to circulate, and resulted in a story on a BBC online service!

Planning for a discovery announcement will require special consultations with IT security specialists and the implementation of a security strategy. It will also require communications strategies (such as crisis management) to deal with any malicious effects of a cyberattack. As with media relations planning, countermeasures for cyberattacks should be implemented, or at least examined, long before a discovery is made. Security audits of key facilities and infrastructure would be a useful interim step.

17.7 Filling the Air

The discovery of extraterrestrial life will certainly be treated as a major news story by all conventional news sources. Media outlets that are not exclusively dedicated to news coverage will certainly cover the story. They will dedicate a major proportion of their print space or news broadcast time to the discovery, but will probably not change their overall format of news delivery. It will be easy for such channels to find enough material to use.

A recent trend in media has been the growing domination of dedicated 24-hour news channels, mostly on television but also on radio. These news channels need to be addressed as special cases in terms of overall media responses. The rapid nature of "breaking news" on such channels has already been mentioned in this chapter, but this is only one of the challenges they pose. There will be a specific need to "feed" these channels, not only in terms of quality, but quantity.

The need for quality treatment comes from the influence and reach of 24-hour news channels. They command large audience numbers of hundreds of millions of people globally, meaning that they will be used as immediate sources of reference by multitudes. The channels are also used as primary producers of major news stories for regional news sources, who licence content produced by them. Thus, 24-hour news channels represent an effective "choke point" in the media food chain, allowing a single message to be distributed to numerous media sources and audiences.

There will also be a demand for a high quantity of material. Air time must be constantly filled with content, and the discovery of extraterrestrial life will be treated as a story of great significance. There will be repetition of the basic news story on a regular basis, together with any interviews or other material supplied at the time of the announcement. However, this will not be enough to satisfy the appetite of 24-hour news for content.

News channels will probably respond to this gap in material by tapping commentators from various sources. Scientists, astronomers, reporters and social commentators will be hurriedly ushered into television studios, and grilled for their opinions on the breaking news. Some of these commentators will probably be worth hearing, at least for entertainment value. However, the rush to fill air time quickly has the potential for creating problems. Some credible sources will simply be unavailable. News producers could find themselves haplessly recruiting conspiracy theorists or ill-informed commentators.

Some news channels will help to fill their air time by performing "vox populi" broadcasts. They will interview people randomly on the street for their reactions to the news, or read emails sent to the channel. Such a strategy would not be a mistake, either for broadcast networks or for the propagation of a discovery story. This feedback approach is regularly used with other news. However, it will not be enough to fully satisfy the airtime demands that will be placed on a story of this importance. After a dozen people have stated their viewpoints, further comments will seem repetitious and boring.

Budgetary constraints at some dedicated news channels have reduced their ability to deploy reporters on a wide scale. As a result, networks are increasingly using non-professional "citizen reporters" to send reports and audiovisual material to these networks. The networks often provide little or no financial compensation to these contributors. The widespread adoption of mobile phones, digital cameras and Internet connectivity has made the "citizen reporter" technically possible, assuming that someone is in the right place, at the right time, to witness and document a newsworthy event.

This reliance on citizen reporters, potentially coupled with managerial pressure to break big news quickly, leaves dedicated news channels vulnerable to hoaxes and fraud. It is possible that malicious, false news reports will be transmitted to these channels. Editors normally exercise caution in using unverified material, but a false report could still slip through. Once the report had been broadcast, it could find its way into other news sources.

17.8 Persuasion and Agenda-Setting

General media studies, and discussions of public reactions to an extraterrestrial discovery, have both followed a similar course in terms of considering power and effects on the general public. Media studies are fairly young when compared to other disciplines within the humanities and social sciences. The earliest formal studies arose in the 1930s, and attributed almost hypnotic powers of control to the mass media. The scholars behind these viewpoints were influenced by the use of propaganda in regimes such as Nazi Germany and Fascist Italy. Over time, this near-omnipotent view of media persuasion has been diluted. The media is still seen as influential, but its effects are limited by factors such as personal judgement and the influence of non-media entities on opinion.

Scholars considering the potential effects of the discovery of extraterrestrial life have followed a similar course in much of their own reasoning. It was once feared by some scientists that general panic could follow such an announcement, but this is no longer believed to be realistic.

It is ironic that the notorious 1938 Orson Welles radio adaption of the H.G. Wells novel *War of the Worlds* influenced both media scholars and SETI researchers in forming their more alarmist viewpoints! Presented in a "mockumentary" format, with simulated news reports, many listeners were convinced that Earth was being attacked by Martians, and panic did arise in some areas. It is important to disconnect this event from a potential contemporary discovery. In 1938, there were fewer alternative sources of news that could be checked immediately. It should also be noted that this was not a simple discovery announcement, but a description of an attack. Some historians state that the real scope of panic resulting from the broadcast was far more limited than contemporary news reports had claimed. It has been suggested that malice on the part of newspaper journalists for the rapidly encroaching "new electronic medium" of radio also fuelled these exaggerations (Lovgen 2005). Rivalry between old media and new media is hardly new!

A media paradigm that represents the modern influential-but-limited view of media power is known as "agenda-setting." The seminal research paper on "agenda-setting" (McCombs and Shaw 1972) basically suggests that the media cannot necessarily convince the public to believe a specific viewpoint, but can persuade people that the topic encompassing the viewpoint is itself important, and on the public "agenda." It is almost certain that the media will judge an extraterrestrial story to be important, but its prominence on the public "agenda" will depend on the style and strength of coverage. If coverage drops off rapidly after the initial announcement, it will not become prominent in the "agenda" of issues being considered and debated within public circles. People will know about it and remember it, but will not extensively talk about it or prominently think about it.

A lengthy, repeated series of stories relating to the discovery, over an extended period, will place the story on the public agenda. This would suggest a more complex, deliberative style of media coverage, as the issues surrounding the discovery were explored. The story will be prominent, but the media will be unable to universally convince people of how to interpret or react to the discovery. The influence of agenda-setting on public considerations of a discovery is significant in its own right. However, agenda-setting also has the potential to influence the depth and duration of discussions of a discovery on social media, which draws much of its "raw" material for circulation, forwarding and recommendation from media sources.

17.9 Two-Step Communication

An interface for communications between primary sources (such as governments and institutions) and the general public has been documented in conventional media theory. In the twentieth century, scholars such as Paul Lazarsfeld produced mass communications models that involved the dissemination of information through traditional mass communications channels (such as broadcast media and print), which was then echoed and moderated by talk between individuals (Katz 1957). Various theories and systems have been developed around this concept, but all acknowledge the interplay of mass media messages with personal discussions and "opinion leaders." One simplistic, linear model of this is a "two-step" process. A message, such as the introduction of a new government policy, is announced from a high-level source. The message may or may not directly reach the general public. However, the message will be received by "opinion leaders," who are considered by the public to be trusted authorities on particular topics. The "opinion leaders" will process and relay the message to a wider audience. The success or failure of a particular communications project could depend on the judgement of the "opinion leaders," who may relay the message but also relay their own disapproval of the concept. People who formulate a favourable opinion of a certain announcement through the mass media may change their viewpoint when a trusted "opinion leader" denounces it.

The only consistency among "opinion leaders" is their ability to influence others to adopt their opinions. Even personality traits are not firm indicators. In one personality-based study, Chan and Misra (1990, 53) note that "risk preference, open-mindedness, and mass media exposure, though correlated with opinion leadership, were not found to important predictors of opinion leadership."

Some "opinion leaders" are qualified experts in specific fields, who will be consulted by media sources. Other "opinion leaders" could be the most popular member of a group of friends, or the most prolific users of social media within specific groups. Marlow (2004) notes that "opinion leaders" can be easily detected in online social environments, "by their centrality to a given network, or by their ability to exercise large portions of the population in question by controlling the flow of information." This can be documented through quantifiable means on social media.

The influence of "opinion leaders" on the public is difficult to predict. There will be so many "opinion leaders" from different backgrounds, with different viewpoints, and different levels of influence among different groups of people. Potential types of opinion leaders outside of the media and the scientific community include political figures, clerics, business leaders and celebrities. It may be useful to provide briefings and documentation to opinion leaders with the potential to influence large numbers of people. Such individuals will certainly have high public profiles and will have ready access to the media.

17.10 Viral Transmissions on Social Media

Online and social media allow ideas to be spread to large numbers of people with astonishing speed. The propagation of news or information is frequently described as "viral." This has the potential to generate the semi-chaotic spread of wild stories and rumours.

The potential vulnerability of the mainstream media to hoaxes transmitted by "citizen reporters" was previously mentioned in this chapter. Such hoaxes will almost certainly spread more rapidly on social media, which have no editorial checks. Despite their propagation, hoax stories on social media are likely to be less damaging than a potential report in the mainstream media. This is partially due to the lower credibility of reports on social media. People will generally seek confirmation from the mainstream media for any extraordinary claims, and will not find it.

Another factor is the behavioural modus of social media. As their name implies, they are largely social in their focus. Passing any sort of information between friends is a social act that reinforces social connections, and also helps to boost the social prominence of an individual within a group or clique. The message is essentially social currency, to be exchanged for rewards such as fame, influence or attention. In some cases, the content of the message itself is largely irrelevant. The media scholar Marshall McLuhan was famous for preaching that "The Medium is the Message." In social media, the real message from such communications is simply to acknowledge that physically disconnected people remain connected as friends and family. The "medium" for demonstrating this connectivity is a regular stream of trivia, tidbits and short greetings, forwarded to others.

The transmission of discovery-related material as social currency will often produce short, garbled, silly or hoax messages. It will be a poor substitute for professional communications. However, it is important to understand the context of this activity. Simply forwarding messages about extraterrestrials (spurious or otherwise) will be just another tweet in an ongoing flow of trivia. It will be subsumed for the more important goal of simply staying in touch with others.

It is possible that bogus Twitter or Facebook accounts will be established by pranksters, hoping to impersonate scientists or institutions connected to the discovery. These will probably be detected and discredited fairly quickly, but they will probably fool some people for a short time.

17.11 Online Collaborative Media

Some online collaborative information sources (such as Wikipedia) have a higher credibility level than social media, but are still regarded with suspicion by some media sources and academics. The open nature of such projects has produced a wealth of useful information, but has also rendered these sites open to abuse. The discovery of extraterrestrial life will produce a flurry of contributions to Wikipedia and similar forums. The popularity of these Web sites will certainly help to disseminate news of a discovery, and could serve as an effective clearinghouse for updates. However, it is almost certain that the sites will be targeted by malicious users hoping to spread rumours, or well-meaning but poorly informed users. Malicious or inaccurate content on Wikipedia would have the potential to be more damaging than social media, given the relatively high popularity and credibility of the site.

Content on extraterrestrials will be in a state of flux after a discovery, and "editorial wars" between contributors could quickly surface. Content could be posted, then erased or rebutted, then changed again. Wikipedia notes that "vandalism" to some of its contents is often corrected within minutes, thanks to the diligence of its user base (Wikipedia 2012). Pages subjected to repeated malicious editing attacks can also be "locked" to prevent further editorial changes. Administrators on Wikipedia, who have more editorial power than conventional users, would be called upon to manage the situation.

It seems likely that an extraterrestrial discovery would prompt a high degree of diligence from the large user base of Wikipedia, who would police the site for spurious content. This would not render Wikipedia (or other collaborative media sites) completely free of errors, but it would probably keep the contents fairly accurate. There would arguably be fewer errors on a well-managed collaborative media site than in the conventional media as a whole.

17.12 Shared Video

Video hosting sites such as YouTube are very popular, and have a social media dimension to their usage patterns. An extraterrestrial discovery would probably generate a stylized series of responses on YouTube. News and documentary clips from television networks would be posted to the site, sometimes unofficially, sometimes under the control of their producers. YouTube also regularly hosts videos from self-styled "soapbox" orators, who voice their opinions on current events. Commentaries on the discovery would also appear in this format. Such pieces will attract limited attention.

More attention will focus on "mashups," or video collages on the subject. Snickars and Vonderau (2009, 24) define mashups as "individual videos that make use of disparately sourced sounds and images remixed into a new composite." They are already prominent on YouTube. As Snickars and Vonderau (2009, 270) note, "mashups and remixes of commercial film and television often appeared in search results adjacent to the works they'd appropriated, which constitute a kind of recognition of the remixer by the remixed." Mashups may seem semi-anarchic, but they have the potential to relay messages about a discovery to audiences who do not relate well to mainstream media. Staging a hoax or prank on YouTube would be more difficult than simply typing a message on Twitter or Facebook, as it would require media production skills and infrastructure. In general, YouTube is likely to be the least controversial of the more popular social media sites.

17.13 Crisis Management

Although the world in general seems unlikely to descend into panic and chaos in a post-detection scenario, there is still obviously the potential for some people and groups to react badly to a discovery. Thus, crisis management must factor into the design and disclosure of an extraterrestrial discovery.

The premature "leak" of a story is one scenario that would require crisis management. In such a case, media and public relations officers would be required to act quickly, to disseminate a legitimate message and possibly quell rumours. The leakage to the media of the putative discovery of Martian microfossils in 1996 demonstrated how easily this could happen. Correcting spurious rumours (possibly circulated by hackers) is another crisis management scenario.

Calm, rational statements from higher authorities would serve as one effective tool for crisis management. The aforementioned Martian microfossil story was a well-managed example. The announcement was made by NASA officials on

August 7, 1996 (Savage, Hartsfeld and Salisbury 1996). On the same day, following the NASA announcement, a statement was read to the media by US President William Clinton (Clinton 1996). Clinton's speech highlighted the excitement of the research while also noting its speculative nature. His generally sanguine tone also demonstrated that the President was not worried by the discovery. This was highly reassuring to listeners, as it used the power of the President as an "opinion leader." This could have prevented panic or fears among some sections of the public.

The 1999 hoax announcement of a signal discovery from EQ Pegasi demonstrated the need for a rapid response in crisis management. Astronomers reacted slowly and cautiously to the incident, and did not issue an immediate and widespread denial. Consequently, the story propagated to numerous journalists. As Oliver, Sim and Shostak (1999) note, "A simple one or two-paragraph press release...may well have shortened the life of the EQ Peg story considerably." This also highlights the fact that crisis management does not always need to be complex or difficult. Oliver (2010) also notes that the rise of social media has the potential to contribute to a "leak" when a putative signal is being investigated but not yet verified. As Shostak and Oliver (1999) note, "in the best of cases, confirmation will require several days." Rumours can spread globally within hours.

This poses a dilemma for crisis management. In some cases, a rapid denial could be useful, but in others, it could inadvertently draw unwanted attention to an investigation. The appropriate response will depend on the extent of circulation of the story on social media and its possible migration to mainstream media.

17.14 Monitoring the Public

Social media sites are regularly "harvested" by marketing moguls, public relations officers and anyone else interested in measuring public opinion, moods and trends. Methodologies and tools for exploring public opinion, as expressed on the Internet, are fairly mature. It would be logical to adopt such methods to explore public reactions to the discovery of extraterrestrial life. Such information would not be useful for the initial announcement of a discovery or media strategy following an announcement. However, it would be a useful intermediate step in the days and weeks that followed an announcement. Analysis of public opinion and concerns would help in the development of a "second wave" of media and public relations, which would address some of the concerns and questions raised by sampling public opinion.

Major news Web sites regularly run simple opinion polls on newsworthy topics. It is almost certain that many of these sites will run polls relating to an extraterrestrial discovery. Questions such as "Are you excited?," "Do you believe it?" and "Are you scared?" would be expected. Results from these sites would be interesting to observe, but should not be considered to be scientifically accurate reflections of public opinion. Similarly, the monitoring of social media should be treated as a guide, but not a scientific survey.

17.15 Other Cultures

Given their current levels of participation in SETI and space exploration, an extraterrestrial discovery is likely to originate from a socially liberal, technologically advanced nation such as the United States of America, Canada, Australia or a European state. These nations generally conform to the media paradigms described in this chapter. However, the aforementioned models do not precisely apply to all parts of the world. Mass communications operate very differently in states with authoritarian control of the media and fewer social liberties. In some countries with widespread Internet connectivity, access to social media is heavily regulated. Much of the world also remains disconnected from online services and has poor access to mass media. Cultural factors can also influence communications and opinions, as Weigel and Coe (2013) observed. Exactly how the communication of a discovery scenario would unfold in other socio-political cultures is difficult to generalize. No two places are exactly alike. Some governments may fear that news of extraterrestrials could be socially destabilizing, or a potential threat to the credibility of certain regimes. This could provoke attempts at censorship or denouncement of a discovery as a hoax.

17.16 Conclusion

We live in a media-saturated world. News of an extraterrestrial discovery would travel quickly and somewhat haphazardly through the mainstream and social media. Dubious information would certainly appear amid the flotsam of messages on the subject. It would be circulated extensively, but would not necessarily have much influence in the face of more credible reports. A fairly accurate message could still reach the majority of audiences if an appropriate media strategy is enacted.

References

Capova, Klara A. 2013. "The Detection of Extraterrestrial Life: Are We Ready?" In *Astrobiology, History, and Society: Life Beyond Earth and the Impact of Discovery*, ed. Douglas A. Vakoch. Heidelberg, Springer.

Chan, Kenny and Shekhar Misra. 1990. "Characteristics of the Opinion Leader: A New Dimension." *Journal of Advertising* 19 (3): 53–60.

Clinton, William. 1996. "President Clinton Statement Regarding Mars Meteorite Discovery." *White House, Office of Press Secretary*. August 7, 1996. http://www2.jpl.nasa.gov/snc/clinton.html. Accessed on February 10, 2012.

Denning, Kathryn. 2013. "Anticipating the Alien: Historical Analogy and the Work of Prognostication." In *Astrobiology, History, and Society: Life Beyond Earth and the Impact of Discovery*, ed. Douglas A. Vakoch. Heidelberg, Springer.

Department of Homeland Security. 2010. "Computer Network Security & Privacy Protection." *Privacy Cybersecurity White Paper*. February 19, 2010. http://www.dhs.gov/xlibrary/assets/privacy/privacy_cybersecurity_white_paper.pdf. Accessed on March 26 2012.

Dick, Steven J. 2013. "The Societal Impact of Extraterrestrial Life: The Relevance of History and the Social Sciences." In *Astrobiology, History, and Society: Life Beyond Earth and the Impact of Discovery*, ed. Douglas A. Vakoch. Heidelberg, Springer.

Katz, Elihu. 1957. "The Two-Step Flow of Communication: An Up-To-Date Report on an Hypothesis." *Political Opinion Quarterly.* 21 (1): 61–78.

Louis, Godfrey, and Kumar A. Santhosh. 2006. "The Red Rain Phenomenon of Kerala and its Possible Extraterrestrial Origin." *Astrophysics and Space Science* 302 (1–4): 1–18

Lovgen, Stefan. 2005. "'War of the Worlds': Behind the 1938 Radio Show Panic." *National Geographic News*. June 17, 2005. http://www.pkwy.k12.mo.us/homepage/jmcmullen/File/War_of_the_Worlds_Behind_the_Panic.doc Accessed on 20 February 2012.

Lowrie, Ian. 2013. "Cultural Resources and Cognitive Frames: Keys to an Anthropological Approach to Prediction." In *Astrobiology, History, and Society: Life Beyond Earth and the Impact of Discovery*, ed. Douglas A. Vakoch. Heidelberg, Springer.

McCombs, Max, and Donald Shaw. 1972. "The Agenda-Setting Function of Mass Media." *Public Opinion Quarterly* 36 (2):25–45.

Marlow, Cameron. 2004. "Audience, Structure and Authority in the Weblog Community." *International Communication Association Conference* New Orleans.

Marks, Jonathan. 1996. "The Anthropology of Science Part II: Scientific Norms and Behaviors." *Evolutionary Anthropology* (5):76.

Oliver, Carol, Helen Sim, and Seth Shostak. 1999. "The Case of EQ Peg: Challenge and Response." *International Astronautical Congress*, Amsterdam 1999.

Oliver, Carol. 2010. "Social Media: Implications for Post-Detection Strategies." *International Astronautical Congress*, Prague 2010.

Peters, Ted. 2013. "Would the Discovery of ETI Provoke a Religious Crisis?" In *Astrobiology, History, and Society: Life Beyond Earth and the Impact of Discovery*, ed. Douglas A. Vakoch. Heidelberg, Springer.

RIA Novosti. 2012. "Life Spotted on Venus" *RIA Novosti* January 20, 2012. http://en.rian.ru/science/20120120/170865269.html.

Savage, Donald, James Hartsfeld, and David Salisbury. 1996 "Meteorite Yields Evidence of Primitive Life on Early Mars." *National Aeronautics and Space Administration* (*NASA*) *Media Release*. August 7, 1996. http://www.nasa.gov/home/hqnews/1996/96-160.txt.

Shostak, Seth. 2002. "SETI and the Media." *Bioastronomy 2002 conference*. Sydney, Australia.

Shostak, Seth, and Carol Oliver. 1999. "An Immediate Reaction Plan for SETI." *Bioastronomy 1999 conference*, Hawaii.

Snickers, Pelle, and Patrick Vonderau. 2009. *The Youtube Reader*. Stockholm: National Library of Sweden.

The Economist. 2012. "The Wow Factor." *The Economist*, March 10, 2012, p. 77

Tuchman, Gaye. 1978. *Making News: A Study in the Construction of Reality*. Ann Arbor, MI: The Free Press.

Wainwright, Milton, N. Chandra Wickramasinghe, Jayant Narlikar, and P. Rajaratnam. 2002. "Microorganisms Cultured from Stratospheric Air Samples Obtained at 41 km" *FEMS Microbiology Letters* 10778:1–5.

Weigel, M.Margaret, and Kathyrn Coe. 2013. "Impact of Extraterrestrial Life Discovery for Third World Societies: Anthropological and Public Health Considerations." In *Astrobiology, History, and Society: Life Beyond Earth and the Impact of Discovery*, ed. Douglas A. Vakoch. Heidelberg, Springer.

Wikipedia. 2012. "Wikipedia: Editorial Oversight and Control." *Wikipedia* http://en.wikipedia.org/wiki/Wikipedia:Editorial_oversight_and_control

Chapter 18
Christianity's Response to the Discovery of Extraterrestrial Intelligent Life: Insights from Science and Religion and the Sociology of Religion

Constance M. Bertka

Abstract The question of whether or not extraterrestrial life exists and its potential impact for religions, especially Christianity, is an ancient one addressed in numerous historical publications. The contemporary discussion has been dominated by a few notable scientists from the SETI and astrobiology communities, and by a few Christian theologians active in the science and religion field. This discussion amounts to scientists outside of the faith tradition predicting the demise of Christianity if extraterrestrial intelligent life is discovered and theologians within the tradition predicting the enrichment and reformulation of Christian doctrine. Missing from this discussion is insight drawn more broadly from the science and religion field and from the sociology of religion. A consideration of how possibilities for relating science and religion are reflected in the US public's varied acceptance of the theory of evolution; the growth of Christianity in the Global South; and a revised theory of secularization which inversely correlates religiosity to existential security, gives credence to the proposal that the response from those outside of academia would be much more varied and uncertain.

18.1 Introduction

From its inception, those charged with launching and nurturing the field of astrobiology have noted its potential for inspiring thought on humanity's place in the universe, "…the payback to the American public and the worldwide public is a continuing new perspective on ourselves, on our role, how our environment shapes us and we shape the environment. The impact is more of a philosophical impact than a practical impact…" (DeVincenzi 1997). The first astrobiology roadmap,

C. M. Bertka (✉)
Science and Society Resources, Potomac, MD, USA
e-mail: cbertka@gmail.com

D. A. Vakoch (ed.), *Astrobiology, History, and Society*, Advances in Astrobiology and Biogeophysics, DOI: 10.1007/978-3-642-35983-5_18,
© Springer-Verlag Berlin Heidelberg 2013

released in 1999, also included an operating principle that highlighted societal interest in the search for extraterrestrial life and led to the following question to be posed, "How will astrobiology affect and interact with human societies and cultures" (Dick and Strick 2004, 218)? In trying to answer this question it is reasonable that we are curious about how astrobiology will impact world religions, as they also reflect societies' efforts to understand "humanity's place in the universe." The question of whether or not extraterrestrial life exists, and its potential impact for religions, especially Christianity, is an ancient one previously explored under the title "plurality of worlds."

A predominant theme that emerges from a historical review of this work through the 19th century is both the "readiness with which extraterrestrial life was accepted by an array of religious figures" and concern over the implications for specific Christian doctrines like the incarnation—the belief that God became man on Earth and died to redeem humanity (Crowe 1986, 557). The discussion around this topic was extensive with publications numbering in the thousands. By the 1700s it was largely assumed that the discovery of extraterrestrial life would be supportive evidence of God's benevolence and omnipotence (Crowe 1986; Crowe and Dowd 2013). It should be also noted that the historical record of this discussion is dominated by Christian perspectives (Crowe 2008). The same is true of the contemporary discussion.

The focus on Christianity in this work is pragmatic. Over 82% of US respondents to the Baylor Religion Survey (Baylor Institute for Studies of Religion 2006) identify themselves as Christians, and Christianity remains the world's largest religion accounting for over one-third of the world's religious belief (The Association of Religion Data Archives 2010). Overall interest in this question may be more general in nature, but of necessity the question of impact will need to be posed for individual religious traditions, for each brings its own perspective to reflection on "humanity's place in the universe." As this work will highlight, even the range of beliefs within Christianity precludes a single generic answer. However, this reality is not readily apparent from the contemporary discussion that has been dominated on one hand by a few notable scientists from the SETI and astrobiology communities and on the other by a few Christian theologians active in the science and religion field. Missing from the contemporary discussion is insight from the sociology of religion and one goal of this chapter is to shed light on how work from this field might influence ideas about the potential impact of the discovery of extraterrestrial intelligence (ETI) on Christian worldviews.

18.2 Conventional Wisdom of the SETI and Astrobiology Community?

The question of the impact of the discovery of extraterrestrial intelligent life on religious belief is not a question that many scientists are likely to address. Although it may be studied from a social scientific perspective, there is no evidence that it is a research driver for the majority of those interested in exploring natural explanations for the origin, extent, and future of life. Individual scientists, like any individual, may ponder these implications from the perspective of their own philosophical leanings, but the various answers they arrive at need not hold any authority within their community or the public at large. That said, through the writings of a few notable members of the SETI and astrobiology communities it has been proposed that the conventional wisdom of these communities predicts a crisis for terrestrial religion if extraterrestrial intelligent life is discovered (Peters 2009 and 2011).

Predicting the demise of religion is not a new idea. The secularization theory, popular with sociologists of religion throughout much of the 20th century, held that as societies become increasingly modernized and secular, religious belief, and the influence of religious institutions would decline. In his seminal critique of the secularization theory, Rodney Stark (1999, 249) reminds us that predicting the decline of religion has deep historical roots, "For nearly three centuries, social scientists and assorted western intellectuals have been promising the end of religion. Each generation has been confident that within another few decades, or possibly a bit longer, humans will 'outgrow' belief in the supernatural." Stark argues that despite its popularity, the secularization theory has never been consistent with empirical data. Peter Berger (2001, 445), earlier a major proponent of the theory, has now abandoned this idea, as have most sociologists of the 21st century, because "the theory seemed less and less capable of making sense of the empirical evidence from different parts of the world, not least the United States." The world today has more people with traditional religious views than ever before (Norris and Inglehart 2011).

Jill Tarter, a prominent figure in the SETI community, has expressed a view of religion that is reminiscent of the traditional secularization theory. To paraphrase from Jill Tarter's (2000) essay "SETI and the Religions of the Universe," the argument put forward is that technologically advanced aliens will have no religion, either because they never had it or because they outgrew it; or if they do have religion it will be a single religion that is superior to any found on Earth because it has enabled a more stable social structure than ours to exist. Implicit in this logic, as highlighted by Peters (2009) as part of what he refers to as the "ETI myth," is the idea that extraterrestrial intelligence will not only be technologically advanced, but morally advanced as well. These assumptions led Carl Sagan and Frank Drake (1997, 8) to argue that contact would "inevitably enrich mankind beyond imagination."

Likewise Paul Davies (1995), acting as a public spokesperson for astrobiology, expresses the same view in his popular book *Are We Alone?* He assumes that extraterrestrial intelligent life will have "discarded theology and religious practice long ago as primitive superstition" or "If they [ETI] practiced anything remotely like a religion, we should surely soon wish to abandon our own and be converted to theirs (Davies 1995, 37)." Steven Dick (2000), a historian of science who has written extensively about astrobiology and the search for extraterrestrial life, presents a more nuanced but similar view suggesting that acceptance of a biological universe will call for a "wrenching" adjustment for Christianity and likely the ultimate acceptance of a natural rather than supernatural God. Just how widely this view may be representative of the SETI or astrobiology communities at large is by no means certain, but it is a view that has elicited the response of contemporary Christian theologians.

18.3 The Response from Contemporary Christian Theologians

In the late 1900s astrobiology reinvigorated the question of "plurality of worlds" with progress towards addressing theoretical probabilities with empirical data. Not withstanding the Vatican study week on astrobiology, which was a gathering of astrobiologists focused on the science of astrobiology and not a philosophical or theological discussion of its implications (Pontifical Academy of Sciences 2009), the interests of contemporary Christian theologians in the implications of astrobiology has been limited to a few theologians prominent in the science and religion field. Ted Peters is one significant example.

Within the public sphere, the rise in space consciousness in the 1900s lead one scholar to suggest that the search for extraterrestrial intelligent life had become a popular surrogate religion (Guthke 1985). Steven Dick (2000, 205) has also conceded that "SETI may be science in search of religion." This is a theme that resonates with Peters. Not only is he critical of the "Conventional wisdom [which] seems to suggest that terrestrial religion would collapse under the weight of confirmed knowledge of extra-terrestrial intelligence (Peters 2011, 644)," he suggests that the idea of a "celestial savior," a civilization advanced in religious belief and morality, is the basis of "assumptions at work in the field of Astrobiology" (Peters 2011, 649). He documents evidence for this view in the speculations of those intrigued by UFOs and in the writings of prominent SETI figures (Peters 1995, 2009, 2011). In contrast to the notion that Christianity would face a crisis upon the discovery of extraterrestrial intelligent life, Peters (2011, 654) contends that "In fact, theologians might relish the new challenges to reformulate classical religious commitments in light of the new and wider vision of God's creation."

Peters'negative critique concerning "conventional wisdom" was shared by theologian Ernan McMullin (2000). He was wary of speculations that arise from an ignorance of the flexibility of Christian doctrine, including the range in their interpretation and how they evolve over time. Regardless of whether or not extraterrestrial intelligent life is discovered, theologians' understanding of key Christian doctrines concerning original sin, the soul, and the incarnation are wide ranging. Another theologian who has published on the topic, David Wilkinson (1997), contends that Christian beliefs are grounded in biblical revelation, not ideas about the natural evolution of religion, and as such will not be threatened by religions from potentially more advanced extraterrestrial societies. He also argues that God's love for humanity can be unique without being exclusive.

While other contemporary theologians have briefly commented on the implications of extraterrestrial intelligent life for Christianity, what is significant about the work of Peters, McMullin, and Wilkinson is that it has a detailed focus and responds specifically to generalities espoused by those outside of the Christian tradition. Each of these authors carefully considers the potential impact of discovery on key Christian doctrines and they do so within the context of an understanding of the general aims of the science of astrobiology.

The contemporary dialogue amounts to scientists outside of the faith tradition predicting a famine for Christianity if extraterrestrial intelligent life is discovered, whereas theologians within the tradition are suggesting the possibility of a relative feast of ideas. Insights from the fields of science and religion and the sociology of religion, however, give credence to the proposal that the response from those outside of academia may be much more varied and uncertain.

18.4 Insights from Science and Religion

Peters, McMullin, and Wilkinson are scholars who incorporate what science has learned about the natural world into their theology, where theology can be defined as "a systematic and rational reflection on faith." Their approach to science and religion, one of "integration," or at minimum, "dialogue" is common among scholars active in the science and religion field. One could argue that this field, despite its deep historical roots, began as a contemporary effort in the year 1966. This year saw both the publication of Ian Barbour's (1966) *Issues in Science and Religion* and the first issue of a non faith-based scholarly journal for science and religion, *Zygon*. At the time Ian Barbour was a physics professor at Carleton College with graduate work in theology. The founding editor and driving force behind the journal *Zygon* was Ralph Wendell Burhoe, a Unitarian Universalist professor of Theology and the Sciences at Meadville Theological School. Burhoe's leadership was also instrumental in the founding of an early institution in science and religion, The Center for Advanced Study in Theology and Science (Breed 1991). Today the field also boasts an International Society for Science and Religion (ISSR, since 2002), a European Society for the Study of Science and Theology (ESSSAT, since 1997), as well as another scholarly journal devoted to science and religion, *Theology and Science*, the journal of the Center for Theology and the Natural Sciences, first published in 2003.

In addition two other publications are available. *Perspectives on Science and Christian Faith* is the journal of the American Scientific Affiliation, a voluntary group of Christian scientists, first published in 1949. *Science and Christian Belief,* founded in 1989, is sponsored by Christians in Science, a voluntary group of Christian scientists in the UK, and by the Victoria Institute. These latter two publications are products of organizations with whom full membership requires agreement with a statement of faith and whose membership is popular with evangelical Christians. However, even a cursory review of the scholarly societies and journals will reveal the large degree to which issues of significance to Christianity in general dominate this dialogue as well, though their continues to be a concerted effort to widen participation to include other religious traditions.

Pertinent to our discussion is the assumption about the relationship between science and religion which permeates this scholarly community and I would argue, that like the approach taken by Peters, McMullin, and Wilkinson, it is one that encourages "dialogue" and "integration," one where theologians, in the words of Peters (2011, 654), "might relish the new challenges to reformulate classical religious commitments in light of the new and wider vision of God's creation." These approaches to relating science and religion do not necessarily reflect the approaches more commonly taken by those outside of this community, namely approaches of "conflict" or "independence." Those who hold that science and religion are in conflict maintain that both are speaking about the same subject matter with claims that cannot coexist. An example of this approach is illustrated by the biblical literalist's quarrels with the scientific community over the theory of evolution. Though the media attention that this topic receives may lead us to believe that "conflict" is the dominant approach towards the relationship between science and religion, a more common approach, especially in mainstream Christian communities, is one of "independence." This view maintains that science and religion deal with different subject matters, one nature the other God, and respond to different questions, science tells us how, religion tells us why, and therefore they cannot be in conflict. This view was eloquently expressed in Stephen J. Gould's (1999) book, *Rock of Ages: Science and Religion in the Fullness of Life* and is the view commonly espoused by the scientific community at large, particularly in response to public concerns in the US over the teaching of evolution.

This typology for the relationship between science and religion (conflict, independence, dialogue, and integration) was originally suggested by Barbour (1997), and whereas subsequent elaborations and modifications have been proposed (e.g., Stenmark 2010), Barbour's original typology serves to highlight the broad range of approaches to relating science and religion. It can be used to categorize the general differences we intuitively recognize between the scholarly science and religion community and the public at large. Also, viewed through the lens of this typology perhaps the "famine" for Christianity that advocates of the conventional wisdom view predict stems from the failure to envision any but a conflict approach if Christianity is faced with the discovery of ETI, and a conflict in which the presumed loser is Christianity.

Ted Peters has attempted to test the validity of the conventional wisdom view with an "ETI Religious Crisis Survey" (Peters 2009 and 2013). He reports that the survey reveals that the majority of religious believers, of all faiths, did not view the discovery of extraterrestrial intelligent life as a threat to their personal beliefs. Peters' survey asks general questions about religious belief and the discovery of ETI life. For example, respondents are asked to "agree," "neither agree or disagree," or "disagree" with the following, "Official confirmation of the discovery of a civilization of intelligent beings living on another planet would so undercut my beliefs that my beliefs would face a crisis (Peters 2009, 21)." The majority, 83–94%, of respondents from different religious traditions (Catholic, Protestant, Mormon, Jewish, Buddhist and Non-Religious) "disagree" or "strongly disagree" with this statement.

Respondents to this survey do not believe the discovery of ETI will impact their religious beliefs. The survey, however, gives us no measure of whether or not the respondents have considered the implications of discovery for the doctrines of their religious traditions in any depth, or if at the conclusion of that exercise they are confident that they can successfully "integrate" the implications of ETI into their existing religious tradition. At least as likely is the possibility that they are operating from an "independence" view of science and religion, which does not require any reflection on implications for specific doctrines.

Peters (2009, Counterbalance Website) does provide detailed written comments received from some of the respondents. Interestingly, missing from these responses, with minor exceptions, is any mention of a recognized connection between the discovery of ETI and the theory of evolution. As Peters (2009, 10) highlights, ideas about evolution are a key foundation to the search for life elsewhere:

> The first and salient feature of the ETI myth is the imaginary exportation of the theory of evolution to other planets or possible habitats in space. There is nothing unscientific about this imaginary exportation, to be sure; it is the most reasonable thing to advance a hypothesis regarding what is not known based on what is known. If we know that life could originate here on earth and could speciate through evolution, then it is reasonable to project that these processes might have occurred more than once in this vast universe.

Of the seven Christian respondents who indicate in their comments that they recognize this connection, two object to the idea of a more highly evolved civilization implying an advanced morality, three do suggest the discovery of ETI will require modifications to their religious tradition, and two oppose the idea of evolution. If the majority of respondents see any connection between the theory of evolution and the discovery of ETI, or any concern about this connection, it is not apparent from their comments. Would survey results be different if they did see a connection?

From 2003 through 2006 I organized a series of workshops through the AAAS Program of Dialogue on Science, Ethics and Religion in which Christian theologians were invited to share their perspectives on astrobiology with a specific focus on the scientific exploration of the origin of life, and the search for and possible discovery of extraterrestrial microbial life (Bertka, Roth, and Shindell 2007; Bertka 2009). These theologians were also confident in Christianity's ability to incorporate a discovery of extraterrestrial life, microbial or intelligent, but there was less optimism over the impact such a discovery would have on current public engagement issues over the validity of evolution. For example, Cynthia Crysdale (2009, 221) states that "Questions about how God is related to an emerging world are at the heart of current religious politics in North America, and confusions in this regard will only be exacerbated with the discovery of extraterrestrial life."

Peters' survey deals with generalities of discovering ET life for religious belief, not specific doctrinal interpretations that this discovery might encourage and the resulting religious politics that might be incited. Popular religious beliefs are not necessarily cognizant of theological rationalizations and this is a particular problem for theological reflections on the significance of nature (theology of nature). This is perhaps nowhere better illustrated than in the lack of awareness of advances in the field of science and religion among seminary students as the subject is far from a routine one in seminary education. In the words of Celia Deane-Drummond (2009, 102), a theologian active in the science and religion field, "Contemporary theologians, reluctant to 'burn their fingers,' have avoided dealing with the subject of nature." Despite its growth among liberal theologians since the 1960s, I would argue that the contemporary science and religion field is a scholarly interest whose impact is yet to be significantly encountered in the larger public sphere of religious understanding .

That acceptance of the theory of evolution continues to be an area of concern is evidenced in the US by numerous surveys over the last fifty years. Notably one-third of US adherents of mainline Protestantism and Roman Catholicism, Christian traditions whose scholars have accepted the theory of evolution, believe that "humans and other living things have existed in their present form only" (Pew Forum on Religion and Public Life 2005). Among evangelical Christians this belief rises to 70%. These individuals may find evidence of extraterrestrial life to attest to "God's benevolence and omnipotence" but perhaps see no reason that this discovery should change their ideas about evolution and inspire reformulation of their particular Christian beliefs. In addition, mainline Protestants and Roman Catholics who accept evolution are fairly evenly split between those who believe that organisms evolved over time "guided by a supreme being," and those who believe organisms evolved over time through "natural selection" (the Pew survey forces them to make a choice between these options, or select "don't know"). The Pew survey allows us to clearly identify those who adopt a conflict approach to science and religion, but it is not as helpful to probe views of integration or independence. Whether, these respondents are incorporating evolution into their worldview, or not, there is no reason to predict that for them another example of evolution in the universe would result in a religious crisis. The contrasting views between the Christian theologians and the SETI and astrobiology scientists who have commented on this topic can be attributed to different assumptions about the possibility of integrating ETI into Christianity, but for the public at large in the US this topic may be a non sequitur because integration is not highly regarded as a necessary option. How might this topic fare outside the US? The majority of respondents to the Peters survey—1,000 out of 1,300 (Peters 2012, personal communication)—were from the US and those surveyed from other parts of the world participated through a web survey largely by invitation. The Peters survey is typical of other surveys conducted on beliefs and assumptions about extraterrestrial life- these surveys sampled educated persons in North American and European countries (Weigel and Coe 2013). Would survey conclusions be different if a broader world demographic was included? Put another way, "who speaks for Christianity?"

18.5 Insights from the Sociology of Religion

A challenge for sociologists of religion today is to explain the large variation in religious belief worldwide and the change in the religious landscape over time. The Pew Forum on Religion and Public Life (2011) recently published a study on the change in the global distribution of Christianity from 1910 to 2000. In 1910 the majority of the world's Christians were in Europe, 66%, today only 26% of Christians reside in Europe. The majority of Christians, 61%, now reside in the Global South (sub-Saharan Africa, Asia–Pacific, and Latin America) versus 39% in the Global North (North America, Europe, Australia, Japan and New Zealand). There has been a rapid growth of Christianity in the developing countries of the Global South with the largest growth in sub-Saharan Africa and Asia–Pacific.

Over half of the world's Christians are Roman Catholic but the segment of Christianity experiencing the greatest growth is Protestant Pentecostal denominations and the Charismatic movement (Pew Forum on Religion and Public Life 2006). These are groups for whom post conversion experiences are central, including speaking in tongues, physical healings and the ability to communicate messages from God. There is a focus in these groups on individual experience. Pentecostal is a term used to describe those who belong to specific Pentecostal denominations, whereas Charismatic is used more generally to define those who describe their religious experiences with a focus on "gifts of the Holy Spirit," though they may be members of mainstream Protestant, Catholic and Orthodox denominations. Worldwide these two groups account for 27% of Christians and in some countries in Latin America, Africa and Asia they are the majority Protestant voice. For example, in Latin America, Pentecostals account for 73% of all Protestants and the Charismatic movement is also widespread in the Roman Catholic Church.

The Pew Forum on Religion and Public Life study, *Spirit and Power* (2006), specifically explored the beliefs of Pentecostals and Charismatics in the US and nine other countries in Latin America, Africa and Asia where these groups are numerous. They find that these groups tend to be more conservative than other Christians with a large majority, for example, believing that "The Bible is the actual word of God and is to be taken literally, word for word." Pentecostals and Charismatics are committed to very traditional views of Christian doctrine. Unlike mainline Christian theologians versed in the scholarship of the science and religion field, these Christians may not relish opportunities to reformulate traditional doctrines. In asking "Who speaks for Christianity?" can we be confident that Christians in the Global South would see their views reflected in those of North American and European mainline theologians? If not does this discrepancy in views give support to the conventional wisdom thesis that contact with ETI will result in a collapse for Christianity? Not necessarily, the revised theory of secularization proposed by Norris and Inglehart (2011) hints at an alternative to both the "famine" and "feast" predictions.

In contrast to the traditional secularization theory which predicted a universal decline of religion correlated to the spread of scientific knowledge with modernization in all industrial socieites, Norris and Inglehart (2011, 27) argue that "secularization is most closely linked with whether the public of a given society has experienced relatively high levels of economic and physical security." The distribution of wealth across the society matters as well, such that a society with a high level of national wealth may still exhibit high religiosity if that wealth is unevenly distributed. In this revised theory of secularization religiosity is inversely correlated with existential security. Using survey data from the 2007 Gallup World Poll, Norris and Inglehart (2011) demonstrate that as an individual's lived poverty index increases, the importance they place on religion in their daily lives, as well as their attendance at religious services, also increases. The lived poverty index includes measures of access to food and shelter, experience with health problems, and access to basic facilities in the home. As confidence in the idea that survival can be taken for granted decreases, religiosity increases. In this revised secularization model the growth of traditional religious views worldwide is attributed to the greater birth rates in poorer countries compared to richer countries.

Norris and Inglehart (2011, 245) do not claim that existential security is the only factor influencing religiosity, but they suggest a stark contrast between theologians and the laity, "Theologians may have been primarily concerned with the meaning and purpose of life, but for most ordinary people, the sense of reassurance that religion provides, that's one fate is in the hands of a benevolent higher power even when it is uncertain that one's family will have enough to eat, has been the most prominent factor drawing people to religion." How would the discovery of extraterrestrial intelligent life impact feelings of existential security and how might Christianity, particularly in the Global South respond? If the answer is a decrease in existential security then we should be wary to predict either a famine for Christian beliefs or a feast for liberal theology, if anything more likely would be an increase in traditional Christian views, particularly in the Global South were the growth of conservative, personal, experience-based Christianity is the current trend.

18.6 Conclusion

The difficulty in trying to generalize about Christianity's response to the discovery of extraterrestrial intelligent life cannot be overstated. The variety in Christianity worldwide, both at the denominational level as well as at the level of individual experience, and the variety of options for relating science and religion, will combine to insure that integrating what SETI or astrobiology learns about the universe into Christian worldviews will at minimum be a long and convoluted process with more than one likely outcome. The answers for the academic community, both astrobiologists and theologians, may bear little resemblance to the response of a public whose religious beliefs are nurtured by more than systematic rationalization. If the question of potential impact of the discovery of ETI on Christianity is raised with the goal of preparing societies for this possibility, then preparation should focus on the reality of a variety of responses.

The question as applied to Christianity in the Global North might be further explored with studies designed to provide insight into what relationship between science and religion individuals have adopted toward the question of ETI, and specifically whether they recognize a connection between the search for extraterrestrial life and the theory of evolution. A renewed effort should also include a deliberate focus on exploring the responses of Christians from the Global South. Contemporary astrobiologists and North American and European theologians have initiated a conversation that would benefit from a broader audience.

References

The Association of Religion Data Archives. 2010. *The World.* http://www.thearda.com/internatio naldata/regions/profiles/Region_23_1.asp.

Barbour, Ian. 1966. *Issues in Science and Religion.* New York: Harper and Row.

Barbour, Ian. 1997. *Religion and Science: Historical and Contemporary Issues.* San Francisco: HarperSanFrancisco.

Baylor Institute for Studies of Religion. 2006. "American Piety in the 21st Century: New Insights to the Depth and Complexity of Religion in the US." *The Baylor Religion Survey.* http://www.baylor.edu/content/services/document.php/33304.pdf.

Berger, Peter. 2001. "Reflections on the Sociology of Religion Today." *Sociology of Religion* 62 (4): 443–454.

Bertka, Constance M. 2009. "Astrobiology in a Societal Context." In *Exploring the Origin, Extent and Future of Life*, ed. Constance M. Bertka, 1–18. Cambridge: Cambridge University Press.

Bertka, Connie, Nancy Roth and Matthew Shindell, eds. 2007. *Workshop Report: Philosophical, Ethical and Theological Implications of Astrobiology.* Washington, DC: American Association for the Advancement of Science.

Breed, David R. 1991. "Ralph Wendell Burhoe: His Life and His Thought." *Zygon* 26 (3): 397–428.

Crowe, Michael J. 1986. *The Extraterrestrial Life Debate 1750–1900: The Idea of a Plurality of Worlds from Kant to Lowell.* Cambridge: Cambridge University Press.

Crowe, Michael J., ed. 2008. *The Extraterrestrial Life Debate: Antiquity to 1915, A Source Book.* Notre Dame: University of Notre Dame Press.

Crowe, Michael J., and Matthew F. Dowd. 2013. "The Extraterrestrial Life Debate from Antiquity to 1900." In *Astrobiology, History, and Society: Life Beyond Earth and the Impact of Discovery*, ed. Douglas A. Vakoch. Heidelberg: Springer.

Crysdale, Cynthia S.W. 2009. "God, Evolution, and Astrobiology," In *Exploring the Origin, Extent and Future of Life*, ed. Constance M. Bertka, 220–241. Cambridge: Cambridge University Press.

Davies, Paul. 1995. *Are We Alone?* London: Penguin.

Deane-Drummond, Celia. 2009. "The Alpha and the Omega: Reflections on the Origin and Future of Life from the Perspective of Christian Theology and Ethics," In *Exploring the Origin, Extent and Future of Life*, ed. Constance M. Bertka, 96–112. Cambridge: Cambridge University Press.

DeVincenzi, Donald. 1997. Oral History Interview by Steven J. Dick, May 12, 1997, p. 16, deposited in the NASA History Office, NASA HQ, Washington, D.C. Quoted in Dick, Steven J. and James E. Strick. 2004. *The Living Universe: NASA and the Development of Astrobiology.* New Brunswick, NJ: Rutgers University Press.

Dick, Steven J. 2000. "Cosmotheology: Theological Implications of the New Universe." In *Many Worlds: The New Universe, Extraterrestrial Life and the Theological Implications*, ed. Steven J. Dick, 199–205. Radnor, PA: Templeton Foundation Press.

Dick, Steven J. and James E. Strick. 2004. *The Living Universe: NASA and the Development of Astrobiology*. New Brunswick, NJ: Rutgers University Press.

Gould, Stephen J. 1999. *Rock of Ages: Science and Religion in the Fullness of Life*. New York: The Ballantine Publishing Group.

Guthke, Karl S. 1985. "The Idea of Extraterrestrial Intelligence." *Harvard Library Bulletin* 33: 196–210.

McMullin, Ernan. 2000. "Life and Intelligence Far From Earth: Formulating Theological Issues." In *Many Worlds: The New Universe, Extraterrestrial Life and the Theological Implications*, ed. Steven J. Dick, 151–175. Radnor, PA: Templeton Foundation Press.

Norris, Pippa and Ronald Inglehart. 2011. *Sacred and Secular: Religion and Politics Worldwide, 2nd ed.* Cambridge: Cambridge University Press.

Peters, Ted. 1995. "Exo-Theology: Speculations on Extraterrestrial Life." In *The Gods Have Landed: New Religions from Other Worlds*, ed. James R. Lewis, 187–206. Albany, NY: State University of New York Press.

Peters, Ted. 2009. "Astrotheology and the ETI Myth." *Theology and Science* 7 (1): 3–30.

Peters, Ted. 2011. "The Implications of the Discovery of Extra-Terrestrial Life for Religion." *Phil. Trans. R. Soc.* A 369: 644–655.

Peters, Ted. 2013. "Would the Discovery of ETI Provoke a Religious Crisis?" In *Astrobiology, History, and Society: Life Beyond Earth and the Impact of Discovery*, ed. Douglas A. Vakoch. Heidelberg: Springer

Pew Forum on Religion and Public Life. 2005. *Public Divided on Origins of Life* http://www.pew forum.org/Politics-and-Elections/Public-Divided-on-Origins-of-Life.aspx.

Pew Forum on Religion and Public Life. 2006. *Spirit and Power: A 10-Country Survey of Pentecostals.* http://www.pewforum.org/Christian/Evangelical-Protestant-Churches/Spirit-and-Power.aspx.

Pew Forum on Religion and Public Life. 2011. *Global Christianity: A Report on the Size and Distribution of the World's Christian Population.* http://www.pewforum.org/Christian/Global-Christianity-worlds-christian-population.aspx.

Pontifical Academy of Sciences. 2009. *Study Week on Astrobiology.* http://www.vatican.va/roman_curia/pontifical_academies/acdscien/2009/booklet_astrobiology_17.

Sagan, Carl and Frank Drake. 1997. "The Search for Extraterrestrial Intelligence." *Scientific American* January. http://www.sciam.com/article.cfm?id=the-search-for-extraterre.

Stark, Rodney. 1999. "Secularization, R.I.P." *Sociology of Religion* 60 (3): 249–273.

Stenmark, Mikael. 2010. "Ways of Relating Science and Religion." In *Science and Religion*, ed. Peter Harrison, 278–295. Cambridge: Cambridge University Press.

Tarter, Jill Cornell. 2000. "SETI and the Religions of the Universe." In *Many Worlds: The New Universe, Extraterrestrial Life and the Theological Implications*, ed. Steven J. Dick, 143–149. Radnor, PA: Templeton Foundation Press.

Weigel, Margaret M. and Kathryn Coe. 2013. "Impact of Extraterrestrial Life Discovery for Third World Societies: Anthropological and Public Health Considerations." In *Astrobiology, History, and Society: Life Beyond earth and the Impact of Discovery*, ed. Douglas A. Vakoch. Heidelberg: Springer.

Wilkinson, David. 1997. *Alone in the Universe?* Crowborough, UK: Monarch.

Chapter 19
Would the Discovery of ETI Provoke a Religious Crisis?

Ted Peters

Abstract Noting how some prophets of crisis forecast that traditional religious traditions are vulnerable to challenge if not collapse upon confirmation of the existence of extraterrestrial intelligent beings, this chapter subjects this claim to examination. Citing findings from the *Peters ETI Religious Crisis Survey*, we find evidence that those who affirm religious belief have no difficulty affirming the existence of ETI and incorporating ETI into their respective worldviews. This applies to Orthodox Christians, Roman Catholics, mainline Protestants, Evangelical Protestants, Jews, Mormons, Buddhists, and to those who self-identify as non-religious. Surprisingly, the self-identified non-religious respondents are the only ones who fear a religious crisis precipitated by contact with extraterrestrials, a crisis expected to happen to others but not to themselves. Turning to the new field of *Astrotheology*, the question of de-centering both geocentrism and anthropocentrism is raised in light of the prospect of discovering intelligent celestial neighbors.

Like a thirsty desert traveler surveying the horizon in hope of sighting an oasis, we Earthlings are surveying the heavens in hope of sighting a new neighbor. Just as a speck of green would excite a desert hiker, even the hint of a biofriendly habitat excites the astrobiologist. In muted and dispassionate language, the scientists who identified two Earth-sized planets orbiting Kepler-20 added: "theoretical considerations imply that these planets are rocky, with a composition of iron and silicate. The outer planet could have developed a thick water vapour atmosphere" (Fressin et al. 2012, 195). Could this be an oasis of extraterrestrial life, maybe even intelligent life? Are all Earthlings thirsty for confirmation that we share our universe with other beings, with extraterrestrial intelligence (ETI)? Not according to some

T. Peters (✉)
Center for Theology and the Natural Sciences and Pacific Lutheran Theological Seminary, Berkeley, CA, USA
e-mail: tedfpeters@gmail.com

D. A. Vakoch (ed.), *Astrobiology, History, and Society*, Advances in Astrobiology and Biogeophysics, DOI: 10.1007/978-3-642-35983-5_19,
© Springer-Verlag Berlin Heidelberg 2013

observers who fear that Earth's religious believers will suffer a crisis of faith. There are prophets of crisis among us who predict that confirmation of ETI will shatter traditional religious beliefs. Can this be true? In what follows I will examine the claims by some seers that terrestrial religions are vulnerable to a crisis because their alleged geocentrism and anthropocentrism are out of date, rendering them unable to adapt to a large universe shared with other intelligent races. I will subject this claim to analysis. In doing so, I will summarize some of the findings of *The Peters ETI Religious Crisis Survey*, which confirms that—though some religious believers tend toward geocentrism and anthropocentrism—a significant majority welcome the prospect of intelligent neighbors in space. This survey conclusion can be buttressed by a brief review of the history of thought, noting an amazing openness on the part of theological minds to accept if not embrace intelligent creatures on other worlds.

19.1 The Prophets of Crisis

Why would we raise the question: would ETI provoke a religious crisis? Because there are social prophets among us who predict it. Some prophets-of-crisis tell us to worry about our religious traditions because of certain assumptions they make: they assume that religions born in a pre-modern age are geocentric and anthropocentric and vulnerable to disappointment at new scientific findings. Some of these crisis prophets rely upon survey data that seem to connect resistance to belief in ETI with conservative religious views. Let us look a bit more closely.

Physicist and astrobiologist Paul Davies is among those who worry about Earth's religions. Religions seem to be fragile, breakable. "The existence of extra-terrestrial intelligences would have a profound impact on religion, shattering completely the traditional perspective of God's special relationship with man" (Davies 1983, 71). Davies worries particularly about the Christian religion, because of the vulnerability of its Christology. "The difficulties are particularly acute for Christianity, which postulates that Jesus Christ was God incarnate whose mission was to provide salvation for man on Earth. The prospect of a host of 'alien Christs' systematically visiting every inhabited planet in the physical form of the local creatures has a rather absurd aspect" (Davies 1983, 71). Christians face a theological choice: either the single divine incarnation on Earth is efficacious for all sentient beings on all planets or, alternatively, each planet receives its own species-specific incarnation. The latter alternative, according to Davies, would be "absurd." This absurdity makes Christianity vulnerable to "shattering."

SETI's Jill Tarter supposes that the ETI who contact us will be more advanced than Earthlings. ETI will have either avoided religion altogether or have outgrown whatever organized religion they once had. This will precipitate a crisis on our home planet. "An information-rich message from these extraterrestrials will, over time, undermine our own world's religions" (Tarter 2000, 147).

Some crisis prophets believe they can support this worry with data. Based upon a 2005 survey by the Center for Survey Research and Analysis, George Pettinico infers that "devout Christians in America are more likely than other Americans to hold very traditional views of humanity as the single culmination of God's creation and Earth as the one divinely chosen place for this culmination of creation to live and prosper. For many conservative Christians, the universe—despite its size and scope—exists for the benefit of humanity alone" (Pettinico 2011, 104). The survey asks, *"Do you believe that there is life on other planets in the universe besides Earth?"* Note how the survey question employs a word with a religious overload, *believe*, rather than a term one might expect here such as *think* or *surmise*. Among those who stated they do not "believe" in ETI, 45% attend religious services weekly; 57% monthly; and 70% rarely. This climb in percentage from apparently more devout to less devout seems to support Pettinico's conclusion that religious devotion is correlated with doubt about the existence of ETI. How does Pettinicio know that the explanation lies in conservative Christian theology that espouses geocentricity, with Earth existing for "the benefit of humanity alone?" Did he test for this belief? No, at least not according to the data he cited for his conclusion.

William Sims Bainbridge references a 1981 study of University of Washington students and concludes: "a major factor discouraging people from supporting attempts to communicate with extraterrestrial intelligence was their religion—in this case evangelical Protestants, exemplified by the Born Again movement" (Bainbridge 2011, 119). He then analyzes in detail a 2001 survey conducted by the National Geographic Society along with the National Science Foundation. The survey contained many questions about UFOs and New Age beliefs. The survey did not zero in on religious devotion in relation to the existence of ETI. Bainbridge concludes, "although geocentrists were somewhat rare in the *Survey* 2001 dataset, they are probably more common in the general population… In the history of western civilization, the geocentric viewpoint reflected the religious belief that human beings were central to God's plan for the universe, and this prejudice retarded the development of science" (Bainbridge 2011, 137). In other words, we draw a conclusion about religion not based upon the data but rather based on what is not in the data. Curious.

Bainbridge does note fittingly a reliable study performed by Douglas Vakoch and Yuh-Shiow Lee in 2000 which suggested that anthropocentric and religious individuals are less likely to affirm the existence of ETI (Vakoch and Lee 2000). In this case, the conclusion corresponds to the evidence. One must grant that some survey material does exist which suggests an inverse correlation between traditional religious devotion and affirmation of the existence of extraterrestrial life.

In contrast, however, a survey conducted by Victoria Alexander suggested welcome acceptance on the part of religious believers to the prospect of someday meeting ETI (Alexander 2003). So, it appears we have conflicting data. Matthew Shindell, who surveyed a large number of surveys "doubts that public reaction to the announcement of the discovery of extraterrestrial life can be predicted" (Shindell 2007, 95). He adds, "I don't think these surveys necessarily ask the best questions" (Shindell 2007, 100). What all of this warrants, in my judgment, is further investigation to see whether prophecies of doom are realistic and just what might be transpiring in religious belief systems.

19.2 The Peters ETI Religious Crisis Survey

With the prophecies of crisis in mind, I constructed a survey instrument that zeroed in directly on the relationship between religious beliefs and the possibility of contact with ETI. Along with my Berkeley research assistant, Julie Louise Froehlig, I devised a survey: the *Peters ETI Religious Crisis Survey* (Peters and Froehlig 2008, Peters 2009, 2011). The *Peters ETI Religious Crisis Survey* received more than thirteen hundred responses worldwide from individuals in multiple religious traditions. It became clear that the vast majority of religious believers, regardless of religion, see no threat to their personal beliefs caused by potential contact with intelligent neighbors on other worlds. When we asked respondents to distinguish between their own personal beliefs and the beliefs of the religious tradition to which they adhere, anxiety rose just slightly that their religious leaders might face a challenge. Still, religious adherents overwhelmingly registered confidence that neither they as individuals nor their religious tradition would suffer anything like a collapse.

We then asked respondents to forecast what would happen with religion in general, with religious traditions other than their own. What is startling, is that respondents who self-identify as non-religious are far more fearful (or gleeful?) of a religious crisis than are religious believers.

In Fig. 19.1 (Question 3), note the consistency of the dominance of the third bar, "disagree/strongly disagree." The short bars are "strongly agree/agree" and "neither agree nor disagree." This shows how Roman Catholics, evangelical Protestants, mainline Protestants, Orthodox Christians, Mormons, Jews, and Buddhists right along with the non-religious fear no threat to their personal beliefs.

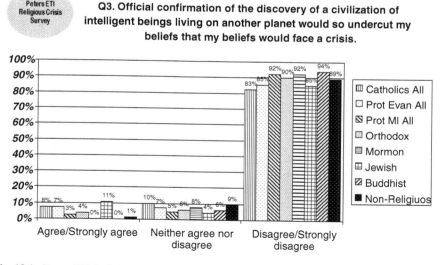

Fig. 19.1 Peters ETI Religious Crisis Survey, Question 3

The survey instrument permitted respondents to offer additional comments. Some reflected just what the crisis prophets would predict. One self-identified evangelical Protestant admits that confirmation of ETI would provoke a crisis but, curiously, not if ETI were a long distance away. "The actual interaction with advanced extraterrestrial life would create a crisis for my belief system. Finding some extraterrestrial life form in a far away planet would not." Another respondent, self-identified as a non-denominational Protestant, associates alien beings with Lucifer. "I believe that all extra-terrestrial beings are fallen angels (demons, if you will). And whatever traits they have can be traced back to Lucifer." These comments came from the Agree/Strongly agree pillars.

As already mentioned, the vast majority spoke positively about potential interactions with ETI. Among those who Disagree/Strongly disagree we hear a Muslim state, "Islamically we do believe that God created other planets similar to Earth." An evangelical Protestant sanguinely reports: "I can't see why the discovery of other life would affect our belief at all. God has made our world—and can make millions more, I suppose. Is Jesus the savior for all of them too, or did God do things very differently in those places? I'd be fascinated to find out, but not at all disturbed by it." A Roman Catholic foresees no crisis. "My religious viewpoints and practices would remain unchanged. The same God who made me is also capable of making extraterrestrials. His message of faith, hope and love of neighbor goes beyond the borders of the known universe."

Some comments suggest more than mere grudging adaptation. They put out a welcome sign for space visitors. A mainline Protestant hopes that ETI would actually strengthen faith. "If life were discovered elsewhere in the universe, I think my faith in the absolutely mysterious and grace-giving God would actually be more confirmed than it is now. I would have to believe that God is involved not just on our planet, but in the universe in its entirety." So also does an evangelical Protestant: "Traditional Christian understanding teaches there are other intelligences in the universe who are more powerful than humans. Discovery that this teaching is confirmed strengthens Christianity, not weakens it." Another mainline Protestant said, "I'd gladly share a pew with an alien."

Mormons already incorporate extraterrestrial entities in their theology; so it is not surprising to read one Mormon comment: "I believe that God, however he did it, created other worlds with other beings." Another Mormon foresees ETI as confirmation of beliefs already held. "First of all, my religion (LDS, Mormon) already believes in extra-terrestrials, and official doctrine and scripture even discusses names of extra-solar planets that are habitable for sentient life forms. If anything, an extra-terrestrial might even be looked at as confirmation of religious beliefs rather than something which would be though of as something to avoid."

Eastern Orthodox Christians seem split. One comments, "I strongly disbelieve in the possibility of other intelligent life other than on earth. I think Christ came to release us from our sins on this planet and that is exclusive.... But, if I were wrong and Christ can redeem other races, it would not change what I believe at all." In contrast, another Orthodox respondent comments, "I am constantly amazed at the ridiculous idea among some (a minority, I hope) in the scientific community that people of faith are ignorant and of low intelligence and therefore must have a 'God-of-the-gaps' theology (i.e., God is used to explain all things which we cannot at this time explain scientifically).... I am a person of DEEP faith with a genius-level IQ, two doctorates, a hope that we will encounter intelligent life from outside of Earth, and a hope (but not necessarily an expectation) that any extraterrestrial intelligent life humanity encounters will be benevolent."

Numerous respondents mentioned they had read the works of British classicist and theologian, C.S. Lewis. One response is typical of a dozen or so: "There's an essay by C. S. Lewis (unfortunately published under various titles in various anthologies) that strongly argues against any particular significance for Christianity of any conceivable type of extraterrestrial. Whether or not one agrees, it is clear at any rate that extraterrestrials would not be disruptive to all religious belief."

Our survey did not test directly for geocentrism or anthropocentrism. Yet, these two items appear repeatedly among the voluntary comments offered by respondents. A Buddhist takes a stand against geocentrism: "I believe that anything is possible including life on other worlds. To think that in the infinity of the universe that we are the only intelligent life form in existence is ludicrous. I would only hope those beings would exhibit more wisdom than humans have in how they relate to their world and fellow beings." Another respondent self-identified as non-religious says almost the same thing. "I believe that we are not unique in the universe (it would be sheer hubris on our part, not that we are not a completely narcissistic species) but the universe is so large that contact among advanced civilizations is limited to neighboring planetary systems; and we may not have very advanced neighbors."

Opposition to geocentrism and anthropocentrism is common to respondents regardless of tradition. A Mormon exclaims: "Our universe is huge. So astonishingly huge that I find it absurd to think we are alone in this universe as a sentient life form." A Roman Catholic trumpets, "The world is too vast and wonderful and God's power is so limitless, that there must be more than little old us." One mainline Protestant explicitly rejects anthropocentrism. "God is God of all creation and all that is within it. The only way this should be a religious problem is if the true (though unstated) center of our worship is humankind." A Muslim similarly chastises anthropocentrism: "Only arrogance and pride would make one think that Allah made this vast universe only for us to observe."

One evangelical Protestant thinks out loud, so to speak, about Christology and Soteriology. "From an evangelical Christian perspective, the Word of God was written for us on Earth to reveal the creator. We were created to bring glory to God. Why would we repudiate the idea that God may have created other civilizations to bring him glory in the same way? Christ as our Savior may be the method he chose to redeem us on Earth, but he could have used similar methods in other galaxies if he desired." A Roman Catholic follows the same thought experiment. "I believe that Christ became incarnate (human) in order to redeem humanity and atone for the original sin of Adam and Eve. Could there be a world of so called "extraterrestrial's? Maybe. It doesn't change what Christ did."

In parallel fashion, a Buddhist thinks out loud about the path to enlightenment. "As a Buddhist, it is clear that ALL sentient beings are subject to birth, old and death and are, therefore, impermanent, subject to various forms of suffering and have no separate self. E.T's would be, essentially, no different from other sentient beings i.e. they would have Buddha Nature and would also be subject to karmic consequences of their actions. We might or might not be able to learn from them."

Finally, from an evangelical Protestant we read: "I don't think they are out there. But if they are, that's cool."

When we turn away from one's own personal beliefs and ask about the beliefs of the respondent's religious tradition (Fig. 19.2, Question 4), we notice a slight shift. The Disagree/Disagree strongly pillars still rise high, to be sure; yet not quite as high. Might this indicate that for some religious individuals who welcome confirmation of ETI a worry about their own religious tradition is evoked?

Some of the voluntary comments are illuminative. One Roman Catholic is not worried about his or her own faith; but the beliefs of the Catholic Church are in jeopardy because of its alleged anthropocentrism. "The foundations of my religion (Catholic) and many others may be shaken by such a discovery because most human religions view human beings as special or privileged beings on the earth and in the cosmos. In Christian traditions humans seem to hold special favor with God. However, I see no reason God could not (or has not) create(d) other beings. It does not imply a lessening of God's love for us. In fact, come to think of it, it would be good for us to discover other beings–especially ones equally or more intelligent than ourselves–because it might knock down human arrogance towards other species right here on earth." We find the same denominational anxiety in an evangelical Protestant. "I think the religious tradition with which I identify (Protestant Evangelical) is not prepared for the day we do make contact, but we need to start thinking this out and become prepared."

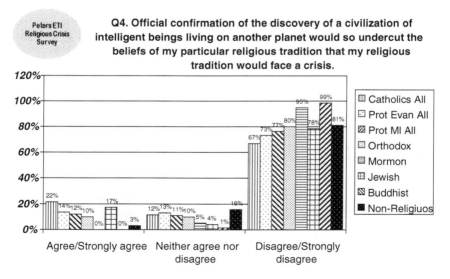

Fig. 19.2 Peters ETI Religious Crisis Survey, Question 4

A few Roman Catholic laypersons seem to be suspicious of their church's leadership, worrying that the hierarchy might find contact with ETI difficult to accept. This is curious, because the Holy See through the Jesuits sponsors the Vatican Observatory which, among other tasks, actively searches for extraterrestrial life (Ariel 2008; Coyne 2000).

Next, we turn away from one's own personal beliefs and the beliefs of one's own tradition; we ask about forecasts for religions other than one's own. The *Peters ETI Religious Crisis Survey* asked respondents to forecast what will happen to the world's religions, those holding beliefs other than one's own. Here, something startling is revealed. See Fig. 19.3 (Question 5). Note how those self-identifying as non-religious are the ones who forecast a crisis in the world's religions. To say it in the first person: "my non-religious beliefs will not suffer a crisis, but other religious believers will have a problem."

How should we interpret this graph in Fig. 19.3? First, we can distinguish between one's own religious beliefs from the beliefs of others who differ. One Buddhist respondent expects easy acceptance of ETI by Asian religions but difficulty for the Abrahamic traditions. "Lumping together all the world's religions is a conceptual error. The religions of the book (the Abrahamic traditions) would have a very different set of reactions than the Asian traditions." Surprisingly, a Roman Catholic agrees. "I think Buddhism and Hinduism would be better equipped to face any encounter with other civilizations. The three monotheistic traditions, however, would enter into a serious crisis."

Second, and more dramatically, non-religious respondents are more likely to prophesy a crisis impending for those who are religious. The Agree/Strongly agree towers indicate this. According to the respondent comments, those who self-identify

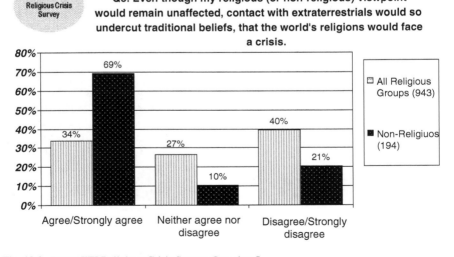

Fig. 19.3 Peters ETI Religious Crisis Survey, Question 5

as non-religious or atheist tend to welcome confirmation of ETI on the grounds that they are open-minded. Presumably, religious people are not open-minded. "I hope we do find life beyond our planet. I have an open mind as to what shape it may take and what influence it will have." An atheist comments, "Discovery of ET would not affect my personal belief system because I am a stone atheist. I do search for ET and think we will come across some one day." Another hopes that ETI will mark a victory over religious anachronism. Contact with ETI is "bound to happen sooner or later, whether they're more advanced than us is uncertain, but it will certainly get rid of the notion that we're God's special little creatures made in his image…. man I can't wait :)."

What might all this signify? (Bertka 2013) It might signify that the prophecies of crisis are primarily the construction of non-religious persons who are making predictions about what will happen to religious persons. A decade prior to this survey Steven Dick made this observation: "In general, for Christians as well as for other religions, indigenous theologians see little problem, while those external to religion proclaim the fatal impact of extraterrestrials on Earth-bound theologies" (Dick 1998, 247). The *Peters ETI Religious Crisis Survey* tends to disconfirm the prophets of crisis and confirm Dick's judgment.

This is a judgment already filed by David Wilkinson, a hybrid physicist and theologian at Durham in the UK. He asks: "would the discovery of life elsewhere in the Universe so contradict the central beliefs of Christianity that it would bring it crashing back to the grave?" He answers: "for the vast majority of the Christian church, the existence of extraterrestrial intelligence is not a big deal" (Wilkinson 1997, 116). Or, in the words of Notre Dame's Michael J. Crowe, "It is sometimes suggested that the discovery of extraterrestrial life would cause great consternation in religious denominations. The reality is that some denominations would view such a discovery not as a disruption of their beliefs, but rather as a confirmation" (Crowe 2008, 328–329).

In summary, such survey evidence requires us to acknowledge that religious believers themselves do not fear that contact with ETI will undercut their beliefs or precipitate a religious crisis. It suggests that what passes for prophecies of crisis may be the product of what non-religious people say about religious people.

19.3 The Question of Geocentrism and Anthropocentrism

Recall what one Buddhist said above: "To think that in the infinity of the universe that we are the only intelligent life form in existence is ludicrous. I would only hope those beings would exhibit more wisdom than humans have in how they relate to their world and fellow beings." This anonymous Buddhist seems to assume that he or she has an opponent, someone who would defend geocentrism and perhaps anthropocentrism. Is this a serious issue? Let us turn to this question.

Even without a telescope, Nicholas Copernicus (1473–1543) of Krakow redesigned the universe for human consumption. He switched the Sun and the Earth. He re-oriented the planets, now orbiting the Sun. Earth lost all her satellites, save the Moon. He dismantled the vault of Earth's heaven and scattered myriads of stars into the vastness of space. His 1543 work on revolutions—*De revolutionibus orbium coelestium* (*On the Revolutions of the Celestial Orbs*) became itself a revolution in science. What about culture? Whether accurately or inaccurately, we think of the Copernican revolution as the one that de-centers our planet, de-centers the human race, and compels us to adopt a new humility in the face of our immense cosmos. Johannes Kepler set the theme for what some believe to be the cultural form of the Copernican Revolution. "If there are globes in the heavens similar to our earth, do we vie with them over who occupies a better portion of the universe? For if their globes are nobler, we are not the noblest of rational creatures. Then how can all things be for man's sake? How can we be the masters of God's handiwork?" (Dick 1998, 246). Or, more tersely said by Mark Twain in 1875, "How insignificant we are, with our pigmy little world!" (Crowe 2008, 463).

The standard story we today tell ourselves is that Copernicus along with Kepler and Galileo fomented a revolution not merely in astronomy but also in culture (Crowe and Dowd 2013). Allegedly, pre-Copernican Europeans believed that the Earth took central place in the universe, with the Sun and stars circling Earth in homage and respect. Also allegedly, religious beliefs had so baptized this geocentrism with rigid dogmas that theologians fought tooth and nail to refute heliocentrism. Science has de-centered the Earth and, thereby, marginalized the importance of our planet and our race. So we assume.

Andrew Dickson White provides our working definition of *geocentrism*. The "geocentric doctrine" is "the doctrine that the earth is the centre, and that the sun and planets revolve about it" (White 1896, 1:115). Nobel prize winning biologist Christian de Duve presumes the standard de-centering story and draws out its significance for culture and religion. "Thus, if life exists elsewhere in the universe, the likelihood that it may produce intelligent forms, some perhaps more advanced than the human form, is far from negligible.... this possibility... threatens one of our most cherished beliefs, a cornerstone not only of many religions, but also of humanism...the conviction that humankind occupies a central position in a universe somehow constructed around it, if not for it" (Dick 2000, 8). Accordingly, modern scientists should feel sorry for out-of-date and dogmatically stubborn religious atavists.

Not only scientists, but prophets of crisis within the theological camp even tell the standard de-centering story. They assume their own religious tradition is geocentric and vulnerable to a devastating shock. Cynthia Crysdale, for example, contends that contact with extraterrestrial intelligence would not change our notion of God but it would demote the place of humanity. "We have faced this dilemma before: Copernicus and Galileo dethroned the human. Darwin made us mere coincidences of evolution. Slowly the human race is discovering that we're not the center of the universe, but that both space and time are so fast that we are mere blips on the screen. This... won't go down lightly" (Crysdale 2007, 201). If Crysdale provides an example of what the *hoi polloi* think they are hearing from their religious leaders, then this helps explain why so many survey respondents seemed to rebel against the standard story of de-centering.

But, we ask: does this standard story accurately reflect history? No, not precisely. More nuances are present than the simple words 'geocentrism' and 'anthropocentrism' imply. Let us look into the matter in a bit more detail.

The prevailing pre-Copernican cosmography was like a vegetable soup, a mixture of Greek cosmology with justifying biblical passages. The Platonic Principle of Plenitude along with the Aristotelian principle of centering were simmered together with the Gnostic and Ptolemaic model of concentric spheres to produce the worldview within which Christians and other Europeans placed the human drama. Imagine globes within globes like Russian boxes (Crowe and Dowd 2013). No pre-Copernican intellectual believed the Earth was flat; everyone assumed a round planet. Above the Earth's curved surface we could see rotating transparent spheres at differing levels–rotating heavens energized by angels–each carrying one or more of the heavenly bodies. The heaven nearest the Earth carries the Moon. Above that, Mercury. Next, Venus. Next the Sun. Mars, Jupiter, and Saturn are borne by the next three heavenly spheres. The eighth heaven holds the stars, the fixed stars. The ninth heaven is the *primum mobile*; and the tenth, enclosing all that is below, is the immovable Empyrean, the boundary between our circumscribed universe and the mysterious void beyond. The music of the spheres would rise up to the Empyrean, where the enthroned Trinity could enjoy it in full divine majesty.

Having dined on this basic Hellenistic cosmology, Christians spiced it with Bible passages that seemed to fit. The opening chapter of Genesis describes the sky as a firmament, perhaps a roof from which to hang the Sun, Moon, and other heavenly bodies. The Psalms describe God as enthroned above the stars, the stars being the work of God's fingers (Psalm 8). God made the Sun stand still at the battle of Gibeon (Joshua 10:12). A mystical experience took St. Paul to the third heaven (2 Corinthians 12:2). Romantics of our own day find their love in the seventh heaven.

Would such a geocentrism all by itself make Earthlings feel good about themselves? After all, the heavenly spheres and even the enthroned Trinity look to Earth for their centering! But, geocentrism did not produce anthropocentrism. Those of us who live on the surface of the Earth look downward toward a more centered center, namely, hell. The center of the Earth is the location of hell. We on the surface of the terrestrial ball lie somewhere between hell below and various heavens above. As members of the human race, we are not in fact the center of this geocentric cosmology.

Historian of ideas Arthur Lovejoy tries to correct the widespread misunderstanding of pre-Copernican cosmology. The standard story tells us that pre-Copernicans thought of themselves as at the center and as centrally important; "but the actual tendency of the geocentric system was, for the medieval mind, precisely the opposite. For the centre of the world was not a position of honor; it was rather the place farthest removed from the Empyrean, the bottom of the creation, to which its dregs and baser elements sank. The actual centre, indeed, was Hell; in the spatial sense the medieval world was literally diabolocentric" (Lovejoy 1936, 101–102). Pre-Copernican believers sought to escape the center and soar to the heavens.

The centeredness of Earth in relation to the celestial bodies did not count toward human pride. Geocentrism by itself did not lead to anthropocentrism. A variant of anthropocentrism did reign, to be sure. But, this human self-importance is lodged in our capacity for reason, our intelligence, not in our geocentric location. Our human intellects belong to the *imago Dei*—the image of God in us—medieval Christians thought. So, as the new breed of modern scientists proceeded to gain increased knowledge of the natural world, they were celebrated as reading the mind of God. Lovejoy adds to our historical precision here. "It was not the position of our planet in space, but the fact that it alone was supposed to have an indigenous population of rational beings whose final destiny was not yet settled, that gave it is unique status in the world and a unique share in the attention of Heaven" (Lovejoy 1936, 102–103). If Lovejoy is correct, then the capacity for reason on the part of Earthlings becomes the focal concern as we consider the possibility of sharing our universe with other intelligent beings, maybe even beings higher on the intelligence ladder than we.

The nineteenth century witnessed advances in both astronomy as well as cultural and theological discussions of astronomy's implications. Camille Flammarion (1842–1925) simply assumed that traditional religions were compatible with belief in ETI. In his pamphlet of 1862, *La pluralité des mondes habités*, he wrote, "All peoples, most notably the Indians, Chinese, and Arabs, have conserved down to our own day theogonic traditions which recognize among ancient dogmas that of the plurality of human inhabitations in the worlds which shine above our heads... either in a religious context, as concerning the transmigration of souls and their future state, or in an astronomical context, as concerning simply the inhabitability of heavenly bodies" (Crowe 2008, 408; Crowe and Dowd 2013). Flammarion de-centers, challenging both geocentrism and anthropocentrism. "Let us assert that no preeminence has been bestowed on Earth in the solar system to make it the only inhabited world" (Crowe 2008 418).

Poet Alfred Lord Tennyson (1809–1892) accepts the loss of geocenrism while holding a grasp on anthropocentrism. On the one hand, he feels puny in the face of the vastness of creation. "When I think of the immensity of the universe, I am filled with the sense of my own utter insignificance, and am ready to exclaim with David [Psalm 8]: 'What is man that Thou art mindful of him?'" Yet, on the other hand, Tennyson could not completely surrender his regard for the human race. "A certain amount of anthropomorphism must, however, necessarily enter into our conception of God, because, though there may be infinitely higher beings than ourselves in the worlds beyond ours, yet to our conception man is the highest form of being" (Crowe 2008, 449). Despite the de-centering of Earth and the admission that some extraterrestrials might beat us in a quiz show, this did not result in a refusal or even a resistance to believe the facts regarding heliocentrism or ETI.

Rabbi Norman Lamm developed what he called "a Jewish exotheology," in which he argued that the Copernican Revolution required acceptance. So also must we accept sharing our universe with ETI. He added that the "nonsingularity of humanity did not mean insignificance" (Dick 1998, 251).

Today, some within the theological community positively extend the principle of de-centering. New Testament historian N.T. Wright, for example, states as
emphatically as euphemistically that *"We are not the center of the universe.* God
is not circling around us. We are circling around him"* (Wright 2009, 23). And,
also in theological circles, de-centering provides the leverage for releasing the
grip of Western culture on the indigenous cultures and religions of the colonized
world. The movement we know as *Postcolonialism* relies upon occasions "when
Christianity's central position is replaced with multiple centers…. These are reve-
latory moments when religions are judged not by the standard set by the Christian
Bible" (Sugirtharajah 2012, 123). A momentum is already in place that carries theological thinking in the direction of welcoming another de-centering precipitated
by sharing our celestial neighborhood with extraterrestrial friends.

We are now working in the domain where astrobiological science overlaps with
culture and society. Michael Michaud reminds us to look beyond science when
speculating about social impact. "Many speculations about the societal implications
of contact are outside science: they tread on other sensitivities" (Michaud 2007, 6).
Steven Dick thinks the standard de-centering question is still culturally significant.
"As Darwinism placed humanity in its terrestrial context, so exobiology will place
humanity in a cosmic context. That context—a universe full of microbial life, full
of intelligent life, or devoid of life except for us—may to a large extent determine
both humanity's present worldview and its far future" (Dick and Strick 2005, 9).

SETI Institute's Douglas Vakoch looks forward to the religious value of contact because it will yield for us "a more humble, more realistic, and yet paradox-
ically more complete and more extensive understanding of our own place in the
universe" (Vakoch 1999, 21). This suggests we might become intentional about
interdisciplinary cooperation, even cooperation between science and theology. A
Boston University systematic theologian, John Hart, foresees that "the collaboration of scientists, ethicists, and theologians will enhance both reflection on Contact,
and terrestrial-extraterrestrial interaction when Contact occurs" (Hart 2010, 390).
With such thoughts in mind, Robert John Russell, physicist and theologian, forecasts a healthy mutual growth for both terrestrials and extraterrestrials through our
future interaction. "I predict that, against those voices who say life in the universe
is meaningless, or that human life is absurd, we'll be able to recognize the common
journey of life everywhere, and we will finally be able to understand our place in
the Universe. Welcome home humanity! Welcome home, ET?" (Russell 2000, 66).

19.4 Conclusion

We are ready for serious theological speculation on astrobiology and the search for
intelligent life off-Earth. Formerly, I called this *Exotheology*; but now I believe the
appropriate term should be *Astrotheology* (Peters 2008, 2009). Today's astrothe-
ologian can safely forecast that any new knowledge regarding the immensity
and grandeur of our barely fathomable universe will influence if not determine

our expanding worldview. No doubt confirmation of contact with extraterrestrial neighbors will expand our sense of the community of intelligence. But, it would be misleading to forecast that our already de-centered religious traditions would suffer from a second loss of geocentrism. With regard to our sense of self-importance as intelligent human beings, we will simply have to wait and see what happens when we engage the aliens in conversation. Such inter-planetary conversation will only energize the emerging new breed of *astrotheologians* (Hart 2010; O'Meara 1999, 2012; Russell 2000).

References

Alexander, Victoria. 2003. "Extraterrestrial Life and Religion." In *UFO Religions*, ed. James R. Lewis, 359–370. Amherst, NY: Prometheus Books.

Ariel, David. 2008. "Vatican: It's OK to Believe in Aliens." Associated Press, May 13, 2008. http://news.yahoo.com/s/ap/20080513/ap_on_re_eu/vatican_aliens.

Bainbridge, William Sims. 2011. "Cultural Beliefs about Extraterrestrials: A Questionnaire Study." In *Civilizations Beyond Earth: Extraterrestrial Life and Society*, ed. Douglas A. Vakoch and Albert A. Harrison, 118–140. New York: Berghahn Books.

Bertka, Constance. 2013. "Christianity's Response to the Discovery of Extraterrestrial Intelligent Life: Insights from Science and Religion and the Sociology of Religion." In *Astrobiology, History, and Society: Life Beyond Earth and the Impact of Discovery*, ed. Douglas A. Vakoch, Heidelberg: Springer.

Coyne, George V., S.J. 2000. "The Evolution of Intelligent Life on Earth and Possibly Elsewhere: Reflections from a Religious Tradition." In *Many Worlds: The New Universe, Extraterrestrial Life and the Theological Implications*, ed. Steven Dick, 177–188. Philadelphia and London: Templeton Foundation Press.

Crowe, Michael J. 2008. *The Extraterrestrial Life Debate from Antiquity to 1915: A Source Book*. Notre Dame, IN: University of Notre Dame.

Crowe, Michael J. and Matthew F. Dowd. 2013. "The Extraterrestrial Life Debate from Antiquity to 1900." In *Astrobiology, History, and Society: Life Beyond Earth and the Impact of Discovery*, ed. Douglas A. Vakoch, Heidelberg: Springer.

Crysdale, Cynthia. 2007. "God and Astrobiology." in *Workshop Report: Philosophical, Ethical, and Theological Implications of Astrobiology*, eds. Connie Bertka, Nancy Roth, and Matthew Shindell, 196–208. Washington, DC: American Association for the Advancement of Science.

Davies, Paul. 1983. *God and the New Physics*. New York: Simon and Schuster.

Dick, Steven J. 1998. *Life on Other Worlds: The 20th Century Extraterrestrial Life Debate*. Cambridge, UK: Cambridge University Press.

Dick, Steven J., ed. 2000. *Many Worlds: The New Universe, Extraterrestrial Life and the Theological Implications*. Philadelphia and London: Templeton Foundation Press.

Dick, Steven J., and James E. Strick. 2005. *The Living Universe: NASA and the Development of Astrobiology*. New Brunswick, NJ: Rutgers University Press.

Fressin, Francois, et al. 2012. "Two Earth-Sized Planets Orbiting Kepler-20." *Nature* 482(7384): 195–198.

Hart, John. 2010. "Cosmic Commons: Contact and Community." *Theology and Science* 8(4): 371–392.

Lovejoy, Arthur O. 1936. *The Great Chain of Being*. New York: Harper.

Michaud, Michael A.G. 2007. *Contact with Alien Civilizations: Our Hopes and Fears About Encountering Extraterrestrials*. New York: Springer, Copernicus Books.

O'Meara, Thomas F., O.P. 1999. "Christian Theology and Extraterrestrial Life," *Theological Studies* 60(1): 10–30.

O'Meara, Thomas F., O.P. 2012. *Vast Universe: Extraterrestrials and Christian Revelation.* Collegeville, MN: Liturgical Press.

Peters, Ted. 2008. *The Evolution of Terrestrial and Extraterrestrial Life.* Goshen, IN: Pandora Press.

Peters, Ted, and Julie Louise Froehlig. 2008. "Peters ETI Religious Crisis Survey" http://www. counterbalance.org/search/search.php?query=Peters+Religious+Crisis+Survey&search=1.

Peters, Ted. 2009. "Astrotheology and the ETI Myth." *Theology and Science*, 7(1): 3–30.

Peters, Ted, 2011. "The Implications of the Discovery of Extra-terrestrial Life for Religion." The Royal Society, *Philosophical Transactions A*, 369 (1936): 644–655. http://rsta.royalsocietypu blishing.org/content/369/1936.toc.

Pettinico, George. 2011. "American Attitudes about Life Beyond Earth: Beliefs, Concerns, and the Role of Education and Religion in Shaping Public Perceptions." In. *Civilizations Beyond Earth: Extraterrestrial Life and Society,* eds. Douglas A. Vakoch and Albert A. Harrison. 102–116. New York:Berghahn Books.

Russell, Robert John. 2000. "What are Extraterrestrials Really Like?" In *God for the 21st Century,* ed. Russell Stannard. 65–67. Philadelphia and London: Templeton Foundation Press.

Shindell, Matthew. 2007. "The Public Response: Reviewing Public Policy Survey Data on Science, Religion, and Astrobiology." In *Workshop report: Philosophical, Ethical, and Theological Implications of Astrobiology*, eds. Connie Bertka, Nancy Roth, and Matthew Shindell, 95–104. Wasnington DC: American Association for the Advancement of Science.

Sugirtharajah, R.S. 2012. *Exploring Postcolonial Biblical Criticism.* Oxford: Wiley-Blackwell.

Tarter, Jill Cornell. 2000. "SETI and the Religions of the Universe." In *Many Worlds: The New Universe, Extraterrestrial Life and the Theological Implications*, ed. Steven Dick, 143–149. Philadelphia and London: Templeton Foundation Press.

Vakoch, Douglas A. 1999. "Framing Spiritual Principles for Interstellar Communications: Celestial Waves." *Science and Spirit* 10: 21–22.

Vakoch, Douglas A., and Yuh-Shiow Lee. 2000. "Reactions to Receipt of a Message from Extraterrestrial Intelligence: A Cross-Cultural Empirical Study." *Acta Astronautica* 46: 737–744.

Vakoch, Douglas A., and Albert A. Harrison. eds. 2011. *Civilizations Beyond Earth: Extraterrestrial Life and Society.* New York: Berghahn Books.

White, Andrew Dickson. 1896. *A History of the Warfare of Science with Theology in Christendom.* 2 Volumes: New York: Dover.

Wilkinson, David. 1997. *Alone in the Universe?* Crowborough UK: Monarch.

Wright, N.T. 2009. *Justification: God's Plan and Paul's Vision.* Downers Grove, IL: IVP.

About the Authors

Constance M. Bertka, M.T.S., Ph.D., is an independent scholar and consultant for Science and Society Resources. She is the Co-Chair of the Broader Social Impacts Committee of the National Museum of Natural History's Hall of Human Origins, and she teaches on contemporary issues in science and religion at Wesley Theological Seminary. From 2002 through 2008 she directed the Program of Dialogue on Science, Ethics, and Religion at the American Association for the Advancement of Science. Prior to that she was a Senior Research Associate at the Carnegie Institution of Washington's Geophysical Laboratory where she also served as the Program Director of the Deep Carbon Observatory from 2009–2011. In addition to her research in planetary sciences, Dr. Bertka has had a long-term scholarly and pragmatic interest in the relationships between science and religion and their influence on public understanding of science. She is the editor of *Exploring the Origin, Extent, and Future of Life*: *Philosophical, Ethical, and Theological Perspectives*.

Danielle Briot, Ph.D., is an astronomer at the Paris-Meudon Observatory. She first specialized in the spectroscopy of Be stars, i.e., hot stars with emission lines originating in a circumstellar envelope. She has also studied binary stars. Since soon after the discovery of the first extrasolar planets, she has been studying these celestial objects. She investigates the search for life in universe, researching indicators of chlorophyll in Earthshine spectra. Dr. Briot is also working on the history of astrobiology, and she is specially interested by the discovery of unknown pioneers in this science. Because she thinks that the diffusion of science is important and she wants to share her enthusiasm about astronomy, a part of her time is dedicated to conferences, talks, and articles in popular astronomy magazines. Moreover, for over a decade, she has gone into prisons to speak about astronomy by giving courses and lectures for inmates.

Klara A. Capova is a doctoral candidate in sociocultural anthropology at Durham University. She is specializing in the anthropology of science and is currently working on her ethnographic dissertation on the scientific search for other life in

D. A. Vakoch (ed.), *Astrobiology, History, and Society*, Advances in Astrobiology and Biogeophysics, DOI: 10.1007/978-3-642-35983-5,
© Springer-Verlag Berlin Heidelberg 2013

the universe and current concepts of other life as understood, perceived, and interpreted by the scientific community and popular culture. Ms. Capova completed her masters degree in general anthropology at the Charles University in Prague, Czech Republic, in 2008. She has studied anthropological aspects of interstellar messages and worked on content analysis of the Pioneer Plaque and Voyager Interstellar Record. Her recent research interests include sociocultural dimensions of space exploration, attempts to detect life beyond Earth, and scientific entrepreneurship.

Kathryn Coe, Ph.D., is Professor of Public Health in the Department of Public Health at Indiana University-Purdue University Indianapolis (IUPUI). She is the author of many papers on health interventions and cultural tailoring, as well as *The Ancestress Hypothesis: Visual Art as Adaptation.* Her research examines the evolution of social behavior and culture, as well as chronic and infectious diseases. Dr. Coe has over thirty years of experience conducting health research among African Americans, Hispanics/Latinos, and American Indians using a community-based participatory methodology.

Michael J. Crowe, Ph.D., is the Rev. John J. Cavanaugh, C.S.C. Professor Emeritus in the Program of Liberal Studies and Graduate Program in History and Philosophy of Science at the University of Notre Dame. Among his nine books, the two most relevant to this volume are *The Extraterrestrial Life Debate 1750–1900: The Idea of a Plurality of Worlds from Kant to Lowell,* which has appeared in a three volume Japanese translation, and his *The Extraterrestrial Life Debate, Antiquity to 1915: A Source Book.* In 2010, Crowe received the biennial LeRoy E. Doggett Prize from the American Astronomical Society for his lifetime contributions to the history of astronomy. He has developed a course titled "The Extraterrestrial Life Debate: A Historical Perspective." He is the founder of Notre Dame's Graduate Program in History and Philosophy of Science and of its Biennial History of Astronomy conferences.

Dennis Danielson, Ph.D., Professor of English at the University of British Columbia in Vancouver, is an intellectual historian with interests in the literature and cultural meaning of science, past and present. His anthology *The Book of the Cosmos: Imagining the Universe from Heraclitus to Hawking* was named to Amazon.com's "Editor's Choice" top 10 science books for the year 2000. He is the author of a biography of the man who discovered Copernicus, *The First Copernican: Georg Joachim Rheticus and the Rise of the Copernican Revolution,* and he has written articles for *Nature, Journal for the History of Astronomy, American Journal of Physics, American Scientist,* and *Spektrum der Wissenschaft.* His work explores the insights of both science and the humanities, especially as these are enhanced and engaged by the enduring legacy of Copernicus's attempt to re-imagine the universe.

Kathryn Denning, Ph.D., is Associate Professor in the Department of Anthropology at York University. Her research examines scholarly and popular ideas about Others, their relationships to us, and how we can know them. The Others she studies include the ancient (in archaeology), the animal (in zoos), and the alien (in the Search for Extraterrestrial Intelligence, or SETI). In SETI, Dr. Denning studies scientists' reasoning processes, the technology and sites used to search the sky for signals, and ideas about how one might communicate with a radically different intelligence. She is a member of the International Academy of Astronautics' SETI Committee and has research projects with the NASA Astrobiology Institute.

Steven J. Dick, Ph.D., is the Charles A. Lindbergh Chair in Aerospace History at the National Air and Space Museum. He served as the NASA Chief Historian and Director of the NASA History Office from 2003 to 2009. Among his books are *Plurality of Worlds*: *The Origins of the Extraterrestrial Life Debate from Democritus to Kant*; *The Biological Universe: The Twentieth Century Extraterrestrial Life Debate and the Limits of Science*; *Life on Other Worlds*; *The Living Universe*: *NASA and the Development of Astrobiology*; and *Many Worlds: The New Universe, Extraterrestrial Life, and the Theological Implications*. In 2006, Dr. Dick received the LeRoy E. Doggett Prize from the American Astronomical Society for a career that has significantly influenced the field of the history of astronomy. In 2009, minor planet 6544 Stevendick was named in his honor.

Matthew F. Dowd, Ph.D., received his doctorate in the history and philosophy of science from the University of Notre Dame in 2004. His dissertation examined the astronomical and computistical works of Robert Grosseteste, an English scholar of the twelfth and thirteenth centuries. He has since worked as an editor at the University of Notre Dame Press. He has co-taught with Michael Crowe a course titled "The Extraterrestrial Life Debate: A Historical Perspective." He has been instrumental in organizing Notre Dame's Biennial History of Astronomy conferences, and in 2011 received the Adler-Mansfield Prize for contributions to the history of astronomy for his work on the conferences.

Aaron L. Gronstal, Ph.D., is an author, illustrator, and independent researcher in the field of astrobiology. He is a contributing editor and illustrator for *Astrobiology* magazine (www.astrobio.net), as well as the artist and co-author behind the Astrobio Comics series, including *The Amazing Adventures of AstrobioBot* and the upcoming *Abominable Snowaliens of Europa*. Dr. Gronstal has worked on numerous projects with NASA, and he is currently acting as author and artist for the graphic history series Astrobiology: *The Story of our Search for Life in the Universe,* produced by the NASA Astrobiology Program. His research is in the field of geomicrobiology, focusing on life in extreme environments and the microbial biosphere of the deep subsurface. His recent work includes studies as part of the International Continental Drilling Program (ICDP)–US Geological Survey (USGS) deep drilling of the Chesapeake Bay Impact Structure.

Chris Impey, Ph.D., is a University Distinguished Professor and Deputy Head of the Department of Astronomy at the University of Arizona. His research is on observational cosmology, gravitational lensing, and the evolution of galaxies. He has over 170 refereed publications and 60 conference proceedings, and his work has been supported by $20 million in grants from NASA and the NSF. Dr. Impey is past Vice President of the American Astronomical Society and has been NSF Distinguished Teaching Scholar, a Phi Beta Kappa Visiting Scholar, and the Carnegie Council on Teaching's Arizona Professor of the Year. He has written thirty popular articles on cosmology and astrobiology, two introductory textbooks, and three popular books: *The Living Cosmos*, *How It Ends*, and *How It Began*. He was a panel co-chair for the Astronomy Decadal Survey of the National Academy of Sciences. In 2009, he was elected a Fellow of the American Association for the Advancement of Science.

Morris Jones, Ph.D., has qualifications in science and journalism. He has worked as a freelance journalist and advisor on scientific matters to the media for more than two decades. Dr. Jones is the author of five books on space exploration: *Out of This World*, *The Adventure of Mars*, *The New Moon Race*, *When Men Walked on the Moon*, and the children's book *Is There Life Beyond Earth?* He has written more than 100 articles for the popular media on spaceflight. In addition to his interests in astrobiology and SETI, Dr. Jones is also known for his investigations into the Chinese astronaut program and spaceflight activity in Asia. He has appeared in broadcast sources such as Voice of America, Bloomberg, Al Jazeera (English), ABC Australia, Sky News Australia, CCTV China, Radio Television Hong Kong, and Phoenix TV Hong Kong.

Ian Lowrie received his B.A. in anthropology from Reed College in Portland, Oregon, and he is currently a graduate student in the Department of Anthropology at Rice University, pursuing a Ph.D. in sociocultural anthropology. Working at the borders between continental philosophy, social theory, and the study of science and technology, his research has touched on topics such as the development of adaptive optics and US-Soviet collaboration on the Search for Extraterrestrial Intelligence (SETI). Also interested in the anthropology of religion, he is the book review editor for the journal *Magic, Ritual, and Witchcraft*. His dissertation fieldwork is a multi-sited ethnographic investigation of contemporary discourses on and of intelligence in the United States and Russia. His research looks toward the concrete scientific communities where models of intelligence are built and tested, and it examines the circulation of these models within both the scientific and public spheres. Ultimately, the aim of this project is to understand better the articulation of notions of intelligence with our increasingly post-human understanding of life in the evaluation of pressing ethical, scientific, and political problems.

Ted Peters, Ph.D., is Professor of Systematic Theology at Pacific Lutheran Theological Seminary and the Graduate Theological Union. He co-edits the journal *Theology and Science*, published by Routledge for the Center for Theology and the Natural Sciences. For three decades he has researched the religious implications of natural science, giving special attention to the cultural impact of genetics and evolutionary theory. He currently serves on the Scientific and Medical Accountability Standards Working Group of the California Institute for Regenerative Medicine, which funds stem cell research. He is author of *Science, Theology and Ethics*; *The Evolution of Terrestrial and Extraterrestrial Life*; *Anticipating Omega*; and *The Stem Cell Debate*. He is co-author of *Theological and Scientific Commentary on Darwin's Origin of Species* and co-editor of *Bridging Science and Religion*.

Florence Raulin Cerceau, Ph.D., is Associate Professor in the Department *Men, Nature, Societies* at the National Museum of Natural History (MNHN) of Paris, France. From 1988 to 1997, she was in charge of the preparation of several exhibitions at the MNHN. In particular, she collaborated on the preparation of the *Grande Galerie de l'Evolution* which opened in 1994. Since 1997, she has been a researcher at the Centre Alexandre Koyré (UMR 8560 – EHESS/CNRS/MNHN), a Center for the History of Science and Techniques partner of the MNHN. Her research is focused on the history and epistemology of exobiology and astrobiology. Her current interests deal with the concept of planetary habitability and the first attempts to contact planets beginning in the nineteenth century. Among her books are *A l'écoute des planets*, *Les origines de la vie*: *histoire des idées*, and *D'autres planets habitués dans l'univers?*

Joseph T. Ross, M.A., M.T.S., M.L.S., is Rare Books Cataloger in the Department of Special Collections at the University of Notre Dame. His master's thesis in the history and philosophy of science at the University of Notre Dame treated Kant's and Hegel's use and critique of analogical arguments for life in the solar system beyond the Earth and life outside the solar system. German *Naturphilosophie* of the eighteenth and nineteenth centuries, especially that of Kant, Goethe and Hegel, continues to be a field of great interest to him. His work in theological studies for the degree of Master of Theological Studies at Harvard Divinity School (1977) focused on theological aspects of nature. His other research interests include the development of orthography and diacritics in the Greco-Roman and Slavic alphabets, especially in relation to the history of typography and codicology.

Woodruff T. Sullivan, III, Ph.D., is Professor in the Department of Astronomy and Adjunct Professor in the Department of History at the University of Washington. He is also past Director of that university's graduate interdisciplinary Astrobiology Program. His research in astrobiology has centered on the Search for Extraterrestrial Intelligence (SETI), for instance as co-founder of the pioneering SETI@home project. With John Baross, he was co-editor of the graduate textbook *Planets and Life: The Emerging Science of Astrobiology* and in history of science he published *Cosmic Noise: A History of Early Radio Astronomy*. In 2012 he was awarded the Leroy E. Doggett Prize for lifetime achievement in the history of astronomy by the Historical Astronomy Division of the American Astronomical Society. Sundials are one of his passions (Google to learn more).

Stéphane Tirard, Ph.D., is Professor in History of Science and Director of the Centre François Viète d'Epistémologie et d'Histoire des Sciences et des Techniques (EA 1161) at the University of Nantes. His research interests include the epistemology and history of theories of the origins and limits of life. He is particularly interested in the epistemology of interdisciplinary fields. Tirard served as the field editor in epistemology for the *Encyclopedia of Astrobiology*, and he has published *Histoire de la vie latente*. He is currently Vice-President of the Societé Française d'Histoire des Sciences et des Techniques.

Douglas A. Vakoch, Ph.D., is Professor in the Department of Clinical Psychology at the California Institute of Integral Studies, as well as Director of Interstellar Message Composition at the SETI Institute. He serves as chair of both the International Academy of Astronautics (IAA) Study Group on Interstellar Message Construction and the IAA Study Group on Active SETI: Scientific, Technical, Societal, and Legal Dimensions. Vakoch's books include *Communication with Extraterrestrial Intelligence (CETI)*; *Civilizations Beyond Earth: Extraterrestrial Life and Society*; *Extraterrestrial Altruism: Evolution and Ethics in the Cosmos*; *Psychology of Space Exploration: Contemporary Research in Historical Perspective*; *On Orbit and Beyond: Psychological Perspectives on Human Spaceflight*; *Ecofeminism and Rhetoric: Critical Perspectives on Sex, Technology, and Discourse*; and *Feminist Ecocriticism: Environment, Women, and Literature*.

M. Margaret (Meg) Weigel, Ph.D., a biological anthropologist, is Professor of Public Health Sciences and founding Director of the Master of Public Health Program in the Department of Public Health Sciences at The University of Texas at El Paso. She has over 25 years of experience conducting mixed-methods research on the health, nutrition, and food security issues of ethnic minority and other communities in Ecuador, Mexico, and the United States.

Index

D. A. Vakoch (ed.), *Astrobiology, History, and Society*, Advances in Astrobiology
and Biogeophysics, DOI: 10.1007/978-3-642-35983-5,
© Springer-Verlag Berlin Heidelberg 2013

Printed by Publishers' Graphics LLC
DBT140604.23.34.183